Metalworking in Bronze Age China

Metalworking in Bronze Age China

The Lost-Wax Process

Peng Peng

Cambria Sinophone World Series
General Editor: Victor H. Mair

CAMBRIA PRESS

Amherst, New York

Requests for permission should be directed to
permissions@cambriapress.com, or mailed to:
Cambria Press
100 Corporate Parkway, Suite 128
Amherst, New York 14226, USA

Front cover image is the openwork ornament portion of the lid of a bronze vessel dated early sixth century BCE, Metropolitan Museum of Art, Charlotte C. and John C. Weber Collection, Gift of Charlotte C. and John C. Weber, 1992, https://www.metmuseum.org/art/collection/search/44734.

Library of Congress Cataloging-in-Publication Data

Names: Peng, Peng, 1987- author.

Title: Metalworking in Bronze age China : the lost-wax process / Peng Peng.

Description: Amherst, New York : Cambria Press, 2020. |
Series: Cambria Sinophone world series |
Includes bibliographical references and index.
| Summary: "This is the first study that adopts a comprehensive, thorough, and interdisciplinary approach toward early Chinese lost-wax castings. With more than 80 images, this book provides a study on the "norms," which are seldom questioned. By examining the reasons why Chinese founders often chose not to use the lost-wax process they had clearly mastered, the book refutes the idea that lost-wax technology is the only "right way" to cast bronzes. This study demonstrates that a "norm" is in many ways an illusion that twists our comprehension of art, technology, civilization, and history"-- Provided by publisher.

Identifiers: LCCN 2019031639 (print) | LCCN 2019031640 (ebook) |
ISBN 9781604979626 (hardcover) | ISBN 9781621964803 (epub)

Subjects: LCSH: Bronzes, Chinese--To 221 B.C. |
Precision casting--China--History. | Bronze founding--
Technological innovations--China--History.

Classification: LCC NK7983.22 .P445 2019 (print) | LCC NK7983.22 (ebook) |
DDC 739.5/120951--dc23

LC record available at https://lccn.loc.gov/2019031639
LC ebook record available at https://lccn.loc.gov/2019031640

To my parents 徐峰 *and* 彭培讓

To my teachers Robert Bagley, 李水城 *and* 蘇榮譽

To Hye-shim Yi and Leo Peng

for the beautiful years

Table of Contents

List of Figures

ACKNOWLEDGMENTS

I am indebted to many individuals for their help in the preparation of this book. In particular, I must thank my academic advisor, Robert Bagley, for his unfading inspiration and enduring encouragement, as well as his always insightful comments on the manuscript. To him, I owe more than I can say.

I am also grateful to W. Thomas Chase, Alain Thote, Mei Jianjun, Victor Mair, and Robert Murowchick, for their valued advice and support. Friends, colleagues, and teachers who have helped me bring this book to fruition also include: Li Shuicheng, Jerome Silbergeld, Hua Jueming, Tan Derui, Zhang Changping, Li Yung-ti, Michael Notis, Qian Wei, Chen Jianli, Lian Haiping, Ma Jinhong, Wu Laiming, Li Yuxin, Huang Fengchun, Shao Anding, Sammy Li, Cao Dazhi, Cui Jianfeng, Robert Silberman, Michael Gaudio, Anna Lise Seastrand, Matthew Canepa, Sinem Casale, Frederick Asher, Liu Yang, Wang Cheng-hua, and Bryan Just (to avoid the list from being endless, I have to omit many significant names here).

Special thanks should go to Su Rongyu and Wang Yong, for many illustrative photographs and brilliant ideas they provided for my research, Kimberly Wishart and Allison Vandev, for their painstakingly careful proofreading of my work, David Armstrong and Toni Tan (Cambria Press), for their perfect arrangement of everything related to this publication, as well as the UMN (University of Minnesota) Wilson Library, for its warm assistance in acquiring certain permissions of copyrighted items for the current book.

I would also like to express my gratitude to Sidney Cheung, Pedith Chan, Fanny Chung, Isaac Leung, and all my current colleagues in the Faculty of Arts at the Chinese University of Hong Kong. Without the most encouraging, engaging, and exciting intellectual environment they offered to me, this book would not have come into being so successfully.

While working on this book I have presented my research at a series of conferences and symposia. I should like particularly to thank audiences at the 105th College Art Association annual conference (New York, February 2017) and the Art History Faculty Symposium of University of Minnesota (Minneapolis, April 2018), for their constructive feedback. It should also be noted that a version of chapter 3 has been previously published in *Orientations*, January/February 2018, pp. 103–11. An old version of chapter 8 can be found in *Sino-Platonic Papers*, edited by Department of East Asian Languages and Civilizations, University of Pennsylvania, January 2017 (Number 265), pp. 1–48. These two sections have both been modified and revised for this book.

Last but not least, words cannot express my deepest gratitude to my family. My mother, a beloved doctor, and my father, a determined scholar, are my model of being a whole person. Their names are written in Chinese on the dedication page. I am also blessed to have a wonderful sister, Peng Li. My wife, Hye-shim Yi, has shared with me every joy and frustration of my research, and most important of all, the unparalleled bliss of having our first kid, Leo Peng (彭勵豪)—Leo, you know, without you, your father's book would not have been finished so cheerfully.

Metalworking in Bronze Age China

Introduction

An overwhelming majority of Western scholars used to believe that most early Chinese bronzes were cast using the lost-wax process.[1] The section-mold[2] theory proposed by Orvar Karlbeck[3] received little attention because it seemed unbelievable that some of the particularly intricate shapes (e.g., fig. 116) could have been cast integrally within piece molds.[4] Later, Rutherford J. Gettens's study verified that certain visually complex Chinese bronzes could have been cast within the section molds, but not necessarily in one piece, because individual castings could have been mechanically or metallurgically joined together.[5] Studies on the relationship between casting technique and vessel design[6] won wide acceptance for the section-mold theory in the manufacture of most bronzes from early China. Careful research carried out on thousands of examples also led two Chinese scholars, Guo Baojun and Hua Jueming, to agree that most early Chinese bronzes, if not all, had invariably been cast by the section-mold process.[7]

Recently the question of whether the lost-wax technique was used at all during Bronze Age China has been hotly debated.[8] The bronze *zun-pan* set buried with Marquis Yi of Zeng (d. 433 BCE, figs. 15, 16) from Suizhou of Hubei[9] and certain other bronzes are the focus of this debate. In response, the following questions motivate this book: 1) Was the lost-wax process ever practiced in Bronze Age China? 2) If the lost-wax process existed in Bronze Age China, how many known artifacts can be confidently identified as lost-wax castings? How was the lost-wax process adapted to each individual artifact? What are the burial contexts, chronology, and geographical distribution of these lost-wax castings? 3)

Was the lost-wax process independently developed in China, or was its use stimulated by contact with casting traditions outside China? If the latter, when, where, why, and how was lost-wax casting transmitted?

OUTLINE OF THE BOOK

My book explores these questions. In chapters 1 and 2, I introduce the technical basis of lost-wax casting and review the research history of the lost-wax problem in Bronze Age China, respectively. Chapter 3 conducts a case study of the rim appendage of the bronze *zun* (of the *zun-pan* set) from the tomb of Marquis Yi of Zeng. This study has a twofold purpose. On the one hand, it contributes to the recent debate on whether the lost-wax technique existed in Bronze Age China. On the other, it seeks to identify lost-wax castings in early China. My study of this artifact and other relevant objects proves that the two objecting arguments are incorrect. Because the Zeng *zun-pan* is stylistically not foreign, it is reasonable to advocate that the lost-wax technique emerged in China no later than the Eastern Zhou period (Late Bronze Age; see Table 1 of this book for the chronology of Bronze Age China).

Chapters 4–7 examine additional lost-wax castings from Bronze Age China with the aim of discovering the driving force of the designs behind them. On this point, two main clues suggest that an object was cast by the lost-wax process. One clue is an intricately shaped object: in such a case, any means of fabrication apart from lost wax is difficult to imagine. The other clue is the archaeological/cultural/artistic context of the object: although the shape of the object may not be enough to be convincing, if it is stylistically related to other confirmed lost-wax castings, it was likely manufactured by a caster who was familiar with lost wax.[10] Because neither of these clues is clear-cut; uncertainty must be tolerated in some cases.

Chapter 8 looks for the technical origin of the lost-wax process used in Bronze Age China. Specifically, it asks whether this technique was independently developed in China or was introduced from outside. Eastern Zhou lost-wax castings make it clear that early Chinese foundries still relied on section molds even when they already knew how to use the lost-wax process with sophistication.[11] This means it is possible that Chinese casters mastered lost-wax or similar processes considerably earlier. Today there are three main hypotheses: (1) that the lost-wax process was developed locally in China from earlier "burn-out" casting methods, (2) that the lost-wax process was introduced from the southwest, and (3) that the lost-wax process was introduced from the north. This chapter tests these hypotheses with known archaeological materials. One question worth pondering is how important the origin of the lost-wax process was. Was its advent an earthshaking development or an event far from revolutionary or consequential? In either case, it is clearly wrong to consider the lost-wax technique the only right way to cast bronzes, implying that Chinese casters used the wrong way only because they did not have the right way at their disposal.[12]

Chapter 9, as the concluding chapter, steps back to assess several unresolved issues, especially those related to other possible lost-wax traditions and technological variations from Bronze Age sites in China. It also carefully examines some uncritically accepted "lost-wax castings" that deserve particular discussion. Even as I uncover possibilities, many questions remain unanswered. Further research and future discoveries will enrich the picture I have drawn here and deepen the present understanding of art, technology, and invention.

NOTES

1. For a description of the lost-wax process, see Hemingway, "Bronze Sculpture," 37–46; Mattusch, *Greek Bronze Statuary*, 10–30; Mattusch, *Classical Bronzes*, 1–34.
2. For a description of the section-mold process, see Bagley, "Shang Ritual Bronzes," 7–20.
3. Karlbeck, "Anyang Moulds," 39–60.
4. Bagley, *Shang Ritual Bronzes in the Arthur M. Sackler Collections*, 37, 44.
5. Gettens, "Joining Methods in the Fabrication of Ancient Chinese Bronze Ceremonial Vessels," 205–217.
6. E.g., Fairbank, "Piece-mold Craftsmanship and Shang Bronze Design," 18–32; Smith, *A Search for Structure*, 128–135, 258–270.
7. See Guo, *Shangzhou Tongqiqun Zonghe Yanji*, 124; Hua, "Shilafa de Qiyuan he Fazhan," 63–81.
8. For a comprehensive review of this debate, see Peng, "Was Lost-wax Casting Adopted in Pre-Qin Chinese Bronze Art?" 103–109.
9. For convenience, the *hanzi* characters of the provincial-level administrative divisions in contemporary China (e.g., Hubei) shall not be provided in the Character Glossary of the book. For locations of Chinese provincial divisions, see figure 1.
10. Robert Bagley, personal communication, September 2014.
11. Bagley, *Shang Ritual Bronzes in the Arthur M. Sackler Collections*, 44.
12. Robert Bagley, personal communication, August 2014.

CHAPTER 1

WHAT IS LOST-WAX CASTING?

PRINCIPLES OF THE LOST-WAX PROCESS

Casting (or founding) is the process of pouring molten metal into a refractory mold with a hollow cavity and allowing it to solidify into the desired shape.[1] The earliest-known castings in the global archaeological record were made in open stone molds.[2] Stone, as a durable material, allows for the repeated casting of a relatively simple shape from the same open mold. Metalworkers in time devised the closed mold, which was cut in stone or, more frequently, fashioned in clay.[3] Of these closed molds, the bivalve or two-piece mold, consisting of two abutting halves, was decidedly the most rudimentary, capable of casting only simple tools and weapons, such as axes, adzes, daggers, knives, and spearheads.[4]

Stone molds, whether open or bivalve, were manufactured by carving or cutting directly in stone, likely with only a conceptual model in mind. The clay bivalve mold, in contrast, seems to have been taken in two pieces from a physical model and subsequently pressed together. The use of a tangible model undoubtedly provided better control of the shape and dimensions of the final casting. In fact, such a model was essential when the desired shape of the finished object became too complex to be visualized in the negative. A model with a complex shape that would not permit the withdrawal of the mold in only two pieces would render the bivalve mold insufficient. How could this problem be resolved?

One solution to this problem was to increase the number of mold pieces. The shape of the design, if not exceedingly intricate, would permit the clay mold to be removed from the model in a controllable number of pieces, which would be reassembled later for casting. This method is called "section-mold" (short for "multi-section-mold") or "piece-mold" (short for "multi-piece-mold") casting. Referring to a diagram in figure 7, we could briefly illustrate how a three-legged vessel similar to that shown figure 9 might be made by the section- or piece-mold casting process in principle: the round tripod is modeled in clay or similarly manageable material; the mold is constructed against the model in multiple fitted sections; then the mold is peeled off and finally reassembled around a core, waiting for the pour of molten metal.[5]

An alternative solution to the problem exists. The complex shape of the design requires a model, but a mold is not easily separated from this model in two pieces; an ideal solution would allow for the creation of a casting cavity without detaching the mold. In this case, the model would necessarily disappear, whereas the undetached mold would have to remain in place to keep the casting cavity intact. How could this be achieved? The model would have to undergo a physical transformation that would allow it to be easily discharged from the enclosed mold-- something like a fluid or gas. Once the model was "lost," a hollow casting matrix would be formed within the mold to await the pour of the molten metal. Wax, which is easy to shape and carve and which melts away under the proper conditions, has been the most commonly used material for this purpose since antiquity. As a result, this method has been termed the "lost-wax" (or *cire perdue*) process; the idea of a "lost model" lies at the heart of this method. The clay mold, as an investment onto or enveloping of the model, need not be constructed in pieces, removed from the model, and reassembled. This method is also referred to as "investment casting," which is a more inclusive term with regard to other expendable materials (e.g., bitumen, tallow, resin, or modern foam) that in some instances replaced wax; nevertheless, the term "lost-wax process" remains in common usage.

This is the basic rationale behind the use of lost-wax (or investment) casting methods. For convenience, I describe the section-mold and lost-wax processes as two solutions to the same problem. Historically, it is conceivable that the lost-wax process developed in multiple regions for different reasons. It is possible that in some locations, wax was only occasionally used as a model, and casters developed the process without having been forced to determine how to deal with a complicated design. However, the section-mold and lost-wax processes, though differing essentially in principle, were by no means two parallel, unrelated techniques. They could be, and often were, freely combined with each other in the same process.

The Lost-wax Process and Its Variations

Though the reasons for using the lost-wax process appear straightforward, casting itself is by no means simple. The layman's understanding of "the lost-wax process" is a highly unrealistic abstraction, far removed from the realities of the workshop. As Robert Bagley has pointed out, "the term 'the lost-wax process' lumps together all sorts of different things. The bronze caster does not work from one textbook diagram; he does whatever it takes to make a particular object."[6] For this reason, it is useful to introduce the lost-wax processes, or rather one process with multiple variations, with specific well-studied objects. Ancient Greek casting provides ample case studies for this purpose.

The Direct Process

The diagram in figure 2 explains one of the simplest forms of lost-wax casting. To cast a solid horse figurine, the founder begins by making a wax model of the horse. A system of gating or venting, composed of wax rods for the pouring of metal and venting of gases, is attached to the model.[7] Subsequently, an investment mold is made by packing clay all around the wax model, presumably in multiple layers.[8] The investment mold is then baked, during which process anything made of wax (i.e., the model and its sprues) melts away from the mold. The now-hollow mold is inverted to receive the molten metal, which lodges in the space previously occupied by the wax, in this case the horse figurine model and its gate system. After the poured metal solidifies, the investment mold is broken open, the gate system removed, and the bronze horse revealed, cleaned, and perhaps further refined.[9] Because metal shrinks during solidification, cracks can easily emerge on thick, solid castings. To prevent shrinkage and to conserve metal, one or more clay cores held in position by metal chaplets can be added within the casting to create a hollow form.[10]

Compared with section-mold casting, the primary benefit of the direct lost-wax process lies in the stage of the mold's construction.[11] Because mold sections do not have to be removed from the model, the lost-wax process is theoretically capable of casting objects of any shape, even those with formidably deep undercuts.[12] Meanwhile, behind the "theoretical elegance and simplicity" of the lost-wax process is "much skill in its application."[13] As Bagley pointed out, the lost-wax caster is not completely free from shape constraints: to avoid trapping air when the mold is filled with metal, the gate system becomes correspondingly more complex the more intricate the shape to be cast.[14] Compared with section-mold casting, the lost-wax procedure, moreover, has some apparent disadvantages. For example, founders are unable to inspect the interior of the enclosed investment mold to detect potential defects before casting.[15] More importantly, the wax model, as the artist's original, is "lost" so a new model must be made if another product is wanted, or if an accident happens before or during casting.[16]

The Indirect Process

To avoid the permanent loss of the original model, the "indirect" lost-wax casting process was invented by borrowing technical elements from the section-mold process. This method maintains the original model through the use of a permanent section mold, which can yield any desired number of wax models that are to be cast in a given shape.[17] In this way, the metal handles of classic Greek vessels were produced as multiples after a single model. The working models for these handles are thought to have been reproduced in wax from a permanent section mold (the "master mold") that retained the impression of the original model.[18] The use of the indirect lost-wax process in this way facilitated reproductions.[19]

Alternatively, wax models after the same original could be extensively detailed, retouched, and individualized before being cast.[20] The renowned Riace bronzes (fig. 3) provide a good illustration of this point. Almost identical in height, body outline, and posture, the two nudes were apparently based on a single original model, yet they differ in appearance, spirit, and musculature, apparently due to extensive individual work on the wax-working models (especially on the heads).[21] The manufacture of the Riace bronzes freely combined the solid casting, indirect lost-wax, and direct lost-wax processes. As Bagley commented, the procedure for making the Riace bronzes "raises the possibility of a sort of customized mass production in which contributions to the finished work might have been divided between different workshops and even different cities."[22] They also provide inspiration for the study of lost-wax casting in Bronze Age China.

Before studying the lost-wax casting of Bronze Age China, it is important to dispense with the idea that "the lost-wax process" is a simple thing. The problematic part of this term is "the," which betrays a presumption that the process is uniform, when in fact a variety of methods were used. The bronze caster does not work from a single textbook diagram; he does whatever it takes to make a particular object. As will be discussed, the layman's idea of "the lost-wax process" is an unrealistic abstraction, far removed from the realities of bronze foundries.

Notes

1. Rao, *Zhizao Jishu: Zhuzao, Chengxing he Hanjie*, 55.
2. This may have been inspired by the phenomenon in which smelting copper, if accidently splashed on a paved floor during the transfer, cooled and hardened into the shape of the stone. See Knauth, *The Metalsmiths*, 44.
3. Knauth, *The Metalsmiths*, 44.
4. Ibid.
5. For a comparable description, see Bagley, "Shang Ritual Bronzes," 13.
6. Robert Bagley, personal communication, August 2014. For detailed discussion, see Mattusch, *Greek Bronze Statuary*, 15–30.
7. The gate system usually comprises a funnel (or sprue) to introduce the liquid metal, runners to permit its flow inside the mold, gates to regulate the flow, and vents to allow the escape of gasses. See Mattusch, *Greek Bronze Statuary*, 17; Rao, *Zhizao Jishu: Zhuzao, Chengxing he Hanji*, 117; Meyers, "Casting Technology in Cambodia and Related Southeast Asian Civilizations," 33.
8. The innermost layer of the investment had to be the finest in consistency to guarantee that all the surface details of the wax were completely enveloped; the external layers were made coarser and harder with sandy and organic inclusions, for the purpose of keeping the clay from cracking and minimizing its shrinkage during baking. See Mattusch, *Greek Bronze Statuary*, 17; Mattusch, *Classical Bronzes*, 15.
9. See Mattusch, *Greek Bronze Statuary*, 31; Mattusch, *Enduring Bronze*, 55. Hemingway terms this method "solid lost-wax casting," but the term "solid casting" could be applied to both the direct and indirect lost-wax processes ("Bronze Sculpture," 39). The method illustrated here, to use the most accurate, specific terminology, is "solid casting by the direct lost wax process."
10. Mattusch, *Greek Bronze Statuary*, 16, 25; Mattusch, *Classical Bronzes*, 15.
11. Bagley, *Shang Ritual Bronzes in the Arthur M. Sackler Collections*, 64, note 248.
12. Ibid., 44; So and Bunker, *Traders and Raiders on China's Northern Frontier*, 59–60.
13. Sandars, *Prehistoric Art in Europe*, 160.
14. Bagley, *Shang Ritual Bronzes in the Arthur M. Sackler Collections*, 44.

15. Bagley, "The Beginnings of the Bronze Age," 70.
16. Bagley, *Shang* Ritual *Bronzes in the Arthur M. Sackler Collections*, 44.
17. Ibid., 20, note 2.
18. Mattusch, *Enduring Bronze*, 76.
19. As Mattusch describes, "from a single original model, more than one bronze can be cast" by the indirect lost-wax process. Mattusch calls this operation "serial production"; it routinely produced utilitarian objects in ancient Greece. See Mattusch, *Classical Bronzes*, 18.
20. See Mattusch, *Greek Bronze Statuary*, 20; Mattusch, *Classical Bronzes*, 10, 18.
21. As Mattusch points out, Statue A has a wide, vigorous face framed by a great mass of loose and imposing curls. By comparison, the head of statue B is longer and narrower, the face older and less aggressive, with the hair and moustache shorter and more contained, and the beard longer but less curly. Statue A is more robust with distinct blocks of muscles and prominent veins, exhibiting vigor and strength. Statue B's body, on the other hand, is slightly more slender and flatter, with less compelling musculature. See Mattusch, *Greek Bronze Statuary*, 202–205; Mattusch, *Classical Bronzes*, 64, 194, 221.
22. Robert Bagley, personal communication, August 2014.

Chapter 2

Metalworking in Bronze Age China

Documentation of lost-wax casting in China does not appear until many centuries after the start of the Common Era, and ethnography likewise contributes almost nothing to the study of early bronzes, because no casting tradition known to ethnographers approaches ancient Chinese bronze casting in scale or complexity. For Bronze Age China, questions surrounding lost-wax casting can be answered only by artifacts and other related archaeological materials. A brief review of metallurgical history in Bronze Age China is therefore an essential preliminary.

Was Lost-wax Encountered by China's First Bronze Founders?

The origin of metalworking in China has long been debated. Although the "independent invention versus diffusion" argument still remains unresolved in the specific case of China, more and more evidence seems to support the hypothesis that metalworking in early China was introduced or stimulated from external sources.[1] Fitzgerald-Huber's 1995 article, a milestone work on this issue, suggests that Henan Erlitou production, representing the earliest known bronze *industry* (large-scale metallurgy) in the Central Plains region of China, was triggered by the transmission of northern Asiatic metalwork. Contemporaneous metallurgical centers in northwestern China, especially Gansu Qijia and Shaanxi Kexingzhuang Culture II, played an intermediate role in this process.[2] The impact of the renowned Seima-Turbino metalworking tradition[3] to the north has been supported by a series of more recent discoveries from China,[4] namely a number of distinct socketed spearheads with a downward hook and a small loop on the same or opposite sides of the socket that could be confidently associated with the Seima-Turbino visual vocabulary of southern Siberia.[5]

With high mobility and brilliant technology, the Seima-Turbino groups[6] were among the first regular users of a tin-bronze alloy in the north of Central Asia.[7] It is possible that they might have also mastered the lost-wax process, judging by certain tools and weapons with pommels bearing three-dimensional figurines (e.g., fig. 4).[8] This possibility, though unverified, is not unlikely, in light of the flourishing use of lost-wax on the Eurasian landmass from the third millennium BCE. Anthony further identified the source of Seima-Turbino lost-wax casting as the BMAC (Bactria-Margiana Archaeological Complex, c. 2100–1800 BCE),[9] the Late Bronze Age civilization of Bactria and Margiana in Central Asia. Rich materials indicate that the BMAC had close international relationships with the Indus Valley, with the Iranian Plateau, and perhaps even indirectly with Mesopotamia, all of whose civilizations were intimately familiar with lost-wax casting.[10] As might be expected, lost-wax casting was also developed in the BMAC,[11] as evidenced by superb ornaments and delicate figures,[12] in addition to comparatively simple and mass-produced artifacts, such as the compartmented metal seals[13] whose common motifs of crosses and stars are comparable to similar designs from Qijia, Erlitou, and their surroundings.[14]

Because both Seima-Turbino and the BMAC, as Fitzgerald-Huber maintained, display remarkable evidence of cultural transmission to the metallurgical centers of Qijia and Erlitou,[15] it is tempting to suggest that the first metalworkers of Bronze Age China had already encountered the lost-wax process. Against the rapid and long-distance metallurgical expansion model proposed by Fitzgerald-Huber,[16] however, Mei Jianjun has argued that metallurgy was brought to China not directly by the Seima-Turbino groups but through "indirect, occasional, and small scale" interregional communications.[17] According to Mei, numerous interactions occurred between northwest China and Eurasia,[18] and many "intermediaries" acted as both means of passage and "filters" of metalworking.[19] If Mei's theory is applied, then lost-wax casting could have been one technology that had been "filtered out" by the time metallurgy was introduced to China.[20] This is, of course, only a speculation without firm evidence from our current archaeological record.[21] Whatever the case, no clues about the use of lost-wax have ever been associated with Qijia or Erlitou, or any other contemporary sites that fall within modern China. The casting used in early Bronze Age China was almost invariably an alternative to the lost-wax process, as discussed in the following section.

Early Chinese Bronzes: The Lost-wax Problem Reviewed

I first briefly review the research history of early Chinese bronzes centered on the lost-wax problem. Based on accounts in antiquity far later than the Chinese Bronze Age,[22] most modern scholars uncritically believed until the 1960s that early Chinese bronzes were cast by the lost-wax process.[23] For instance, according to *Chinese Art* by Stephen Bushell, published in 1904, "Chinese bronzes have always, as far back as we have any record, been executed by the *cire perdue* process, and finished, when necessary, with the hammer, burin, and chisel."[24] Agreeing with Bushell that the lost-wax process cast all early Chinese bronzes, William

Collins apparently considered the wax models to have been directly and individually shaped, and "in consequence no two could be exactly similar."[25] Bruce Simpson, in a book first published in 1948, still took the delicate decoration on a "Shang" bronze artifact to be the "filigree" cast by the lost-wax process.[26]

W. Perceval Yetts certainly noticed the section-mold cast marks on early Chinese bronzes, but he still staunchly believed in the lost-wax theory. In a catalogue published in 1929, he argued that the fitting of mold sections used in the previously mentioned indirect lost-wax process caused such mold marks, which he problematically called "seams."[27] In 1935, Yetts reiterated his point by criticizing the section-mold theory. Yetts held the opinion that section molds from Anyang were used to shape wax models, and those with scorch marks were fragments of the investment molds.[28] Herrlee Creel also agreed with this notion in 1935,[29] despite having seen the section-mold fragments from Anyang himself.[30] As Creel explained of the mold sections (probably for bronze *jue* goblet),[31]

> Note the smooth edges of the molds and the lugs and notches designed to cause them to fit together properly. The way in which such molds were used is not yet fully clear; it seems plausible that they were used to cast wax models, to be reworked and used for *cire perdue* casting later, rather than for casting the bronze itself.[32]

In the following two decades, most Western students of early Chinese bronzes were still convinced by the lost-wax theory. For instance, in 1954, Herbert Maryon and Harold Plenderleith, knowing of the discovery of section molds from Anyang, still conceived of an unreasonably convoluted and jumbled means of reconstructing what they considered to be a largely lost-wax process (for illustration, see fig. 5).[33]

Mold fragments unearthed from Anyang greatly facilitated the study of the Chinese bronze casting process. In a research paper published in 1931, Li Ji, the director of the Anyang excavations, suspected that the flanges of Chinese bronzes might have originated from the fitting of mold sections.[34] This argument, though not completely appropriate,[35] revealed that Li's understanding of the casting process was moving in a generally more plausible direction. The first substantive step towards an accurate account of the casting process in China was taken by Orvar Karlbeck, who explicitly demonstrated that Anyang bronzes were cast by the section-mold process. Based on traces of soot and color changes likely caused by scorching on the interior of many mold fragments, as well as the occasional bronze remnant left in them, Karlbeck maintained that section molds, rather than the shaping of wax models, were used for the casting of bronzes.[36] This clearly ruled out the possibility of the indirect lost-wax casting proposed by Yetts, Creel, and their followers. Though Karlbeck himself did not completely abandon the theory of direct lost-wax casting,[37] he considered it "extremely unlikely."[38] Karlbeck was almost certain that Anyang bronzes were cast within fitted mold sections. It is hard to imagine any reason for the investment mold to carry the mortises and tenons that Karlbeck had clearly observed.[39]

With the work of Karlbeck as their starting point, later Chinese scholars continued to make progress along this line of inquiry. Many scholars, such as Guo Baojun and Rong Geng, soon used the section-mold theory to explain the casting process of newly unearthed Chinese bronzes.[40] In 1954, Chen Mengjia, with newly acquired excavated materials from Anyang and elsewhere in China, reaffirmed the section-mold theory and rejected lost-wax, including the indirect process proposed by Yetts.[41] The next year, Shi Zhangru published a more exploratory, though problematic, analysis of the Anyang casting process that relied mainly on the section-mold theory. Specifically, Shi assumed that the model was shaved down to serve as the core[42] and the thickness of the layer scraped off corresponding to the thickness intended for the bronze vessel wall.[43] Though Shi's proposal of the use of wax and tallow during mold construction makes no sense,[44] he still contributed some original proposals to the prevailing section-mold theory, including his detailed discussion of how a mold assembly might be constructed in a proper number of fitted sections.[45]

In Western academia, Karlbeck's 1935 publication, though detailed and meticulous, garnered little attention in subsequent years. Leroy Davidson, one of a few immediate followers of Karlbeck, argued in 1937 that a bronze *he* from the Museum für Völkerkunde in Berlin[46] must be one of the earliest decorated Chinese bronzes cast by the section-mold process: with access to the interior of the mold, the section-mold casters could easily create the thread-relief lines and circles.[47] Davidson's conjecture, though eventually verified by later archaeological finds from Henan Erlitou, was representative of a rather weak response by Western academics to the section-mold theory. When Western researchers began to take the section-mold theory seriously, their attention stemmed less from Karlbeck's paper than from the work of Chinese scholars, most notably the aforementioned Shi Zhangru.[48] Shi's discussion of the division of mold sections[49] significantly influenced the Australian scholar Noel Barnard, who in his renowned but controversial 1961 book wholly embraced and wildly elaborated the section-mold theory.[50]

Universal acceptance of the section-mold theory in Western academia was finally realized in the 1960s, thanks to two crucial pieces of research. The first was Wilma Fairbank's discovery of "piece-mold segmentation" in the design of early Chinese bronzes—the compartments of bronze motifs and ornaments that evidently correspond to mold sections.[51] The second was Rutherford J. Gettens's technical study of Chinese bronzes in the Freer Gallery, Washington, D.C., which verified that certain visually complex Chinese bronzes could be cast within section molds, but not necessarily in one piece, because individual castings could be mechanically joined together in an interlocking manner (fig. 6).[52] In addition, Gettens associated the mold join marks on the surface of many early Chinese bronzes with the section-mold process.[53] Meanwhile, significant studies were also being advanced in Taiwan and Mainland of China. Through painstaking efforts to evaluate 168 excavated bronze vessels and over three thousand mold fragments from Anyang, Li Ji and Wan Jiabao concluded that all of the bronzes under examination had been cast by the section-mold process, and the number of mold sections depended on the vessels' shape, size, and level of craftsmanship.[54] Careful research carried out on a large number of bronzes similarly led the leading scholars Guo Baojun and Hua Jueming to believe that most early Chinese bronzes were cast by the section-mold process.[55]

Early Bronze Age: Section-mold Casting and Its Development

Although small-scale metallurgy had appeared sporadically several centuries earlier within the territory now occupied by modern China, the rise of a real bronze industry occurred around the sixteenth century BCE, as manifested by Erlitou metalworking in the Middle Yellow River Valley (the core area of the Central Plains of China).[56] With an exclusive dependence on casting, a technique used with little regard for the cost of metal and usually as a metalsmith's last resort, Erlitou metalworking was immediately distinct from that of the rest of the world.[57] Although simple stone molds and clay bivalve molds have been unearthed from the Erlitou cultural sphere, the archaeological site of Erlitou is most featured by the new-fashioned technique of section-mold casting.[58] This pioneering technique, though experimented at earlier sites in the Central Plains region (e.g., Taosi of Shanxi Province), was substantially explored at Erlitou to produce ritual vessels that had previously been made in pottery.[59] Perhaps owing something to the highly developed pottery-making technique of the east coast cultures during the late Neolithic and early Bronze Age, the making of section molds soon achieved great sophistication.[60]

The section-mold process, by requiring the application of clay to the model, the removal of fitted parts, and the reassembly of those parts around one or multiple cores (see fig. 7, the mold divisions for a *ding* tripod of the Erligang period, which immediately followed the Erlitou culture in this part of the Central Plains), favors symmetrical bronze shapes.[61] The most demanding design among Erlitou bronzes is the slightly odd-shaped *jue* drinking cup with a strap handle (fig. 8), which is also the most frequently seen Erlitou vessel type.[62] The fact that the *jue* cup was cast in a single pour further increases the difficulty posed in the construction of its mold assembly.[63] Aside from many examples of minor repairs on bronzes, no evidence of joins is known in the fabrication of Erlitou bronzes;[64] all known Erlitou bronze vessels seem to have been cast in a single piece by the section-mold process. The earliest bronze decoration known in China, a raised pattern of simple lines and bosses on a *jue* cup from Erlitou (fig. 8), was apparently formed by carving into the interior of one mold section.[65] This mold-carving decorative strategy indicates that the Chinese bronze ornament originated as fundamentally subdivided, with compartmentalized patterns conforming to the divisions of the mold.[66] Even after craftsmen later began to chiefly decorate the model, this characteristic continued as an integral aspect of Chinese bronzes throughout the early Bronze Age.

The Erligang bronze industry (c.1500–1300 BCE?),[67] succeeding the Erlitou tradition, had its own technological breakthroughs, primarily in two respects. The first breakthrough came with interlock casting,[68] a stepped-fabrication process that allowed for the joining together of bronze pieces during successive pours of metal.[69] Significantly, it could sidestep the withdrawal of mold sections from an intricately shaped model and avoid trapping pockets of air, a problem from which even the lost-wax process is not free.[70] This method may have originally been devised to make repairs,[71] as casting-on, a variation of this technique, was already being practiced during the Erlitou period to patch flaws.[72] Interestingly, Erligang founders still had a strong preference for casting vessels in a single pour,[73] as evidenced by many vessels produced by that method (e.g., the round *ding* in fig. 9, with its casting process shown in fig. 7).

Another contribution of Erligang metalworkers was the use of metal spacers, which were scrap-like pieces of bronze fixed between core(s) and their mold, with the purpose of maintaining a consistent gap between mold and core that would result in vessel walls of uniform thickness.[74] Spacers, presumably recycled from casting fragments, would not have necessarily conformed to a particular shape or dimension.[75] As some scholars have observed, if a single spacer was applied, it was most frequently set at the bottom center of the vessel where the mold joins meet; when multiple spacers were used, they often appear around the more angular areas of the finished bronze.[76] Unique to the section-mold casting of early China, the "spacer" cannot be considered identical to the "chaplet," a spike-like device that had to be removed after casting, usually associated with lost-wax casting in the Mediterranean world.[77]

After several centuries of development, section-mold casting was further elaborated in the Anyang period (c. 1200–1050 BCE).[78] With foundry debris excavated on an extensive scale, Anyang workshops are the most thoroughly studied foundry sites of early Bronze Age China. The "consensus view" of Anyang casting, formed through generations of painstaking research, is most precisely summarized by Robert Bagley:

> with rare exceptions, the decoration of Anyang bronzes was transferred to the mold from a decorated model. In this procedure, the designer began by making a clay model of the vessel he wished to cast, complete with all its decoration. The mold-maker, who may or may not have been the same person, then used the model to make the fitted sections of a clay mold...The sections were formed one by one against the model, removed from it, and reassembled around a core for casting. To form a section the mold-maker applied fine clay to the model. Perhaps he began by painting on a few layers of clay slip to ensure that the finest details would transfer. After completing the section, he removed it, cut mortises into its edges, let it dry, and replaced it on the model. Then he formed a new section next to the first. The new section would acquire tenons fitting the mortises in the section next to it, guaranteeing a secure fit when the completed mold was assembled for casting.... When bronze was poured into the space between core and mold, the result was a bronze replica of the decorated clay model.[79]

Though this mainstream view has recently been attacked by proposals asserting that models were never decorated, only molds,[80] and that models were made not of the whole vessel but only of a fraction of it,[81] it seems that this mainstream view cannot be easily shaken based on current evidence.

Late Bronze Age: Innovations in Metalworking

As discussed earlier, in early Bronze Age China, mold making started predominantly with a decorated model of the vessel to be cast, but serial or repetitive production from a single model appears not to have been practiced until quite late.[82] Only around the mid-first millennium BCE was a new decorative technique invented. It is best revealed by the bronzes and workshop remains from Houma of Shanxi Province, the capital of the Eastern Zhou state of Jin. This innovative process has been thoroughly investigated by Barbara Keyser[83] and lucidly described by Robert Bagley. Bagley's description proceeds as follows: to cast a Houma-style *hu* in the Freer Gallery (fig. 11; the two handles were separately cast and will not be discussed here), clay was applied around an undecorated model and removed in two large blank shells. The decoration was carved neither on the model nor in the mold but was manufactured separately from a reusable baked pattern block, similar to the one illustrated in fig. 10. Wet clay was pressed to the block and then peeled off, carrying with it a negative imprint of the decoration. This soft peeled ornament was pared, bent, and plastered piece by piece around the mold in a manner akin to wallpapering. The soft pattern units so planted in the mold were then baked hard, and the bronze cast in the resulting mold bore horizontal registers of ornament, each register comprised of identical, duplicated units.[84]

A remarkable number of Houma products have been discovered at numerous sites in Shanxi, Shaanxi, Henan, Hebei, Shandong, and even more remote areas.[85] With the verified use of reusable decorated models at Houma,[86] it is puzzling that pattern units there were not applied directly to the model, in which case the decorating process would only have had to be carried out once.[87] Even more curiously, Houma casters rarely carved on the model or made mechanical duplications of it, a process they had clearly mastered.[88] A marked change in workshop organization, likely stimulated by the demand for greater efficiency of production, seems to have accompanied this new trend in using pattern blocks.[89] Compared with the traditional section-mold casting inherited from the early Bronze Age, the pattern-block technique made better use of the highly sought-after carvers at the cost of increased difficulty in mold making.[90] Mold makers, in contrast, required no artistic training, and a finer division of labor among them may have been possible—the making of mold sections, reproduction of the pattern units, handling of units and planting them in the mold, and baking of the mold could all have been done individually by different groups. Although it is impossible to reconstruct the exact procedure of mold construction, the work must have been subdivided in a new way.[91]

The mold-making technique of Houma craftsmen was unparalleled in its quality and complexity. This distinctiveness is most clearly manifested in a bronze frontlet of exceptional intricacy and delicacy (fig. 12) and a similar clay model discovered at Houma, demonstrating full three-dimensionality and formidable undercuts (fig. 13a).[92] Were the clay model not known to us, the bronze frontlet might have been unhesitatingly labeled a lost-wax casting. As will be shown, although lost-wax casting was indeed practiced elsewhere during the Houma era, this example, an unmistakable section-mold casting, was created utterly independently of the lost-wax process. It exhibits the amazing transfer skills of the Houma craftsmen, who could do extraordinary things with clay.

No evidence of lost-wax casting has yet been found at Houma or any of the closely related foundries. Additionally, the pattern-block technology best known from Houma was by no means a monopoly of the Houma workshops. It was likely used elsewhere at an earlier date, because the sixth-century BCE burials of the state of Chu at Xichuan Xiasi in southwestern Henan yielded some stylistically quite different bronzes that were nevertheless cast by the pattern-block process.[93] But those same Chu burials also contained virtuosic lost-wax castings. Considering the frequency of interregional communications during the Eastern Zhou period, could it safely be assumed that Houma casters did not know the lost-wax process? It is hard to imagine that they did not, but if they did know it, why has no evidence of the use of lost-wax at Houma been found in the abundant archaeological record? One reason might be that they were already able to cast everything their patrons wanted.[94] Another reason might have been their pursuit of efficiency. Perhaps needing to supply a broad market, they may have focused on mass production and replication, which the direct lost-wax process, at least, does not lend itself to. Houma founders may have been uninterested in the lost-wax process because it was too inefficient for them.[95]

In addition to the pattern-block technique, Late Bronze Age metalworking also made extensive use of mechanical or metallurgical joining. For instance, an intricate table base from Hebei Pingshan (fig. 14) dated to the late fourth century BCE[96] is composed of 78 components that were cast from 188 mold pieces and joined together 22 times by interlock casting and 48 times by soldering.[97] In the same period, more and more bronzes were inlaid with metals of various types, such as gold, silver, tin, or pure copper. The Pingshan table base (fig. 14) and many other bronzes from the same tomb are inlaid with gold and/or silver; the earliest metal inlays in bronze (with cast copper) appeared in China no later than the sixth century BCE.[98] Although not entirely absent from earlier periods, hammering, as a technology distinct from casting, also started to become more common in China at this time, especially with its variants of repoussé and chasing. During the same period, other "alien" techniques arrived in China, such as granulation, gilding, and punching.[99] Lost-wax casting, as another seemingly new metallurgical skill, unavoidably arouses the curiosity of scholars. Did lost-wax casting exist in Bronze Age China? If so, what evidence is there for it? How was it adapted to produce particular castings? Did it revolutionize Chinese metalworking as a whole? The chapters that follow attempt to unravel these mysteries.

Notes

1. For instance, many nationalistic arguments for the independent invention of metallurgy in China (e.g., Tang, "Zhongguo Qingtongqi de Qiyuan yu Fazhan") have been systematically refuted by a dispassionate Chinese scholar, An Zhimin. For details, see An, "Some Problems Concerning Early Chinese Copper and Bronze Artifacts"; An, "On Early Copper and Bronze Objects in Ancient China."
2. See Fitzgerald-Huber, "Qijia and Erlitou," 29–39. One problem with Fitzgerald-Huber's theory, as she herself acknowledges, lies in the "presumed chronological discrepancy" between the Qijia and the apparently later Seima-Turbino cultures. As Fitzgerald-Huber narrated, "whereas the calibrated carbon-14 dates for the Qijia culture cluster around the year 2000 B.C., the Seima-Turbino, more loosely dated on the evidence of their occasional intermingling with the Sintashta-Petrovka cultures of the Tobol River area, have consistently been assigned to a later time, roughly coeval with Shang. Chernykh, for instance, estimates that the consolidation of the Seima-Turbino took place no earlier than the seventeenth century B.C." ("Qijia and Erlitou," 49–50). Yet more and more evidence supports Fitzgerald-Huber's reassessment of the Seima-Turbino chronology already reassigned to between the 22nd and the 18th to 17th centuries BCE by Chernykh (e.g., "Ancient Metallurgy in the Eurasian Steppes and China: Problems of Interactions," 2; "Formation of the Eurasian Steppe Belt Cultures: Viewed through the Lens of Archaeometallurgy and Radiocarbon Dating," 136). This adjusted chronology fits quite well with Fitzgerald-Huber's hypothesis linking the Qijia and the Seima-Turbino.
3. For a detailed description of Seima-Turbino metalworking, see Shaw and Jameson, *A Dictionary of Archaeology*, 517–518.
4. For a review of these discoveries, see Mei, "Qijia and Seima-Turbino," 35–36; Lin, "Saiyima-Tuerbinnuo Wenhua yu Shiqian Sichou zhi Lu," 51–59.
5. To date, more than a dozen such spearheads have been recognized in China; they are distributed from Qinghai and Gansu in the northwest to as far south as Henan Xichuan, which is even some distance to the south of Erlitou. Although some of the samples are certain to be local imitations, the intrusion of the Seima-Turbino impact

seems to have been even deeper than previously imagined. For details, see Chernykh, *Ancient Metallurgy in the USSR*, 227–228; Mei, "Qijia and Seima-Turbino," 35; Lin, "Saiyima-Tuerbinnuo Wenhua yu Shiqian Sichou zhi Lu," 52–59.

6. With small populations, the Seima-Turbino groups consisted of metallurgists and horsemen warriors who seem to have had not only a formidable level of organization, sufficient to allow them to migrate thousands of kilometers in a short span of time, but also technology advanced enough to produce high-level tools and weapons. For details, see Chernykh, *Ancient Metallurgy in the USSR*, 215–216.
7. Anthony, *The Horse, the Wheel, and Language*, 444.
8. Take the Seima-Turbino dagger (fig. 4) from the Rostovka burial near Omsk (the Irtysh Basin) for example: Chernykh stated that this exquisitely executed dagger is "princely" and interpreted the figurines as a horse and a Seima-Turbino skier who holds the horse's reins during travel (*Ancient Metallurgy in the USSR*, 219). Anthony asserted that the dagger pommel was lost-wax cast, but does not provide any technical evidence for this assertion (*The Horse, the Wheel, and Language*, 446, fig. 16.15). In the photo (fig. 4), no mold mark is visible, so this handsome dagger, or at least its highly detailed three-dimensional figurine-pommel, seems quite likely to have been cast by the lost-wax process. Similarly, the lost-wax method might also be applied to other "princely" daggers manufactured in a similar manner, as illustrated by Chernykh (*Ancient Metallurgy in the USSR*, figs. 73.16; 74.21; 75.5, 6; 77.1, 2.). Though Chernykh did not specify the fabrication process of these daggers in his book *Ancient Metallurgy in the USSR*, he did aver that lost-wax casting was mastered by the Seima-Turbino. According to Chernykh, some "unique *cire perdu*-cast figurines of animals and idols" (e.g., see *Ancient Metallurgy in the USSR*, 228, Plate 20), together with typical Seima-Turbino artifacts, were discovered in a hoard at Galich in the Upper Volga region (*Ancient Metallurgy in the USSR*, 203).
9. Anthony, *The Horse, the Wheel, and Language*, 444; Anthony, "The Sintashta Genesis," 65.
10. See Possehl, *The Indus Civilization*, 215–232; Possehl, "The Middle Asian Interaction Sphere," 42.
11. Anthony, *The Horse, the Wheel, and Language*, 435; Anthony, "The Sintashta Genesis," 65. In fact, Ivanova argued that far earlier than the period of BMAC, "lost-wax was practiced at sites in south Turkmenistan dating to the Namazga III period, late fourth millennium BC" (*The Black*

Sea and the Early Civilizations of Europe, the Near East and Asia, 289, note 106).

12. As Anthony stated, "the metalsmiths of the BMAC made beautiful objects of bronze, lead, silver, and gold. They cast delicate metal figures by the lost-wax process, which made it possible to cast very detailed metal objects" (*The Horse, the Wheel, and Language*, 425).
13. According to Chernykh, "complex beads and seals were generally *cire perdu* cast; such artefacts were never subsequently forged" (*Ancient Metallurgy in the USSR*, 182). Similar determinations are also found in Masson, *Altyn-Depe*, 100; Pittman, *Art of the Bronze Age*, 53; and Kuz′mina, *The Origin of the Indo-Iranians*, 211. Because no mold marks are recognizable in this large group of seals, it seems plausible that they were made by the lost-wax process. For pictures of these seals, see Pittman, *Art of the Bronze Age*, 54–57.
14. See Fitzgerald-Huber, "Qijia and Erlitou," 52–62.
15. Ibid., 40–65.
16. Like Fitzgerald-Huber, Chernykh also interpreted the Seima-Turbino phenomenon as the simultaneous movement of several Altaic mobile groups of metallurgists and horse-riders (*Ancient Metallurgy in the USSR*, 226–227). Following the logic of Fitzgerald-Huber and Chernykh, recently Lin has reconstructed how the Seima-Turbino arose in the Altai and Ob river valley area, moved westward to the Urals, shifted southward to the Xinjiang Uighur region and "herded horses" in the territories of the Qijia and Siba cultures around the Qilian Mountains, and penetrated into the Central Plains of China ("Saiyima-Tuerbinnuo Wenhua yu Shiqian Sichou zhi Lu," 50–59). The history of the Seima-Turbino culture group was probably not as neat as Lin has pictured.
17. Mei, "Qijia and Seima-Turbino," 41.
18. As Mei argued, "the early contacts between Northwest China and the Eurasian steppe appear to consist of a number of independent processes carried out during different periods and through various routes and intermediary links, involving not just Seima-Turbino and Qijia, but also other cultural groups associated with Okunev, Andronovo, Siba, Tianshanbeilu, and possibly Afanasevo and Machang" ("Qijia and Seima-Turbino," 40).
19. Mei, "Qijia and Seima-Turbino," 44–45.
20. Equally possibly, Qijia or Erlitou metalworkers could have done the "filtering" if lost-wax was ever known to them.

21. The major testimony of casting from archaeological materials in northwest and northern China is still stone molds, which are in consistence with the finds in the Eurasian Steppes.
22. In the earliest known account, from the first half of the thirteenth century CE, Zhao Xihu was already arguing that bronze vessels "in ancient times" (an imprecise appellation, probably referring to Bronze Age antiquity) were shaped directly from wax models and manufactured by the lost-wax process.
23. For a critical review, see Smith, *A Search for Structure*, 260.
24. Bushell, *Chinese Art*, 61.
25. Collins, "The Corrosion of Early Chinese Bronzes," 37.
26. Simpson, *Development of the Metal Castings Industry*, 24, fig. 20.
27. For details, see Yetts, *The George Eumorfopoulos Collection*, 35–38. The term "seam," which usually refers to a long, narrow space or gap, cannot adequately describe these linear, raised section-mold cast ridges. Seams appear concave, "mold marks" convex.
28. As Yetts asserted, "So far as my experience goes, all archaic Chinese bronze vessels have been cast by this process ('*cire perdue* process'); and I venture to suggest that the excavated moulds (or at least many of them) are really those which were used for shaping the wax models. Some of the mould fragments may have been bits of the fire-clay envelope to the wax model. The envelope is broken away after the metal has been poured and cooled; and, of course, such fragments show signs of burning" ("Recent Finds near An-Yang," 473). However, a careful study published by Karlbeck in the same year clearly showed that many of the mold fragments with scorch marks also had mortises and tenons ("Anyang Moulds"). Yetts's argument is refuted by Karlbeck's mold fragments.
29. Creel, "On the Origins of the Manufacture and Decoration of Bronze in the Shang Period," 68–69.
30. Su, "Ershi Shiji dui Xianqin Qingtong Liqi Zhuzao Jishu de Yanjiu," 405.
31. For fragments of clay molds, see Creel, "On the Origins of the Manufacture and Decoration of Bronze in the Shang Period," Plate 9.
32. Creel, "On the Origins of the Manufacture and Decoration of Bronze in the Shang Period," 69.
33. As Maryon and Plenderleith illustrated in their discussion of casting a bronze *zun*, a wax layer "of the thickness desired for the bronze" with individual edges "joined by heating" was attached in separate pieces over the surface of a wheel-shaped dried core for the lowest section of

the vessel. In a similar manner, clay cores, wax layers, and ornaments were fabricated before the construction of the wax runners and vents. Thereafter, the multi-layered mold was applied, dried, cut with "wedge-shaped mortises" along the edges, and lifted from the model in pieces. After the last-made pieces had thoroughly dried, the mold was opened and every part of the wax was physically *taken* away (not *melted* away). After the final inspection and any necessary repairs, the mold assembly was reconstructed with an additional coarse, thick layer, dried gradually, and baked hard before the form was filled with molten metal ("Fine Metal-work," 628–630). In addition, Maryon and Plenderleith considered their reconstruction to be a variation of the lost-wax process: "That these moulds were piece moulds demonstrates also that the founder wished to lift them from the model and to remove the wax by hand. Had he wished to employ the *cire perdue* process there would have been no need for the piece-mould—the mould could have been made as a whole" ("Fine Metal-work," 630). This excessively complex method, whether a variation of the lost-wax process or not, was impractical and unrealistic. The proposal of such an improbable method was likely due to the authors' deeply entrenched a priori belief in the use of wax models. At the same time, they must have known the already published Anyang materials that included the mold fragments: containing mortises and tenons, these fragments were hardly likely to be pieces broken from a wholly formed investment mold. They must have instead come from molds constructed in fitted sections.

34. Li, "Fushen Zang," 476; Su, "Ershi Shiji dui Xianqin Qingtong Liqi Zhuzao Jishu de Yanjiu," 388.
35. For critical analysis of this issue, see Bagley, *Shang Ritual Bronzes in the Arthur M. Sackler Collections*, 27–28.
36. Karlbeck, "Anyang Moulds," 42–43.
37. Preferring the section-mold theory, Karlbeck, being cautious and circumspect, still did not rule out the possibility of the direct lost-wax process: "There is also the possibility that they [the section molds] were used in the *cire perdue* method, the sections being formed round a wax model which was melted out and replaced by molten bronze" ("Anyang Moulds," 43).
38. Karlbeck, "Anyang Moulds," 44.
39. Ibid., 46.
40. See Guo, *Ruixian Yiqi*, 19; Rong, *Shangzhou Yiqi Tongkao*, 158.
41. Chen, "Yindai Tongqi Sanpian," 36–43.

42. Shi did not distinguish the model and core in many cases. The term Shi originally used in this context was *mo* ("the model"; see "Yin dai de Zhutong Gongyi," 127), although he was, in fact, referring to the core.
43. Shi, "Yin dai de Zhutong Gongyi," 109.
44. According to Shi, the core was applied with tallow or wax so craftsmen could easily tell whether the scraped thickness was consistent and adjust it as needed. The tallow or wax, in Shi's interpretation, could also stabilize the mold assembly; he believed that the dark, shiny layer he observed on the exteriors of some unearthed Anyang cores might have been left by tallow. As Shi presumed, tallow that remained during the casting process could facilitate the flow of the molten metal within the mold, a hypothesis which has been subsequently determined to be incorrect ("Yin dai de Zhutong Gongyi," 127). As Su has pointed out, the tallow would create a large amount of gas when contacting the hot molten bronze, a situation which would have seriously damaged the casting and even endangered the casters ("Ershi Shiji dui Xianqin Qingtong Liqi Zhuzao Jishu de Yanjiu," 409).
45. For details, see Shi, "Yin dai de Zhutong Gongyi," 110–118.
46. The surface ornament of this *he* is stylistically similar to that of the Erligang *ding* in fig. 9, but more "primitive"; its vessel shape also preserves the character of a still earlier stage. For image, see Davidson, "Toward a Grouping of Early Chinese Bronzes," 33, fig. 8.
47. Davidson, "Toward a Grouping of Early Chinese Bronzes," 51. As Bagley commented, "It should be added that thread relief is not a form of decoration likely to arise in lost-wax casting. The lost-wax caster must carve on the wax model or the finished bronze; the mold is closed and he does not have access to its inner surfaces" ("Shang Ritual Bronzes," 10).
48. Bagley, "Shang Ritual Bronzes," 11; 20, note 5.
49. Shi, "Yin dai de Zhutong Gongyi," 110–118.
50. In fact, Barnard elaborated the section-mold theory in an extremely cumbersome manner (a so-called "sectionalism") with the partitioning of an unrealistic number of mold sections (*Bronze Casting and Bronze Alloys in Ancient China*, 112–168). As Wilma Fairbank correctly commented, in Barnard's interpretation, "the number of horizontal mould sections often seems excessive, and the widespread use of detachable insertions seems overcomplicated" ("Review of Noel Barnard's 'Bronze Casting and Bronze Alloys in Ancient China'"). For comment on Barnard's "sectionalism," see also Su, "Ershi Shiji dui Xianqin Qingtong Liqi Zhuzao Jishu de Yanjiu," 400. It is worth mentioning that in his book, Barnard also

proclaimed that *cire-perdue* casting came to China sometime between 200 BCE and 600 CE, and that the section-mold process formed all previous castings (*Casting and Bronze Alloys in Ancient China,* 108–109). This rash argument will be disproved by my research, based on certain lost-wax castings from the Eastern Zhou period (see the following).

51. As Wilma Fairbank also stated, the flanges "along the vertical joins of the mold segments" were expressly meant to "delineate the basic piece-mold assembly of a bronze vessel"; "Evolved from the practical requirements of craftsmanship, they (the 'flanges') became a characteristic and unique aesthetic feature of early Chinese bronze design. But it should be kept in mind that their uniqueness derived from an ingenious technical solution to problems presented by a molding system unique at the time" ("Piece-mold Craftsmanship and Shang Bronze Design," 12–15). As convincingly as Fairbank has argued here, readers should be cautious, for the flanges did not originate as a way to conceal mold marks (for detailed discussion of this issue, see Bagley, *Shang Ritual Bronzes in the Arthur M. Sackler Collections,* 27–28).
52. For details, see Gettens, "Joining Methods in the Fabrication of Ancient Chinese Bronze Ceremonial Vessels," 205–217.
53. Gettens regarded the "evenly and symmetrically spaced narrow ridges of metal" on Chinese bronzes as "characteristic of any object cast in a piece mold and are caused by molten metal running into the narrow crevices marking the joins in the mold or in the build-up of the original model" (*The Freer Chinese Bronzes,* 60, 64). It is far too sweeping a judgement to consider any casting with the "mold-join marks" as indicative of "piece-mold" casting (considered as an alternative the possibility of the *indirect* lost-wax casting); Gettens's argument here applies to almost all early Chinese ritual bronzes.
54. Li, "Yinxu Chutu Qingtong Liqi zhi Zong Jiantao," 526–527.
55. Guo, *Shangzhou Tongqiqun Zonghe Yanjiu,* 124; Hua, "Shilafa de Qiyuan he Fazhan," 63–81.
56. Based on the study of archaeological stratigraphy and ceramic typology, "Erlitou Culture" can roughly be fitted between the Neolithic Henan Longshan culture (c. 2600–2000 BCE) and Erligang culture (c. 1500–1300 BCE). It is commonly divided into four phases. A series of radiocarbon determinations on samples showed that the Erlitou Culture from Phase 1 to Phase 4 should fall within the range of 1900 to 1500 BCE (with each phase roughly lasting 100 years for the sake of convenience). An alternate range of dates recently obtained from Erlitou and the contem-

poraneous site Xinzhai is 1750–1530 BCE, which probably only partly covers Erlitou Culture. For details, see Qiu et al., "Youguan Suowei 'Xia Wenhua' de Tanshisi Niandai Ceding de Chubu Baogao"; Bagley, "Shang Archaeology," 137–139; Liu and Chen, *The Archaeology of China*, 263, 266.

57. As Bagley observed, "elsewhere in the ancient world vessels and other simple metal shapes were invariably made by hammering, a technique far more economical of material. The cast vessels at Erlitou thus represent the first sign of a characteristically extravagant use of metal in China" ("Shang Archaeology," 141). It is understandable that certain weapons and tools were manufactured by casting, inasmuch as hammering, ingenious method that it was for saving metal, sometimes failed to produce objects of sufficient strength. Yet even simple vessels at Erlitou were also exclusively cast; hammering, like needlework, simply requires and even favors a single skilled craftsman (e.g., Benvenuto Cellini) with one or two assistants working from start to finish. But casting processes, which are more akin to an industry, require extensive teamwork and promote the division of labor. As Bagley remarked, "industrial organization, which has some claim to be a distinguishing feature of later Chinese civilization, may well be a legacy of the bronze foundries" ("Shang Archaeology," 141–142).

58. As an elaboration of bivalve mold casting, the section-mold process, known to Bronze Age societies everywhere, was not at all unique to Erlitou. For instance, Near Eastern metalsmiths had used it at a much earlier date, but did not truly rely on it because of their predominant use of hammer work and the complementary use of lost-wax casting, the latter mainly used to fabricate complex shapes that were difficult to raise by hammering. For details, see Bagley, "The Beginnings of the Bronze Age," 70; Bagley, *Shang Ritual Bronzes in the Arthur M. Sackler Collections*, 16.

59. Liu and Chen, *The Archaeology of China*, 272; Bagley, *Gombrich among the Egyptians*, 11. As perhaps the counterpart of Near Eastern statues of deities and kings, ritual vessels in early China must have held considerable symbolic meaning for socio-political purposes; they may have been used to feed the deceased ancestor-deities in exchange for power and protection.

60. For details, see Hwang, "Maixiang Zhongqi Shidai," 582–587. As Bunker inferred, the hollow, bulbous vessel legs of certain types of pottery in prehistoric China "could be individually molded," and this kind of "modular production" could facilitate and is manifested in the "complex sys-

tem of section-mold casting" ("The Beginning of Metallurgy in Ancient China"). Although Bunker here is basically correct, it is important to avoid Ledderose's abuse of the term "modular production" or "module" in describing mass-production in Chinese art (*Ten Thousand Things*).

61. This serves to explain the pervasive symmetry and regularity of bronze shapes following the Erlitou tradition, indicating the difficulty of using section molds for more irregular shapes. See Bagley, "The Beginnings of the Bronze Age," 71.
62. According to Bagley, "The *jue* almost certainly copies wrought-metal prototypes, a suggestion borne out not only by rivets imitated on the handle of a pottery *jue* from Erlitou but also by the rims of early bronze *jue*, which copy the appearance of a rim folded over and hammered down" (*Shang Ritual Bronzes in the Arthur M. Sackler Collections*, 25). For a similar discussion of the "skeuomorph" phenomenon, see also Fitzgerald-Huber, "The Relationship of the Painted Pottery and Lungshan Cultures," 207; and Hwang, "Maixiang Zhongqi Shidai: Zhutong Jishu de Shuru yu Zhongguo Qingtong Jishu de Xingcheng," 600–621. As Fitzgerald-Huber believed, the *jue* vessel originated outside of China and could be linked to the sheet-metal vessel type of Bactria-Margiana ("Qijia and Erlitou," 52–62). Either with the wrought-metal prototype as an antecedent or emulated from a design in another medium, the bronze *jue* vessels must have held great ritual and ceremonial significance for Erlitou elites.
63. Barnard inferred that the handle of the *jue* was precast ("Thoughts on the Emergence of Metallurgy in Pre-Shang and Early Shang China and a Technical Appraisal of Relevant Bronze Artifacts of the Time"), but as Su et al. pointed out, all the *jue* vessels from Erlitou were produced solely through casting (*Zhongguo Shanggu Jinshu Jishu*, 97).
64. For details, see Su et al., *Zhongguo Shanggu Jinshu Jishu*, 97–99.
65. See Bagley, "Shang Ritual Bronzes," 14.
66. Fairbank, "Piece-mold Craftsmanship and Shang Bronze Design," 12–15; Bagley, *Shang Ritual Bronzes in the Arthur M. Sackler Collections*, 18; Bagley, "Shang Ritual Bronzes," 12.
67. The Erligang ceramic culture, associated with two fortified settlements at Yanshi and Zhengzhou, respectively in the Yiluo Basin and the central region of Henan, had already become noticeable during the Erlitou Phase IV. Two levels have been distinguished in pottery assemblages, Lower and Upper Erligang, and in the Central Plains the Erligang-dominant period is roughly assigned to c. 1500–1300 BCE, fixing

it between the Erlitou (c. 1900 to 1500 BCE) and the Erligang-Anyang transition period (13th century BCE). See Bagley, "Shang Archaeology," 165. Alternatively, Liu and Chen have dated Erligang culture to approximately 1600–1400 BCE (*The Archaeology of China,* 278). Because Erligang culture co-existed with the perhaps still-dominant Erlitou culture for a period of time, and the Yanshi walled settlement of Erligang culture supplies a portion of radiocarbon dates "in the same range and some as late as 1400–1100 B.C." (Bagley, "Shang Archaeology," 165, note 53), the Erligang-dominant period in the Central Plains has been temporarily assigned to approximately 1500–1300 BCE.

68. See Gettens, "Joining Methods in the Fabrication of Ancient Chinese Bronze Ceremonial Vessels," 205–209. The bronzes Gettens analyzed in the Freer Gallery of Art were not necessarily Erligang bronzes, but examples of interlock casting that emerged no later than the Erligang period.
69. Two principal schemes of mechanical interlock are distinguished, respectively, by the terms pre-casting and casting on. In the former, one or more previously cast vessel components are inserted into the mold for the body of the vessel. In the latter, the body is cast first and molds for small parts are built against the body. For details, see Gettens, *The Freer Chinese Bronzes,* 76–98; Bagley, *Shang Ritual Bronzes in the Arthur M. Sackler Collections,* 42–43; and Bagley, "Shang Archaeology," 144.
70. As Bagley explained, the more complicated the shape, the more challenging and burdensome it is to withdraw mold sections from the model and construct the fitted assembly. In addition, the challenge of filling the mold with molten bronze without trapping pockets of air multiplies. The lost-wax process could solve the first problem, but not the second; the Erligang casters applied a stepped strategy that overcomes both, in which the joining is almost invariably achieved by a new pour that mechanically locks everything related into place ("Shang Archaeology," 42).
71. Barnard, "Origin of Bronze Casting in Ancient China," 8–9; Bagley, "The Zhengzhou Phase (The Erligang Period)," 99.
72. Su et al., *Zhongguo Shanggu Jinshu Jishu,* 97–99.
73. Bagley, "The Zhengzhou Phase (The Erligang Period)," 99.
74. Bagley, *Shang Ritual Bronzes in the Arthur M. Sackler Collections,* 42; Bagley, "Shang Archaeology," 144.
75. Spacers from Panlongcheng are usually around 5 cm × 9 cm, and irregularly shaped as quadrilateral, triangular, and polygonal. See Liu et al., "Jishu Xuanze he Jishu Fengge de Xingcheng," 208.

76. See Zhang, "Panlongcheng Shangdai Qingtong Rongqi de Chubu Kaocha," 46–47; Liu et al., "Jishu Xuanze he Jishu Fengge de Xingcheng," 208.
77. Meyers, "Casting Technology in Cambodia and Related Southeast Asian Civilizations," 40–41, note 23. As Su noted, Rutherford J. Gettens, an early discoverer of spacers, also inconsistently used the terms "spacer" and "chaplet" to refer to the same thing in *The Freer Chinese Bronzes* ("Ershi Shiji dui Xianqin Qingtong Liqi Zhuzao Jishu de Yanjiu," 433).
78. The Anyang Period (c. 1200–1050 BCE) began when the site known as Yin or Yinxu ("the Ruins of Yin"), located near the later walled city of Anyang, became an important city, corresponding to late Phase 1 and early Phase 2 of Yinxu in pottery seriation. See Bagley, "Shang Archaeology," 181. Compared to the horizon of Erligang, which might be regarded as a "territorial state," decentralization clearly occurred later during the Transition period (13th century BCE) as well as the following Anyang period. Because the king list of the "Shang" or "Shang dynasty" (real or mythical) recorded by the later historian Sima Qian (c. 145 or 135–86/85 BCE) was partially found in the Anyang oracle inscriptions, Anyang is usually associated with the "Shang" or "Late Shang," although the association of archaeological materials and received texts still calls for special caution. For details, see Bagley, "Shang Archaeology," 125; Schaberg, "Texts and Artifacts," 473; and Baines, "Civilizations and Empires," 100.
79. Bagley, "Anyang Mold-making and the Decorated Model," 44. Earlier but more detailed descriptions can be found in Bagley, *Shang Ritual Bronzes in the Arthur M. Sackler Collections*, 37–45; and Bagley, "Shang Ritual Bronzes," 7–19.
80. Against this "consensus view," Nickel argued that "undecorated models" were unchangeably in use "throughout the Bronze Age," and that "decoration was always applied directly to the outer mold and was usually designed in a way that would fit the mold sections." Nickel has mainly provided two reasons, what he calls "imperfect symmetry" in bronze ornament, and what he considers the insurmountable difficulty of decoration transfer from model to mold ("Imperfect Symmetry,"11–35). Bagley is a critic of Nickel's view ("Anyang Mold-making and the Decorated Model").
81. Soon after Nickel's publication, Dong Yawei made a similar argument about the use of undecorated models. Dong further proposed the adoption of "fractional models" in section-mold casting from the beginning of Bronze Age China. For instance, for a *jia* associated with the Erlitou

period, Dong argued that only a one-third model was required. From this fractional model, three identical mold sections could be produced, and later fitted together for the casting of the bronze. Similarly, according to Dong, a one-fourth model was applied to manufacture an Erligang-style square *ding*. This working manner, as Dong believed, was in constant use even during the Late Bronze Age. For details, see Dong, *Fanzhu Qingtong*, 4–6, 41–49. Dong's assertion of the application of fractional models lacks corroboration in archaeological records. As far as I know, none of the foundries in Bronze Age China, including Anyang, Houma, and all others, have ever yielded any physical evidence of the fractional model. Nor do the bronzes themselves offer any proof for this argument, except for the presence of the mold marks central to Dong's assertion. Yet mold marks can only suggest the use of section molds, and do not necessarily indicate that the mold pieces were serially shaped from a single fractional model. At the very least, during the early Bronze Age, none of the bronzes contained exactly identical sections in shape, size, or especially decoration. This negative evidence perhaps could be explained away by Dong's other argument, that decoration was always made directly in the mold. But a similar argument proposed by Nickel ("Imperfect Symmetry") has been convincingly disproved by Bagley ("Anyang Mold-making and the Decorated Model"). Hence, I suspect that Dong's view is simply an anachronistic perspective based on modern practice. As Dong himself admitted, the use of fractional models is prevalent today, but he traced this technical idea back to the Erlitou period by assuming that all early Chinese bronzes had been fabricated in this manner (*Fanzhu Qingtong*, 4). As Cyril Smith pointed out, "The greatest problem that a historian of techniques faces is, perhaps, to avoid jumping to the conclusion that an object was made in the way that he himself would find natural to make it!" (in Doeringer et al., *Art and Technology*, 51). The method that seems natural to modern casters, such as the application of the fractional model, might not have been natural for the ancient craftsmen. A comment by Bagley can be applied to Dong's argument: "For modern observers it is automatic to look for labor-saving ways of doing things, but we should beware of assuming that Bronze Age kings and Anyang casters shared our instincts. Evidence of a concern for *efficient* mass production in antiquity, in China or anywhere else, is very hard to come by" ("Anyang Mold-making and the Decorated Model," 84). *Efficient* mass production is the main motive for Dong's proposal.

82. Bagley, "Replication Techniques in Eastern Zhou Bronze Casting," 238.

83. See Keyser, "Decor Replication in Two Late Chou Bronze *Chien*," 127–162.
84. See Bagley, "Replication Techniques in Eastern Zhou Bronze Casting," 232–234; Bagley, "What the Bronzes from Hunyuan Tell Us about the Foundry at Houma," 217.
85. So, *Eastern Zhou Ritual Bronzes from the Arthur M. Sackler Collections*, 43, 46. As So commented, "The wide distribution of bronzes with Houma décor styles testifies to the foundry's large volume of production and to the penetrating influence of its manufacturing techniques. It is also evidence for an extensive trade network and for active artistic exchange between Jin and the states surrounding it in Shanxi" (*Eastern Zhou Ritual Bronzes from the Arthur M. Sackler Collections*, 43).
86. Because Houma artifacts identical in both shape and decoration are known, such as a bronze quadruped from Shanxi Hunyuan and two other precisely matched examples in Western collections, it is clear that reusable decorated models were used at Houma. For details, see Bagley, "What the Bronzes from Hunyuan Tell Us about the Foundry at Houma," 214.
87. It is evident that this was not the case because matched bronzes manufactured by the pattern-block technique share precisely the same shape and look-like pattern units from the same set of blocks, but not exactly the identical decoration. See Bagley, "What the Bronzes from Hunyuan Tell Us about the Foundry at Houma," 218. Apparently, the pattern units were applied twice individually to the mold instead of the model. One conceivable advantage for this strategy is the ease of withdrawing undecorated mold sections from a plain model. As Bagley explained, "Presumably, then, since there is no obvious reason for making two independent models, two moulds were being decorated. The caster must have begun with an undecorated model, formed two undecorated moulds on it, and then decorated the mould sections" ("What the Bronzes from Hunyuan Tell Us about the Foundry at Houma," 218).
88. Houma casters could proficiently replicate multiple bronzes from a decorated model. This process, with nearly one thousand years of history in China by the time of the Houma operation, also allowed the devising of patterns in accordance with the object and serial production from a single reusable decorated model. Thoroughly versed in using decorated models, Houma casters for some unknown reason preferred the newly developed pattern-block technique, which must suggest some particular

concerns. For details, see Bagley, "What the Bronzes from Hunyuan Tell Us about the Foundry at Houma," 214.

89. Bagley, "What the Bronzes from Hunyuan Tell Us about the Foundry at Houma," 222.
90. Ibid., 221.
91. In this sense, the pattern-bock technique was similar to the "prescriptive" process proposed by Ursula Franklin, composed of a sequence of operations by different specialized groups of workers. According to Franklin, "holistic" and "prescriptive" processes are two different types of organization of production: "the Holistic process is essentially a sequential, linear development" that involves "a single, step-wise approximation toward the final object." The prescriptive process "can be viewed as a sequence of unit processes," with each unit representing "an autonomous skill," and therefore it draws on "different groups of workers." All technologies at the beginning are holistic, but in sophisticated forms "the operational sequence in a prescriptive technology has been rearranged to increase productivity and efficiency," whereas "in a holistic technology, the basic production procedure of the craft is more or less preserved" ("The Beginning of Metallurgy in China," 96–97). See also Franklin, *The Real World of Technology,* 18–20; Li, "The Anyang Bronze Foundries," 290.
92. As Bagley discussed, to make the decorated mold section of the frontlet (where the back half of the ornament is flat and plain), moist clay had to be applied—perhaps by constant dipping in slurry—to the finely carved model and detached from it with great care. Disengaging the mold from the model with the sculptural protruding dragons seen in fig. 13 must have been extremely difficult, because the negative impression had to take the design intact in every detail. See Bagley, "Debris from the Houma Foundry," 51; Bagley, "Anyang Mold-making and the Decorated Model," 50.
93. Bagley, "Debris from the Houma Foundry," 58.
94. As Bagley noted, "[Houma craftsmen's] awesome skill is too easily taken for granted when we see only the finished product and not the means by which it was achieved. The carving on some pieces of the debris is sharper than on the finest bronze, and its effect is not adequately conveyed by even the best photographs" ("Debris from the Houma Foundry," 58).
95. Of course, the archaeological record is too incomplete to determine whether the Houma founders did not use lost-wax casting; it can only

be said that so far there is no direct evidence of its use at this site. Robert Murowchick, personal communication, August 2018.

96. With four intertwined phoenix-like birds and four dragon-like beasts, this table base was unearthed from the Hebei Pingshan Sanji Gongshe M1, a tomb of Cuo, king of the state Zhongshan, who died soon after 310 BCE. See So, *Eastern Zhou Ritual Bronzes from the Arthur M. Sackler Collections*, 60–61.
97. Hua, *Zhongguo Gudai Jinshu Jishu*, 200; Su, "Zhanguo Zhongshan Wang Cuo Mu Qingtong Qiqun Zhuzao Gongyi Yanjiu," 232–233, 262.
98. The earliest known cast inlay process was long thought to have come from Henan Xichuan Xiasi M2, a tomb dated to the second quarter of the sixth century BCE. See So, *Eastern Zhou Ritual Bronzes from the Arthur M. Sackler Collections*, 34. A recent discovery at Shandong Zaozhuang Xulou presents an even earlier example confirming the existence of precast inlays no later than the first quarter of the sixth century BCE. For details, see Zaozhuang et al., "Shandong Zaozhuang Xulou Dongzhou Mu Fajue Jianbao," 26–27; Hu et al., "Zaozhuang Shi Xulou Cun Hongtong Zhuxiang Qingtongqi Wenshi Xiangzhu Tezheng"; and Su and Wang, "Zaozhuang Xulou Chutu Zhuxiang Hongtong Qingtongqi Tanlun."
99. For instance, many of these techniques, including forging, repoussé, chasing, and granulation, are best evidenced in the tomb at Gansu Majiayuan in northwestern China during the late phase of the Warring States Period (475–221 BCE). See Huang et al., "Zhangjiachuan Majiayuan Mudi Chutu Jin Guanshi de Yanjiu"; Huang et al., "Majiayuan Mudi Jinzhipin de Chengfen yu Zhizuo Jishu Chubu Yanjiu"; and Mei et al., "Archaeometallurgical Studies in China," 227–28. A recent discovery from Hubei Zaoyang Caomenwan, dated to the early phase of the Spring and Autumn period (770–476 BCE), reveals the uses of repoussé, punching, and gilding in China centuries earlier. For details, see Su, "Hubei Zaoyang Caomenwan Zengguo Mudi Yihao Mu Chutu Jinshuqi Keji Jiazhi."

Chapter 3

Was Lost-wax Casting Used in Bronze Age China?

Recent intense scholarly debate over the use of the lost-wax process in Bronze Age China has focused on the bronze *zun-pan* set (fig. 15) from the tomb of Marquis Yi of Zeng (d. 433 BCE), discovered at Suizhou in Hubei Province in 1978. Of the more than four thousand bronze artifacts discovered in the Marquis's tomb, the *zun-pan* set is by far the most delicate and complex. The vessel set was likely created for an earlier marquis whose name was erased and replaced by that of Marquis Yi,[1] a fact that underscores the tremendous value of the *zun-pan* in the time of Marquis Yi.

In the *zun* (fig. 15a) of the *zun-pan* set, the rim appendage (fig. 16) has been widely believed to be the only portion made using the lost-wax technique.[2] It is composed of two physical layers connected to each other by irregularly shaped and spaced rods. The inner layer, which was not meant to be seen, is composed of a highly regular network-like structure, whereas the outer layer, the visible part, is composed of two rings, one central and the other distal. Each ring is further composed of thirty-two units, with each unit sitting alternatingly higher or lower than its neighbor. Each individual unit comprises multiple pairs of pattern elements. Small vertical rods support the pattern elements and ultimately connect to the horizontal rods that structurally support this surface pattern (fig. 17).[3] Traditional assessments have held that the rim appendage was composed of four lost-wax cast blocks that were subsequently soldered together. The reason behind this thinking is that the section-mold process could not cast even a quarter-section of such an openwork structure.[4] This traditional view similarly deems the handle-shaped appendages of the *pan* (see fig. 15b) to have been cast by the lost-wax process.[5]

Two objecting perspectives have recently countered this traditional understanding of the manufacturing process of the Zeng *zun-pan* set. *Perspective 1*, proposed by Zhou Weirong et al., argues that the rim appendage of the Zeng *zun* is composed of numerous minute section-mold castings soldered together.[6] The evidence given to support this view includes pronounced "soldering marks" and a series of "mold marks" and other traces supposedly left by the section-mold process in the manufacturing of precast individual segments (as in fig. 18).[7] This evidence has met with vigorous critique.[8] In fact, apart from a few soldering marks left by the joining of the four quarter-sections to the framework, all of the remaining purported soldering points, if they exist at all, seem far more likely to have been caused by the fusing together of pieces of a wax model. Similarly, other so-called "mold marks" that Zhou et al. have cited as evidence of section-mold casting could just as feasibly be the seams created in shaping the components of a wax model. One unanswered key objection to *Perspective 1*, as the scholar Zhang Changping has pointed out, is its failure to explain how one could solder over three thousand precast individual components within a space whose diameter is a mere twenty-five centimeters and whose height is only three centimeters.[9] *Perspective 1* gives no explanation for the mechanics of such an intricate operation, beyond the acknowledgment that "it must have been difficult."

Another objecting opinion, *Perspective 2*, proposed by Wang Jinchao, argues that each quarter of the rim appendage of the *zun* was cast in onc piece using a special kind of section-mold technique.[10] More specifically, manufacturing a quarter of the rim appendage required a preliminary model somewhat simpler than the finished bronze. From the model, five outer mold sections would have been made (fig. 19). The simple pattern elements transferred to these mold sections (fig. 20a) could be carved in the mold with the raised lines and dots that appear on the surface of the finished bronze (fig. 20b). Once the double-layered "inner mold" assembly (fig. 21)[11] was constructed within the outer mold pieces, the outer layer of the inner mold would be overprinted with the embellished design from the outer mold. Before pouring the bronze, space for casting would be reserved between the outer and inner molds, including numerous small casting cavities for pattern elements of the embellished design. At an earlier stage, however, the desired number of small "tunnels" had to be created within the double-layered inner mold through the use of a thin cylindrical tool that could penetrate the outer layer of the inner mold and leave a number of tiny "pits" on its inner layer (fig. 22). By connecting all such "pits," the newly produced horizontal "channels" (fig. 23), together with the vertical "tunnels" achieved earlier, could serve as runners to lead the molten bronze into the numerous casting spaces for the pattern elements. Once the liquid bronze solidified, the quarter of the *zun* rim appendage took its final shape.[12]

A series of details suggests that the method proposed by *Perspective 2*, ingenious though it may be, cannot be historically correct. First, on some deformed pattern elements of the rim appendage of the *zun*, the raised lines become lower and wider (fig. 24).[13] This phenomenon is readily explained if one assumes the use of the lost-wax process: handwork on the soft wax model could easily have led to the deformation of both its basic shape and its decoration. This phenomenon is difficult to reconcile with *Perspective 2*, which holds that the raised linear ornament of the pattern elements was produced by carving on the outer mold pieces. Furthermore, the ripple-like wrinkles on fragments from the handle-shaped appendages of the Zeng *pan* either were left by the liquid wax as it cooled[14] or, more likely, were formed from a series of small, staggered wax slices set in a laminated manner, whose purpose remains unclear (fig. 71). The irregular shapes and wrinkled surfaces of some of the rods within the rim appendage of the *zun* were also likely a byproduct of handwork on the wax model.[15] It is notable that many rods in the famous *jin* (altar table) from the mid-sixth century BCE tomb Xiasi M2 in Xichuan County, Henan Province (fig. 26), another probable lost-wax casting, lack consistent thickness (see figs. 33, 34b), probably because the melting components of the wax model were pressed and fused together.[16] In the method proposed by *Perspective 2*, such rods would have been produced in a tunnel-drilling manner, which cannot explain the inconsistent thickness of the rods. Moreover, an abundance of handmade marks, including rubbing and pinching imprints, are present on these same types of rods in the *jin* table, making the validity of *Perspective 2* still more unlikely.

The lost-wax casting method requires the use of a usually complex gate/venting system (fig. 25) composed of pouring and venting channels that were originally made of wax and attached to the model.[17] Once the casting was complete, these channels took the form of bronze rods that had to be removed from the finished vessel except when they were concealed inside it. In comparison to the section-mold process, the lost-wax process allows for easier construction of a greater number of runners and vents, which were needed to prevent the formation of air pockets in the casting of complex shapes.[18] A number of the bronze stumps that are evenly distributed on the back of the rim appendage of the *zun* are likely the remains of such a venting system, removed after casting. Some of the mistakenly identified "runners" for the "individually precast" pattern elements asserted by *Perspective 1* (fig. 18) are probably simply the scars created during the removal of the bronze strips (the original "vents") from the pattern elements on the finished piece. The Xichuan *jin* contained a similar irregularly shaped inner structure with many bronze rods in mild disarray, an obvious feature of the lost-wax venting system[19] (fig.

34), which would not have been produced had the method proposed by *Perspective 2* been used.[20]

At present, it seems clear that each quarter of the *zun* rim appendage was neither a combination of numerous precast individual components soldered together nor a single piece cast by the section-mold process. Detailed examination has further ruled out the possibility that the artifact was hammered into shape. It appears that the lost-wax process must have been used, and to this particular vessel set, its use might not be limited to the casting of the *zun* rim appendages and the *pan* handle-shaped ornaments (for further discussion, see the final paragraph of chapter 6). Because the Zeng *zun-pan* is not foreign stylistically, it seems likely that the lost-wax technique was used in China no later than the Late Bronze Age. In the next three chapters, this point will be developed with more examples of lost-wax casting from other sites of this period in China.

Notes

1. Hua and Jia, "Zenghou Yu zun-pan he shilafa de qiyuan," 352–359.
2. Zhang, "Guanyu Zeng Hou Yi Zun Pan Shifou Caiyong Shilafa Zhuzao de Shuping," 85–90, 52.
3. Hubei, *Zeng hou Yi Mu*, 228–231.
4. Hua and Jia (1983), 352–359.
5. Hubei, "Zenghou Yu zun-pan he shilafa de qiyuan," 229–231.
6. Zhou et al., "Zhongguo Qingtong Shidai Bu Cunzai Shilafa Zhuzao Gongyi" 80–85.
7. See Zhou et al., "Zhongguo Qingtong Shidai Bu Cunzai Shilafa Zhuzao Gongyi," 80–85; Zhou et al., "Zai Lun 'Shila Gongyi Bushi Zhongguo Qingtong Shidai de Xuanze,'" 41–48, 132; Dong et al., "Zaitan Zenghou Yi Zun-Pan de Zhuzao gongyi," 97–107.
8. E.g., Huang, "Zenghou Yi Zun-Pan Caiyong Shilafa Gongyi Zhuzao Wuyongzhiyi," 109–117; Tan, "Qianyi Gudai Shila Zhujian de Bianren," 118–124; Hua, "Zhongxifang Shilafa zhi Tongyi: Jianping 'Xianqin bu Cunzai Shilafa' yi Shuo," 87–96.
9. Zhang, "Guanyu Zeng Hou Yi Zun Pan Shifou Caiyong Shilafa Zhuzao de Shuping," 85–90, 52.
10. Wang, "Tan Zeng Hou Yi Zun Pan de Zhuzao Gongyi," 70–74; Wang, "Nifanfa Fuyuan Xugong Ning Loukong Panhuiwen Qingtong Shijian Yanjiu," 34–36.
11. Wang believed that the mold into which bronze was poured consisted of three parts, with the two inner parts collectively viewed as the "inner mold" (his original term is *xinfan*, "core mold"). In other words, both the inner and the middle parts of the whole mold assembly belonged to the "inner mold" and were described by Wang as its two constituent layers ("Tan Zeng Hou Yi Zun Pan de Zhuzao Gongyi"; "Nifanfa Fuyuan Xugong Ning Loukong Panhuiwen Qingtong Shijian Yanjiu").
12. For details, see Wang, "Tan Zeng Hou Yi Zun Pan de Zhuzao Gongyi," 70–74; Wang, "Zhongguo Qingtong Shidai Loukong Qingtongqi Nifan Zhuzuo Gongyi Yanjiu," 278–287, 190; Wang et al., "Sanwei Donghua li de Gudai Chuantong Gongyi," 113–122.
13. Zhou et al., "Zai Lun 'Shila Gongyi Bushi Zhongguo Qingtong Shidai de Xuanze,'" 41–48, 132.

14. For details, see Zhang, "Guanyu Zeng Hou Yi Zun Pan Shifou Caiyong Shilafa Zhuzao de Shuping," 85–90, 52; Zhang, *Zeng Guo Qingtongqi Yanjiu*, 262.
15. Hua, "Zhongxifang Shilafa zhi Tongyi: Jianping 'Xianqin bu Cunzai Shilafa' yi Shuo," 87–96.
16. See Li, "Xichuan Chunqiu Chumu Tongjin Shila Zhuzaofa de Gongyi Tantao," 39–45.
17. See Bagley, *Shang Ritual Bronzes in the Arthur M. Sackler Collections*, 44; Hemingway, "Bronze Sculpture," 37–46.
18. Tan, "Qianyi Gudai Shila Zhujian de Bianren," 118–124.
19. Hua, "Zhongxifang Shilafa zhi Tongyi," 87–96.
20. For *Perspective 2*, one question that has not been asked is: why would the Zeng caster have invented such an illogical and confused mold-making process? Wang Jinchao had a motive for inventing it: he was trying to replicate an object he had seen. But what was the Zeng caster's motive? Did his patron tell him: "Do something completely crazy for me?" In fact, as the strongest and most obvious objection to *Perspectives 1 and 2*, both Zhou et al.'s and Wang Jinchao's procedures are intensely complicated; nor do they explain why a Chinese caster would have used such a convoluted procedure. Did they believe that the Chinese caster did this *because he did not know the lost-wax process*? If so, they were trapped in the unrealistic conviction that the Chinese caster followed nonsensical procedures because he simply did not know any better. Robert Bagley, personal communications, August 2014 and December 2017.

Chapter 4

Further Confirmed Lost-wax Castings I

The Xiasi *Jin*

Beyond the Zeng *zun-pan*,[1] castings that demonstrate the use of lost-wax also include vessels from the Xichuan Xiasi cemetery of the sixth century BCE, an especially exquisite example being the bronze *jin* (altar table, fig. 26), which has multiple levels of highly intricate openwork structure on its four profiles and along the peripheries of its top and bottom. In addition, twelve each of two types of precast beast-shaped figurines with a less complex openwork decoration are appended around and underneath the *jin* body in a symmetrical manner.[2] The entire bronze *jin* consists of the table body, twelve laterally appended beasts, and twelve supporting beasts (see fig. 26c).[3] This particular *jin*, the largest one known,[4] measures 131 centimeters long, 67.6 centimeters wide, and 28.8 centimeters high and weighs 94.2 kilograms.[5] The *jin* body alone is 103 centimeters long, 46 centimeters wide, and 14 centimeters high.[6] The sheer size, elegant shape, superb decorations, and unusual fabrication process make this artifact unique and remarkable.

According to Zhao Shigang, the main author of the archaeological report *Xichuan Xiasi Chunqiu Chumu*, the *jin* body comprises twenty-five separate pieces, including one rectangular central panel and twenty-four precast panels (fig. 27): all of these precast panels, after being individually manufactured by the lost-wax process, were subsequently embedded in the two-piece mold assembly for the central panel.[7] The construction of the *jin,* therefore, involved twenty-five separate casting operations, with the twenty-four prepared lost-wax cast panels being locked in place during the casting of the central panel.[8] Yet Zhao believed that the twenty-four beast-shaped figurines were cast solely by the section-mold process and then soldered to the *jin* body.[9] Zhao understood this specific part of the process to start with the soldering together of four section-mold cast parts (head, tongue, tail, and the remainder) to constitute each laterally appended beast, with the two section-mold cast portions (tongue and the remainder) of each supporting beast soldered together.[10]

However, according to Ren Changzhong and Wang Changqing, who were in charge of the *jin*'s restoration, the *jin* body was cast entirely from a single wax model consisting of twenty-five wax pieces, including twenty-four decorated portions and a single undecorated piece for the central panel.[11] In Ren and Wang's conception, all of the beast-shaped figurines were also cast entirely by the lost-wax process and later soldered onto the *jin* body.[12] Li Jinghua also considered the entire body of the *jin* to have been cast from a single wax model that was comprised of twenty-five original pieces. The openwork decorations on the beast-shaped figurines, in Li's estimation, were also products of the lost-wax process, but he shared Zhao Shigang's view that the beast-shaped figurines themselves were cast without resorting to the lost-wax method.[13]

Despite the apparent differences, which will be discussed soon, these opinions have one aspect in common: the belief that the lost-wax process was used to cast the main body of the *jin.* This consensus has been reached for two main reasons. The first is the absence of mold marks typically associated with section-mold casting throughout the *jin* body.

The second, and more important, reason is the extreme intricacy of the multi-layered openwork structure that is integral to most parts of thc *jin* body. Such an openwork structure is commonly composed of five physical layers connected to each other by irregularly shaped and spaced rods. The innermost layer, with the thickest bronze rods (c. 1.0–2.3 centimeters in diameter), serves as the supporting framework for the whole structure. The outermost layer is composed of numerous spirals, often undercut ornamental elements.[14] Between the innermost and outermost layers are an additional three layers of bronze rods of varying thickness (c. 0.2–0.5 centimeters in diameter), which, as the archaeological report describes, constitute a network similar to the bracket-structure *dougong* of traditional Chinese wooden architecture.[15]

This intricate network, as most explicitly illustrated by Li Jinghua, is generally composed of, from innermost to outermost, *kuang geng* ("framework rods"), *gongxing geng* ("arched rods"), *zhi geng* ("straight rods"), *cheng geng* ("supporting rods"), and *huawen geng* ("ornamental elements") (fig. 28).[16] For each minimum grid unit of the framework (see four such grid units in the diagram fig. 29), the thickest *kuang geng* are connected with a number of arched *gongxing geng*, which further support four straight *zhi geng* that lie parallel to each other at a distance of approximately 2 to 2.5 centimeters (fig. 30). The ornamental elements or *huawen geng*, respectively in the shapes of "C" (hereafter referred to as the "人-shaped ornament"), and "X" (hereafter "X-shaped ornament"), are constructed over the *zhi geng*, and they at times require extra support from the slender *cheng geng*, which also connect to the base of the *zhi geng* (figs. 31, 32).[17] Within a space a mere five centimeters in height, this intricate five-layered openwork (fig. 33) is quite unlikely to have been cast entirely by the section-mold process.[18] It also would have been impractical, if following the section-mold casting hypothesis, to solder precast individual pieces together within such a narrow space; furthermore, no corresponding soldering marks have ever been identified.[19] It is thus certainly the case that the lost-wax process was used to create the *jin* body.

The next question is whether the *jin* body was cast as a single piece or in multiple pieces. As mentioned earlier, Zhao Shigang argued in the archaeological report that the *jin* body comprises twenty-five separate pieces, including twenty-four precast pieces individually manufactured by the lost-wax process. Zhao's argument is based mainly on his observations of the bronze rods distributed at the corners of the *jin* body:

> first, *cuo feng* ("seams formed in error") appear inside the *jin* corners, that is, the bronze rods, whether thick or thin, are sometimes unconnected or just barely connected; second, the bronze rods at the *jin* corners are comparatively thin; third, occasionally bronze rods at the *jin* corners are unconnected or are encrusted with impurities.[20]

To assist readers, Zhao provided a photograph illustrating the inner structure of a corner of the *jin* (fig. 34a). Zhao's statement, though partly incomprehensible, can be summarized as follows: bronze rods at the *jin* corners, with noticeable inclusions of other materials, are usually thinner than, and occasionally unconnected to, the surrounding bronze rods. Zhao used this observation as proof to support his assumption that the *jin* body is made up of multiple separately cast pieces. According to Zhao, the inconsistency or disconnection of the bronze rods suggests that the molten bronze could not pass through the corresponding spaces, with the result that the body of the *jin* could not be possibly have been cast as a single piece by filling its entire mold cavity with molten bronze.[21]

Ren Changzhong and Wang Changqing, on the contrary, held that the *jin* body was cast in its entirety from a single model, which was made by joining twenty-five wax pieces together. As Ren and Wang noted, a series of "discontinuous" or "irregularly defined" boundary lines (indicated by the dotted lines in fig. 26c) throughout the *jin* body could divide the whole object into twenty-five well-organized pieces.[22] These boundary lines, obvious to all, provide the common theoretical basis for dividing the *jin* body into twenty-five original pieces. The question now is whether the twenty-five original pieces were in wax or bronze. Zhao

Shigang, as mentioned earlier, upheld the latter view by interpreting these boundary marks as having been produced during the casting of the central panel in a two-piece mold in which precast lost-wax pieces were already embedded.[23] Ren and Wang maintained the former view by arguing that these marks were formed by joining wax pieces together with additional wax.[24] Specifically, as Ren and Wang surmised, when the twenty-five wax pieces were formed, each individual piece was invested fully with fine slurry, except for the two edges. After being air-dried, the twenty-five wax pieces were assembled into a single model by joining neighboring pieces with sticky wax along the edges. Slurry would then be invested again on areas of the wax paste, covering the wax joins just made.[25] While plausible at first glance, this reconstructed procedure seems a bit superfluous in practice, and it does not do a good job of explaining why the boundary lines on the *jin* surface are "discontinuous" or "irregularly defined."[26]

Similarly, Li Jinghua also considered these boundary lines, which differ visually from the section-mold casting marks, to be traces of wax pieces that fused together. The "fusing" procedure, as Li inferred, involved inserting a hot metal sheet into the tiny space between two wax pieces: as the wax melted, the two pieces could be fast-fused together after the metal sheet was extracted.[27] As Li explained, during this procedure, a small amount of liquefied wax might have been pushed out of the interspace area, forming a series of "fusing marks" after the cast bronze had solidified, and these marks could be mistaken for section-mold marks.[28] Compared with Ren and Wang's theory, Li's proposal more reasonably accounts for "discontinuous" or "irregularly defined" boundary marks. According to Li, this wax-fusing operation was not only applied to assemble the twenty-five main wax blocks but was also frequently used to connect numerous rods and ornamental elements within each openwork structure. The wax-fusing marks, which by and large have been polished off the exterior surface of the final piece, are better preserved within the openwork structure.[29] As Li pointed out, they commonly have *yuan leng* ("rounded edges"), which are visually distinct from the section-mold casting marks with sharp *fei leng* ("flying edges").[30]

Li Jinghua also found irregularities in bronze rods in the vicinity of the boundary marks; in particular, as Li noted, ornamental elements are occasionally dislocated, creating a large gap in the décor.[31] In addition, in such cases a new ornamental element was directly added to the model,[32] which would be difficult to explain had the *jin* body not been entirely cast from a wax model. It seems clear, then, that the body of the bronze *jin* was cast as a single piece from wax. But how to answer the aforementioned question raised by Zhao Shigang: why are the bronze rods at the corners of the *jin*, with noticeable impurities, usually thinner than, and occasionally unconnected to, the surrounding bronze rods? If the bronze rods are sometimes unconnected at the *jin* corners, suggesting that the bronze fluid was not able to pass through the corresponding spaces, how could the mold cavity of the entire *jin* body have been successfully filled by the molten bronze?

During their restoration work, Ren Changzhong and Wang Changqing identified twelve pairs of funnels (or sprues) and risers distributed at regular intervals throughout the *jin* body,[33] which was apparently cast upside down. A funnel or a riser occurs at each of the twenty-four joins where the beast-shaped figurines attach to the *jin*; funnels are set at all four profiles of the *jin* body, and risers appear underneath the *jin* body, both deliberately preserved as protruding tenons (see figs. 26c, 27). As Li Jinghua commented, this design is highly ingenious, because the funnels or risers could also be used as devices to discharge the wax from the mold and to facilitate the attachment of the beast-shaped figurines to the *jin*.[34] Most importantly, these funnels and gates made it possible to cast the *jin* body as a single piece, because molten bronze could be poured into the mold simultaneously from twelve separate funnels.[35] Particularly at locations at which the molten bronze would have had to make angular turns, such as at the *jin* corners, the bronze might not have been able to flow through successfully, prompting the phenomenon proposed by Zhao Shigang: the incorporation of multiple funnels and risers that could ensure that the entire mold cavity of the complex *jin* body was filled in a satisfactory manner.

More evidence in support of lost-wax casting has been observed by Li Jinghua, and it reveals further details of the manufacture of the *jin* body. First, Li has observed marks on some ornamental elements of the *jin* surface that were apparently made by round-ended tools on a meltable medium.[36] In addition to the marks left by tools, Li has discovered traces of working the wax with the fingers. For instance, pinch marks are found on the lateral faces of some ornamental elements, either in a half-moon shape when the pinching was comparatively light or in a shape similar to an elongated fish-scale when the action was more forceful.[37] Moreover, both the rods within the openwork and those that constitute the ornamental elements are generally thicker in the neighborhood of joins (see figs. 33, 34b), indicating that pinching and squeezing by hand probably facilitated the fusing of wax pieces.[38] Finally, Li also detected that bronze rods or ornamental elements of the same type always seem to have differed slightly in shape, size, and decoration, which suggests that the wax model was shaped mainly by hand.[39] As reasonable as Li's argument is, one cannot rule out the possibility that wax-shaping molds were applied in the manufacture of the wax model components, such as the aforementioned "C-," "人-, " and "X-" shaped ornamental elements. In this case, mold-shaped wax pieces could have been further worked by hand, with the result that different *jin* components of the same type might not have been produced in completely uniform shapes and dimensions. The possible advantages of using wax-shaping molds would have been to save labor and to help craftsmen produce elements that matched.

The final question is, how were the beast-shaped figurines cast and attached to the *jin* body? As mentioned earlier, Ren Changzhong and Wang Changqing argued that all the beast-shaped figurines were cast entirely by the lost-wax process and later soldered onto the *jin* body. This proposal fails to explain why many similar organic parts, especially the beasts' tongues, had broken off when the *jin* was discovered.[40] The frequent and repeated detachment of similar parts renders Ren and Wang's argument that the beast-shaped figurines were all entirely cast as a single piece insecure. Metal spacers, devices commonly used in section-mold casting, have furthermore been identified on the figurines.[41] This fact further indicates the unreliability of another of Ren and Wang's arguments, namely that all these figurines were manufactured by the lost-wax process.[42]

Zhao Shigang, as mentioned earlier, held a fundamentally opposing view. He believed that the beast-shaped figurines were cast in multiple pieces without the use of the lost-wax process. Specifically, according to Zhao, each laterally appended beast consists of four separately cast parts (head, tongue, tail, and the remainder) soldered together: the head required a core and a three-piece outer mold; the tongue, only a bivalve or two-piece mold; the tail, an inner mold and a two-piece outer mold; the rest an inner mold and a three-piece outer mold.[43] Each supporting beast comprises two section-mold cast parts (the tongue and the remainder) soldered together: the tongue required a bivalve mold for casting; the remaining portion an inner mold and a three-piece outer mold.[44]

Compared with Ren and Wang's argument, Zhao's analysis seems a more reasonable explanation for the fact that many similar essential parts of the beast-shaped figurines had broken off. Gray or black soldering materials have been identified between each figurine and its openwork appendages, as well as between different parts of the same figurine.[45] The traces of solder remain chiefly in the mortises, which with their corresponding tenons could facilitate the effective soldering of separate castings. As mentioned earlier, the twenty-four projections left by the funnels and risers were later used as tenons. These tenons, inserted into mortises dug into the clay cores of the beast-shaped figurines, could join these figurines to the *jin* body through soldering, as appropriate.[46] As time went by, the preserved red-colored clay core within each figurine became damp and loosened, damaging the corresponding mortise-and-tenon join and subsequently leading to the not infrequent detachment of figurine segments.[47]

Zhao's view, that the beast-shaped figurines were cast primarily by section-mold casting, can also be demonstrated to be correct. First, the relatively simple and strongly two-dimensional openwork of the figurines does not introduce any particular difficulty for section-mold casting.[48] Second, marks caused by the assembly of mold pieces for the section-mold castings, although overlooked by Ren Changzhong and Wang Changqing,[49] have been reliably identified on the beast-shaped figurines.[50] A bivalve mold mark is also clearly preserved along each of the detached tongues of these beasts.[51] Finally, as mentioned earlier, spacers commonly used in section-mold casting have also been identified with the figurines.[52] Yet as reasonable as Zhao Shigang's argument is, his statement is still somewhat ambiguous. This is especially apparent where Zhao utterly failed to explain the casting procedure of the openwork appendages of the figurines: the reader is left to wonder if they were cast individually or together with the corresponding figurines, and if they were individually cast, if the lost-wax process might have been used.

Judging by the fact that most openwork appendages of the beast-shaped figurines fell off,[53] they must have been cast separately. Though less complex than the decoration of the *jin* body, they are similarly intricate in structure. Ren Changzhong and Wang Changqing made the following observations: a decorative piece from the tail of one laterally appended beast is shaped like an open umbrella, with the ornamental elements on the surface serving as a "canopy" and the bronze rods beneath it serving as the "ribs" and "handle." The "handle," formed by the thickest bronze rod, supports two comparatively thin "inner ribs," which further sustain a number of curling or coiled slender "outer ribs." These "outer ribs," finally, hold up the "canopy," and the ornamental elements, taken together as the "canopy," are separated from each other without exception, supported only by the "outer ribs" below.[54] As Ren and Wang indicated, these "handles" and "ribs," exhibiting as they do an uneven thickness and traces of handwork left by rubbing and pinching actions, are likely to have been shaped in wax. The intricately structured and spaced "ribs," variously sinuous and twisted, without any identifiable section-mold marks on their surfaces, further support the possibility of lost-wax casting.[55] In fact, the openwork appendages of the beast-shaped figurines mostly have such an umbrella-like structure.[56] This structure, which is little different from the openwork of the *jin* body in constitution and style, was also likely fabricated by lost-wax casting.

In sum, the manufacture of the bronze *jin* involved lost-wax casting, section-mold casting, and soldering: the *jin* body was most likely cast from a single wax model consisting of more than twenty fused wax blocks, including twenty-four decorated panels, in addition to one (or, according to some scholars, four) plain piece(s) for the central panel.[57] The decorated wax panels with the intricately shaped and spaced rods were constructed primarily by hand through the direct manual manipulation of materials and simple tools, such as those used to fuse wax pieces. Wax-shaping molds might also have been used for repeated units of ornament. The twenty-four beast-shaped figurines were all cast by the section-mold process in multiple pieces that were then soldered together, but their openwork appendages were probably fabricated by the lost-wax process. The soldering of separate castings was usually facilitated by mortise-and-tenon devices, wherein a projection (tenon) was fitted into a close cavity (mortise) that had been filled with pasty solder. Through this type of soldering, different parts of the figurines were joined both metallurgically and mechanically. Finally, the figurines were attached to the *jin* body with the same soldering technique.

Notes

1. For the process applied to cast other parts of the Zeng *zun-pan* beyond its rim appendages, see Zhang, "Guanyu Zeng Hou Yi Zun Pan."
2. Henan et al., *Xichuan Xiasi Chunqiu Chumu*, 386.
3. Li, "Xichuan Chunqiu Chumu Tongjin Shila Zhuzaofa de Gongyi Tantao," 39.
4. Hua, *Zhongguo Gudai Jinshu Jishu*, 176.
5. Henan et al., *Xichuan Xiasi Chunqiu Chumu*, 386.
6. Ibid.
7. Ibid.
8. This is, in fact, a logic interpretation from the somewhat confusing description of the archaeological report. The original Chinese text, itself too muddled to translate into English, is "從禁身背後的範痕跡看，它並不是一塊蠟模所鑄，而是由二十四塊失蠟模鑄型和二塊普通泥範組裝澆鑄而成。它們的結構是：中心長方形面板上下範各二塊，面板周圍前後各四塊，左右各二塊，禁體四壁上前後各四塊、左右各二塊。各範在轉角相接處，製成45°角，以便於鑄型的拼合." See Henan et al., *Xichuan Xiasi Chunqiu Chumu*, 386.
9. Henan et al., *Xichuan Xiasi Chunqiu Chumu*, 386.
10. Ibid., 387.
11. Ren and Wang, "Henan Xichuan Xiasi Chunqiu Yunwen Tong Jin de Zhuzao yu Xiufu," 474–475.
12. Ibid., 476–477.
13. Li, "Xichuan Chunqiu Chumu Tongjin Shila Zhuzaofa de Gongyi Tantao," 39, 43–44.
14. See Henan et al., *Xichuan Xiasi Chunqiu Chumu*, 386; Li, "Xichuan Chunqiu Chumu Tongjin Shila Zhuzaofa de Gongyi Tantao," 41.
15. Ibid.
16. Li, "Xichuan Chunqiu Chumu Tongjin Shila Zhuzaofa de Gongyi Tantao," 41–42.
17. Ibid., 43.
18. Ibid., 42.
19. See Ren and Wang, "Henan Xichuan Xiasi Chunqiu Yunwen Tong Jin de Zhuzao yu Xiufu," 474.

20. The original Chinese text is "一是轉角處內部有錯縫，即粗細銅梗有的不相連接或連接很少；二是銅梗在轉角處較細；三是轉角處有的銅梗中間不相連接或混有雜質. " See Henan et al., *Xichuan Xiasi Chunqiu Chumu*, 386.
21. Ibid., 386.
22. Ren and Wang, "Henan Xichuan Xiasi Chunqiu Yunwen Tong Jin de Zhuzao yu Xiufu," 474.
23. As Zhao stated, "except the *fan feng* ('mold-seams') that divide the bronze *jin* into sections, elsewhere the section-mold *fan feng* mark is never discovered." See Henan et al., *Xichuan Xiasi Chunqiu Chumu*, 386. Strictly speaking, it is not proper to call the typically raised or convex section-mold marks *fan feng*, a term that usually refers to concave linear features.
24. Ren and Wang, "Henan Xichuan Xiasi Chunqiu Yunwen Tong Jin de Zhuzao yu Xiufu," 474–475.
25. Ibid.
26. Following Ren and Wang's theory, wax paste could have been applied in a systematic and orderly fashion.
27. Li, "Xichuan Chunqiu Chumu Tongjin Shila Zhuzaofa de Gongyi Tantao," 41, 44.
28. Ibid., 44.
29. Ibid. For visible wax-fusing marks, see Li, "Xichuan Chunqiu Chumu Tongjin Shila Zhuzaofa de Gongyi Tantao," 41, Figs. 5, 6.
30. Li, "Xichuan Chunqiu Chumu Tongjin Shila Zhuzaofa de Gongyi Tantao," 42.
31. Ibid.
32. Ibid.
33. Specifically, if the *jin* body was cast from a wax model that comprised one central board and twenty-four decorated panels, a funnel would be set around the boundary line of each two adjoined wax panels and a riser around the central area of each wax panel. See Ren and Wang, "Henan Xichuan Xiasi Chunqiu Yunwen Tong Jin de Zhuzao yu Xiufu," 474–475.
34. Li, "Xichuan Chunqiu Chumu Tongjin Shila Zhuzaofa de Gongyi Tantao," 44.
35. However, Li did not specify exactly how he envisioned this scene: did twelve casters with twelve separate crucibles crowd around the *jin* table? This does not sound quite practical, but is still possible in operation.
36. Li, "Xichuan Chunqiu Chumu Tongjin Shila Zhuzaofa de Gongyi Tantao," 43.

37. Ibid., 51.
38. Ibid., 43.
39. Ibid., 42–43.
40. Ren and Wang tried to explain this phenomenon by suggesting that the figurines were shaped in wax with multiple original pieces "fused" together. Careless wax-fusing operations led many parts of the figurines to break off. See Ren and Wang, "Henan Xichuan Xiasi Chunqiu Yunwen Tong Jin de Zhuzao yu Xiufu," 476. This phenomenon is too universal, however, to be explained by accidental fusing.
41. Li, "Xichuan Chunqiu Chumu Tongjin Shila Zhuzaofa de Gongyi Tantao," 44.
42. For details, see Ren and Wang, "Henan Xichuan Xiasi Chunqiu Yunwen Tong Jin de Zhuzao yu Xiufu," 476–477.
43. Henan et al., *Xichuan Xiasi Chunqiu Chumu*, 387.
44. Ibid.
45. Li, "Xichuan Chunqiu Chumu Tongjin Shila Zhuzaofa de Gongyi Tantao," 43. Ren Changzhong and Wang Changqing, during their restoration work, also recognized remains of lead-tin alloy in the mortises between some supporting beasts and their corresponding openwork decorations. See Ren and Wang, "Henan Xichuan Xiasi Chunqiu Yunwen Tong Jin de Zhuzao yu Xiufu," 477.
46. Li, "Xichuan Chunqiu Chumu Tongjin Shila Zhuzaofa de Gongyi Tantao," 44–45.
47. Ibid., 43.
48. Ibid.
49. Ren and Wang, "Henan Xichuan Xiasi Chunqiu Yunwen Tong Jin de Zhuzao yu Xiufu," 476–477.
50. Li, "Xichuan Chunqiu Chumu Tongjin Shila Zhuzaofa de Gongyi Tantao," 43. In fact, Zhao Shigang's detailed reconstruction of the section-mold assemblies for all twenty-four figurines should primarily refer to the preserved section-mold marks. See Henan et al., *Xichuan Xiasi Chunqiu Chumu*, 387.
51. Li, "Xichuan Chunqiu Chumu Tongjin Shila Zhuzaofa de Gongyi Tantao," 43.
52. Ibid., 44.
53. Ren and Wang, "Henan Xichuan Xiasi Chunqiu Yunwen Tong Jin de Zhuzao yu Xiufu," 476–477; Su, "Anyang Yinxu Qingtong Jishu Yuanyuan de Shangdai Nanfang Yinsu," 406.

54. Ren and Wang, "Henan Xichuan Xiasi Chunqiu Yunwen Tong Jin de Zhuzao yu Xiufu," 477.
55. Ibid.
56. Ibid., 476–77.
57. Most scholars, such as the aforementioned Ren Changzhong, Wang Changqing, and Li Jinghua, believe that the wax model of the *jin* body consisted of twenty-five pieces, including twenty-four decorated panels and one plain one for the central board. According to Hua Jueming, however, the wax model of the *jin* body consists of twenty-eight pieces, including twenty-four decorated panels and four plain ones for the central board (Hua argued that the uneven surface of the central board was caused by the polishing of the fusion marks). For details, see Hua, "Shilafa de Qiyuan he Fazhan," 63–81; Su et al., *Zhongguo Shanggu Jinshu Jishu*, 311. These two proposals, as Hua later commented, do not differ much in principle. See Hua, *Zhongguo Gudai Jinshu Jishu*, 177.

Chapter 5

Further Confirmed Lost-wax Castings II

Certain "Diatrete" Vessels and Mirrors

In Bronze Age China, the Xiasi *jin* seems to be a unique example that was cast primarily by the lost-wax process. For other confirmed lost-wax castings, this process is proven only in cases of highly ornamental parts or their sculptural appendages. For instance, some of the artifacts termed "diatrete" by Helen Loveday, with a very particular double-walled structure composed of "an outer shell of openwork ornamentation set over a smooth inner shell,"[1] illustrate this point—the lost-wax process was used, but probably only to cast openwork outer shells.

The Xuyi *Hu*

Among the diatrete vessels, the finest and most thoroughly studied piece is the *hu* vessel excavated from Xuyi Nanyaozhuang, Jiangsu Province (fig. 35). The Xuyi *hu*, looted from the state of Yan by the Qi general Chen Zhang after a successful attack, was most likely manufactured in the second half of the fourth century BCE.[2] This exquisite *hu* is 24 centimeters high and 22.2 centimeters in diameter at its shoulder, and weighs 5590 grams.[3]

The vessel has a long-curved neck, a rotund body, a tall circular base, and an outer shell of openwork around its inner wall. This outer shell may have once extended from the vessel shoulder approximately all the way down to the rim of its circular base; the lowest reaches of this décor are damaged over a broad area, revealing a large portion of the inner shell.[4] The outer shell is composed of a multi-layered mesh network of coiled and three-dimensional interlaced serpents punctuated by rows of five-petal flowers, which may represent plum blossoms.[5] A horizontal hoop immediately below the shoulder divides the network into two parts: an upper part surrounding the shoulder (see fig. 35b) and a lower part covering the belly. Both parts of the network have been constructed with more or less the same type of three-layered structure: the vertical rods, representing the undulating "serpent bodies," can be regarded as the middle layer; the internal horizontal rods that connect alternate serpents, also with a series of undulating curves, as the inner layer; and the external ornamental plum blossoms as the outer layer.[6] The question now is how was this network-like structure cast?

With a tricky and heavily undercut multi-layered structure that recalls the Zeng *zun-pan*, the openwork of the Xuyi *hu* could not, in honest estimation, have been cast (or *precisely* replicated) by any imaginable version of the section-mold process.[7] This is especially evident in that the overhanging layers of the openwork do not typically touch each other (e.g., see fig. 35b), and the enclosed spaces trapped in between would not have allowed the withdrawal of mold sections, however skillfully the process might have been carried out. This type of feature, tentatively termed here "absolute undercuts," is usually a reliable sign of lost-wax casting.[8] Such absolute undercuts also increase exponentially the difficulty of using the lost-wax process in terms of the construction of the gate/venting system.[9] This particular piece implies, accordingly, that casting the openwork as one piece would have been a highly formidable task. In an alternative possibility, the bronze openwork could have been cast in separate pieces by the lost-wax process and subsequently joined together. Considering the likelihood of these alternatives raises the question of how many pours of bronze were required to cast this openwork.

Careful observation suggests that both the upper and the lower parts of the openwork can be visually divided into eight sections, each with more or less the same pattern. In places, a small, filled gap exists between two neighboring sections, indicating that not every section equals exactly one eighth of the circumference.[10] The openwork itself is almost perfectly circular,[11] showing the meticulous preparation of a complete wax model, probably with painstaking modifications. However, it is not completely clear how the entire wax model was shaped or whether the model would have been sectioned for casting.

In any case, the openwork, containing as it does intricately entwined and three-dimensionally spaced wax rods, could not have been entirely produced from molds by any means. The fabricators of this basket-like "woven" design must instead have turned to free-hand operations. These rods, of a similarly shaped profile and proximate thickness, might first have been formed in molds before being used to construct the forms, but this is only a possibility.[12] The plum blossoms also seem to have been formed in molds, judging by their almost identical shape consisting of five petals and a central stamen and highly uniform surface decorations.[13] It is quite likely that all these plum blossoms were produced in wax in a single mold or a series of molds. It is possible that a device was used to ensure that all the individually hand-constructed openwork sections were matched in size and pattern.[14] This seems to be the most reasonable process for making the wax model for the openwork, but verification of this conclusion still awaits further evidence.

The wax model, as a whole, may still not have been ready for casting at this stage, because its intricate structure would have required the construction of a correspondingly complicated wax-gate system. Superb craftsmanship would have been essential in any attempt to cast the openwork whole. Instead, craftsmen could have chosen to partition the wax model for casting, perhaps in the same sections (i.e., one eighth of the model) that had been originally constructed. If this were the case, then one would expect to find a series of joining traces, either mechanical or metallurgical, where separately cast sections had been assembled. No such traces have been recognized thus far. In fact, the possibility of casting the whole openwork in a single piece cannot be completely ruled out, considering that the body of the *jin* from Xichuan Xiasi was probably cast whole. The extremely skilled technical work represented by the *hu* from top to bottom also strengthens my estimation that the openwork could have been cast in relatively few pieces. How many parts does this openwork comprise? This key question still awaits further investigation.

Despite the network-like ornamentation, the remaining portions of the Xuyi *hu* do not themselves appear to be difficult candidates for section-mold casting. Among them, the most challenging parts for the section-mold process to cast were the four ring-holder masks. Virtually identical in shape, size, and decoration even in the smallest of particulars,[15] the four masks (see fig. 35c) were almost certainly modeled as a group after the same design. One possibility is that they were all cast within sectional molds, with each set of molds being taken from the same model. Twelve teeth and a long, floating tongue have been identified within each mouth of the two serpents on either side of the main mask motif (fig. 35c).[16] Casting the whole design in a single piece would have been extraordinarily challenging with the section-mold process.[17] A more likely scenario would have been that the tongues were individually cast and then attached to the serpents' mouths, either metallurgically or mechanically. In this case, casting the ring-holder mask in a single pour would not have proven especially difficult with the lost-wax process. One or multiple wax-shaping molds taken from the same original model could have guaranteed that the four ring-holder masks were practically identical. The serpents' tongues (and perhaps also the teeth) could have been separately produced in wax and attached in place before the investment casting. The lost-wax process, of course, also allowed the association of separate castings in metal if each ring-holder mask was not a single piece. No definite conclusion can be drawn here before further examination, except that the four ring-holder masks, as a group, were serially modeled after the same design.[18]

The Metropolitan Tripod

Another diatrete vessel that has received scholarly attention is the unprovenanced spherical tripod purchased in 1947 by the Metropolitan Museum of Art in New York (fig. 36). At 21.9 centimeters high, this vessel is most noteworthy for its oblate openwork with intertwining serpent patterns and its five three-dimensional sculptural birds, one large and four small, all with wings outstretched and heads in a flying posture.[19] Discussion has centered on whether the piece is a hodgepodge,[20] whereas the authenticity of individual vessel parts, especially the double-shelled vessel body, has undergone little scrutiny. Similar to the Xuyi *hu*, the outer openwork shell of the Metropolitan vessel is also composed of two parts, upper and lower. These two parts are of a similar size and of the same hemispherical shape, and they meet at the center of the vessel body.

Close observation of the vessel in the storage room of the Metropolitan Museum provides useful information.[21] In particular, it verifies the fully three-dimensional nature of the openwork, which clearly presents the aforementioned absolute undercuts, separating the overhanging layers and creating an enclosed space between them (fig. 36b). As discussed earlier, this is a trustworthy indication of lost-wax casting because such enclosed spaces, even if they constituted a small portion of the total décor, would have made the withdrawal of section molds impossible. Also visible are the "mold lines" on the exterior of the openwork hemisphere (fig. 36c) identified by Donna Strahan, who cited them as evidence against lost-wax.[22] Yet the Greek example discussed in chapter 1 demonstrates that section-mold lines or marks can also be found on lost-wax castings because the lost-wax process could freely combine with piece molds in the indirect lost-wax process. In fact, such mold lines, irregularly defined and often discontinuous,[23] are distinguishable from those on bronzes cast in section molds. It is more likely that they were created when wax pieces were fused together, in light of a similar case with the Xiasi *jin* (fig. 26) that also has irregular marks on the surface of its main body.[24]

Through observation, the visual evidence also confirms that the interlace openwork is composed of four very similar and perhaps identical ornamental units. Donna Strahan argued that because the four sections of the globe are identical, they, together with the aforementioned mold lines (fig. 36c), appear to suggest that the openwork was formed by the section-mold process.[25] Though Strahan did not provide specifics, her reconstruction of the procedure used to make the openwork shell of this particular vessel seems to be the following: (1) one quarter of the openwork shell was made as a model; (2) this model was used to create four pairs of inner mold pieces and outer mold pieces; (3) the inner mold and outer mold were created by combining the corresponding mold pieces, and the two molds were assembled in the required manner before casting the whole openwork shell. This is the only section-mold process procedure I can imagine that reconciles the two pieces of evidence noted by Strahan. The aforementioned absolute undercuts of the openwork (fig. 36b), however, render step (2) impossible, because the pieces for the inner and outer molds could not have been separated from a multilayered, three-dimensional model.

Alternatively, the four sections of the model could have been individually constructed in wax, and in this case the problem of the absolute undercuts would have been solved. The four wax sections could then have been fused to combine them into a whole; this would likely have created the series of mold lines Strahan identified. Although photographs indicate that the two-dimensional images of the four ornamental units could generally be overlapped, they are not necessarily identical in three dimensions. Indeed, close visual analysis has detected subtle inconsistencies in the exterior patterns, which were probably caused by individual constructional work on different pieces of the wax model.[26] These inconsistencies recall the aforementioned discussion on the Xuyi *hu*, which similarly has eight nearly identical units. Some device, unknown to modern observers, may have been used to guarantee a high degree of consistency in the production of all of the individually hand-constructed openwork sections.

As described earlier, the mesh of the Xuyi *hu* is composed of vertical interlacing designs outside and horizontal rods inside. The Metropolitan vessel's openwork, in contrast, consists of a horizontal interlace in the outer layer and vertical rods in the inner layer (e.g., see figs. 36b, 36c). Compared with the neatly arranged decoration on the exterior, the openwork interior is cruder and less delicate in appearance. This impression is strengthened by the following two phenomena. First, the small, hollow, convex circles that decorate the exterior sides at intervals (figs. 36b, 36c), which may have held an intricate inlay of semiprecious stones (e.g., malachite, turquoise) or glass, are not present on the interior surface (fig. 37). Second, the raised linear surface patterns (see figs. 36b, 36c), now understood as a decoration broadly applied to the exterior of the openwork, are similarly absent from the interior (fig. 37). Most remarkably, the interior has an irregular surface and a rough texture lacking refinement (fig. 37b). From this it can be deduced that the inner side of the openwork was not intended to be seen, and it therefore serves as a better indicator of the technical features of the process used to create the vessel. Compared with the exterior, the interior reveals a less fine appearance with a greater diversity of minute shapes, all of which create a higher degree of three-dimensionality (fig. 37). The third dimension, even if it could have been achieved by the section-mold process, would have tremendously increased the difficulty of its application. In other words, the section-mold casters would have been hard-pressed to attain this three-dimensionality. Why would a section-mold caster work so hard to make a part of the vessel that nobody would see?

Close observation allows the detection of another interesting phenomenon: on the surface of each bronze rod from the interior of the openwork are several groups of small, oval-shaped slices laminated together (fig. 37b). These slices are staggered slightly, and accordingly their arc-shaped edges form a series of ripples. All such slices must have had strong plasticity and great viscosity, because they could be freely folded, bent, and attached to each other. At the same time, many of these slices seem to have been in a molten state, because their edges display a flowing irregular shape (fig. 37b). Hot, viscous, semi-solid wax is the material that best matches all these characteristics. In fact, this feature occurs in a number of other likely lost-wax castings, such as a square openwork vessel lid in the Nanjing Museum (fig. 44c), a bronze finial in the Sackler Gallery, the Smithsonian Institution,[27] and a newly excavated *pan* vessel from Suizhou Wenfengta cemetery, Hubei (fig. 70b).[28] The most remarkable case comes from the *pan* vessel of the famous Zeng *zun-pan* set. By observing the detached pieces within the openwork structure, I, with the help of Yung-ti Li, have identified many similar oval-shaped slices attached to individual rods.[29] Especially in one case, it appears that a craftsman's fingerprints have been left on the rod (fig. 72d); the pinching was surely for the purpose of pressing the slices together for adhesion. In this case as well, everything appears so roughly made because the interior of the openwork structure was not expected to be viewed. It can therefore be concluded with certainty that the oval-shaped slices were not attached for the sake of ornament. In fact, if the interior of more lost-wax castings become accessible, similar additional traces and clues would probably emerge.

As Thomas Chase commented on the Metropolitan vessel,

> Evidence of the use of lost-wax casting is indicated by the break in the band of lappets surrounding the base of the large bird. This kind of damage occurs because of the fragility of the model and because the molds could not be inspected before casting.[30]

Chase's evidence, though not irrefutable, also provides a clue in favor of the lost-wax casting process.[31] The four small birds (see fig. 36a), almost identical in shape and decoration, were surely multiples made from the same model.[32] But it is not certain whether they were shaped in wax from the same mold or cast within sectional molds, with each set of molds being taken from the same model. The same uncertainty occurs also with the four ring-holder assemblies, the four sections of the ring hoop, and the three slightly odd legs.[33] The plain inner shell,[34] with a slightly flat round shape, would not have presented any difficulty for the section-mold process. Indeed, it was, in all likelihood, independently cast by this traditional method. Soldering played an important role in the connection of different segments for the vessel. As Chase plausibly concluded, "Except for the large bird, the various parts of the object were probably separately cast and assembled with soft solder."[35]

The Fengxiang Burner

An incense burner from Shaanxi Fengxiang (fig. 38), with an elliptical openwork sphere and a large bird with open wings standing on its top,[36] most closely resembles the aforementioned Metropolitan tripod in conformation. Discovered around the palace complexes Yaojiagang of the Qin capital from 677 to 383 BCE, the Fengxiang vessel, if it could be reliably dated to the Warring states period[37] (as it seems to have been), would be among the first confirmed Chinese incense burners made in bronze.[38] Taking into account the location of its discovery and the exquisite nature of its workmanship, it might even be a royal artifact used in the Qin court. With an overall height of 35.5 cm and a total weight of around 4 kilograms, the Fengxiang burner is composed of three general parts: an elliptical sphere, a large square base, and a short hollow column that connects them.[39]

The sphere of the Fengxiang burner (fig. 38b) is close in size and form to that of the Metropolitan tripod discussed earlier;[40] the Fengxiang sphere also has a diatrete structure comprising a plain inner shell fitting within an openwork outer shell. It is composed of two hemispherical portions fastened together and soldered with four ring-holder masks encircling the central part of the openwork shell.[41] As Loveday pointed out,

> Apart from the absence of the four small birds on the top half of the sphere and the ring legs, the major difference between the Fengxiang and Metropolitan vessels lies in the type of openwork interlace forming the outer shell. Instead of horizontal rows, the Fengxiang shell is decorated with S-shaped snakes that appear to move diagonally up the sides.[42]

Indeed, in a departure from the basically orthogonal grid of both the Metropolitan vessel (fig. 36b) and the Xuyi *hu* (fig. 35b), the skewed symmetrical mesh of the Fengxiang burner is composed of obliquely inter-crossing rows (fig. 38b). Compared with the tightly packed rows of the Metropolitan openwork (fig. 36b), the interlaces of the Fengxiang burner are much thicker and bolder in appearance.[43] This feeling of boldness is magnified by the low density of the mesh and the extraordinary three-dimensionality manifested in the interlaces. In this respect, the Fengxiang burner is closer to the Xuyi *hu*, which has a similarly bold type of interlace.

As the outer shell of the Fengxiang vessel could not be removed at the time of viewing, the interior of the openwork could not be observed. However, absolute undercuts, which indicate that the overlapping layers of the interlace do not touch each other, could be easily detected by observing the openwork closely (fig. 38b). As a result, the two openwork hemispheres of the Fengxiang burner can be verified as having been cast by the lost-wax process. Though the sculptural bird of the Fengxiang burner (see fig. 38a) is more exquisite than that of its Metropolitan Museum counterpart, its openwork sphere does not seem particularly fine; the interlaces of the Fengxiang burner are uniformly plain, without any decorative lines or spaces for inlays, whereas such ornaments or their corresponding traces can be found on both the Metropolitan tripod and the Xuyi *hu*.

The openwork square base of the Fengxiang burner, however, contains ornamental elements throughout its body (fig. 39). The base, at 18.4 centimeters long and 9.8 centimeters high,[44] consists of four rectangular vertical walls at the lower level, four trapezoidal sloping sides at the higher level, and a flat, square upper surface. The four vertical walls are decorated with the same set of patterns in two registers, composed of warriors, tigers, and birds (see fig. 39a). A stylized tiger in profile is depicted at the center of the upper level. Two men with shields are represented on its sides, both looking down in a defensive posture. Below the tiger is another warrior, more heroic in appearance and holding a shield. To either side of this warrior is an upside-down tiger, stepped on by a corresponding warrior from the upper level. A large, aggressive, center-facing bird is found on the left and right peripheries of the wall. Below the tail of the bird is a small, erect beast with an upturned head and open mouth. As Jing and Wang observed, despite the bird and beast on the two sides, all these motifs are repeated in the four trapezoidal slopes in exactly the same manner.[45] The flat top of the base, like a window, has four small square sections, all decorated with more or less the same one-eyed coiled *kui*-dragon patterns (Fig. 39b).[46]

The decoration of the square base, though called "high-relief openwork" by Jing and Wang,[47] exhibits a very gentle three-dimensionality, especially when compared with the openwork interlace of the sphere. This type of flattened openwork, without any undercuts, would not in fact have posed any casting difficulty for the section-mold process. This point is further strengthened by the completely two-dimensional inner surface of the square bottom (fig. 40). As the interior of the square is quite rough in appearance and was manufactured without a high degree of care, it must not have been expected to be seen. Because of this, it provides a useful perspective for deciding which casting technique may have been used. The technical nature of the casting here differs substantially from that of the lost-wax process. If casters who were familiar with section-mold casting figured out the lost-wax process, they would have used it to make something different and special that the section-mold process could not easily do. The openwork of this type of square base, which would have been made by the section-mold process for a long time before the lost-wax process was used, did not require such a technical substitute. If the casters had used the lost-wax process, why would they have created such a design? Why not use the more familiar section-mold casting process to fabricate it? If the lost-wax process had been used to cast the square base, its interior would not be as flattened as it is in this case. This treatment of the design on the base was deliberate and was clearly intended to make it easier to remove mold pieces from the model--in other words, to use the section-mold process.[48]

OTHER DIATRETE VESSELS

The round-shouldered jar[49] in the gallery of the Harvard University Art Museums (HUAM; later renamed the Harvard Art Museums) (fig. 41), a bequest of Grenville L. Winthrop, exhibits the same stylistic and technological features as the Metropolitan vessel, the Fengxiang burner, and the Xuyi *hu.*[50] Bought by Winthrop in 1935, the artifact was reportedly discovered in 1911 in the vicinity of Xi'an, Shaanxi Province.[51] This anecdotal evidence links the Winthrop jar more closely to the Fengxiang burner, because the geographic distance between Xi'an and Fengxiang, both places within the core territory of the state Qin in the Warring States period, is less than 90 kilometers. With a height of only 9.5 centimeters, this container is very modest in size. From its flat bottom and narrow, cylindrical foot, the vessel expands to a pronounced belly and round shoulder, then contracts to a bowl-shaped neck, opening into a clearly everted lip. Like the diatrete objects discussed earlier, it has a plain inner shell and an openwork outer shell. The outer shell is made up of two individual sections, upper and lower, which join immediately below the shoulder.[52] The openwork shell of the Winthrop jar also retains small plain areas beyond the patterns of the openwork design; "the interlace (of the Winthrop jar) is joined to the foot by a wide plain register and to the neck by a much narrower one."[53]

Compared with the tightly packed serpent bodies of the Metropolitan openwork (fig. 36b), the bold interlace of the Winthrop jar (Fig. 41b) is closer to that of the Fengxiang burner (fig. 38b). Yet the network of the Winthrop jar is generally in an orthogonal form, which also distinguishes it from the skewed symmetrical mesh of the Fengxiang burner. The openwork of the Winthrop jar accordingly most closely resembles that of the Xuyi *hu* (fig. 35b). As Loveday described it,

> The interlace (of the Winthrop jar) follows a pattern similar to that of the Xuyi *hu*, but the almost exaggerated thickness of the snake bodies gives it a cruder and less fine appearance. Instead of the raised flower heads of the *hu*, the points where the snake bodies overlap are marked by a small inlay of turquoise or malachite.[54]

Again, a number of absolute undercuts could be found in the wholly three-dimensional configuration of overhanging layers (fig. 41b), indicating that the lost-wax process was used to cast the openwork shell.

Beyond the three-dimensional openwork with inlays of turquoise and malachite, the Winthrop vessel is also marked by two zoomorphic handles with projecting tongues, back-turned heads, ophiomorphic faces with neutral expressions, and S-shaped curved bodies (fig. 41a).[55] Interestingly, similar S-curved zoomorphic handles also appear on another double-shelled vessel with three-dimensional openwork. Unknown when Loveday was writing, this diatrete vessel was excavated from Hubei Jingmen Zilinggang and published in 2008 (fig. 42).[56] Classified as a *lei* by the report, this vessel was probably also manufactured during the Warring States period, though the tomb that yielded it is dated to the early period of the Western Han (206 BCE–9 CE).[57] With a height of 13.2 centimeters and diameters at the rim and foot of 6 and 7.8 centimeters, respectively,[58] this Jingmen *lei* looks like a slender version of the squat Winthrop jar. Compared with the latter, the Jingmen vessel bears four, instead of two, S-curved zoomorphic handles. The zoomorphic designs of the Jingmen vessel are feline and slightly more imposing than those of the Winthrop jar.[59]

As in the case of the Winthrop jar, the openwork ornamentation of the Jingmen *lei*, which covers a plain inner shell, has upper and lower sections that join in the middle of the vessel. In addition, the interlace ornament of the Jingmen *lei* connects to the foot bordered by a broad, plain register and connects to the neck by a much narrower one. The mesh of the openwork interlace is composed of coiled serpent bodies constructed in a layered, fully three-dimensional assemblage. Although the interior of the openwork ornamentation is not reachable, obvious absolute undercuts can be identified by closely observing its exterior (fig. 42c). It is clear that the two sections of the openwork shell were also cast by the lost-wax process. As the archaeological report infers, the plain inner shell and the two sections of the outer openwork shell were all separately cast. The combination of the three involved placing the inner shell within the lower section of the outer shell, before the other section of the outer shell was fitted to them.[60] The separately cast two shells of the aforementioned diatrete vessels were likely all assembled this way. The serpent interlaces of the outer shell were also cast with small button-like hollow circular forms that might initially have held an inlay material, as on the Winthrop vessel.

In the light of their great similarities in terms of vessel form, openwork structure, and close zoomorphic designs, the Jingmen *lei* (fig. 42) and the Winthrop jar (fig. 41) might have been produced in approximately the same period. If they were manufactured in different workshops, it is also tempting to assume that the workshops were not geographically distant from each other. It is intriguing that these two vessels, one from Jingmen and the other purported to have been unearthed near Xi'an, were buried approximately five hundred kilometers apart. If the workshops that manufactured this type of vessel were not geographically widespread, then these vessels must have been diffused from their production centers over a wide area. Considering the frequent interregional interactions facilitated by trade or military conflict during the Warring States period, the latter possibility seems more likely, especially in the aforementioned case of the Xuyi *hu* (fig. 35), which was looted from one state and taken to another. Except for the Metropolitan tripod (fig. 36), whose contextual information remains unclear, all the aforementioned diatrete artifacts were apparently manufactured, or at least appeared on the market, in North China. This conveys the impression that these diatrete artifacts were products of North China and that the Jingmen *lei* (fig 42) was brought to the south from the north.

It is easy to forget, however, that places in North China were also tremendously far apart. For instance, Fengxiang, the place of discovery of the incense burner (fig. 38), is almost one thousand kilometers away from the core territory of the Yan state, from which location the Xuyi *hu* (fig. 35) was looted. In light of this distance, the Winthrop jar (fig. 41) and the Jingmen *lei* (fig. 42), though one was from North China and the other from South China, were not so far apart. In addition, diatrete vessels other than the Jingmen *lei* have been discovered in South China. For instance, in 1958 a tomb was unearthed at Pixian Liulin, Jiangsu Province, revealing a number of bronzes that could be dated to the Warring States period.[61] Among them, a small openwork jar (fig. 43), termed a *bu* by the report,[62] is a diatrete vessel not mentioned by Loveday. Only 7.1 centimeters high and 5, 8.8, and 4.1 centimeters in diameter through the rim, belly, and foot, respectively,[63] this *bu* is even smaller than the Winthrop jar (fig. 41). With an everted rim, a vertical neck, a bulbous body, and an openwork outer shell set around a plain inner shell, the Pixian *bu* looks like a squat version of the Winthrop jar. Like many vessels of the diatrete type, the Pixian *bu* is attached with four ring-holder masks, which have corroded substantially. The openwork shell of the Pixian *bu* is also made up of two parts, upper and lower, which meet in the middle of the vessel. The interlace network of the Pixian *bu*, like that of most vessels in the diatrete type, also comprises coiled serpent bodies constructed in an entirely three-dimensional intertwined manner. Possibly displaying the effects of deterioration, the thick, rounded, and spaghetti-like individual interlace of the *bu* appears crude when compared with that of most other vessels in the series. Yet this crudeness does not obscure the presence of absolute undercuts (fig. 43b). It was unquestionably the lost-wax process that cast the two sections of the openwork shell of the vessel. Small raised hollow circles can be found on almost every node of the openwork mesh (fig. 43b), indicating that inlays were certainly present and later fell out.

Although it lacks the double-shell structure, a square vessel cover from the same tomb as the Pixian *bu* can also be associated with the group of diatrete vessels in light of its completely three-dimensional openwork interlace (fig. 44). The cover is in the shape of a square cone, with four rectangular vertical walls at the lower level, four trapezoidal slopes at the higher level, and a small square flat top. All of the surfaces, with the exception of the top, are composed of highly realistic serpentine interlaces with vividly three-dimensional movements (fig. 44b diagram and fig. 45a).[64] For instance, as Wan observes, twenty-three serpent bodies and eighteen snake-like heads occur on a single rectangular vertical wall.[65] The interior of the openwork cover (fig. 45b), though without the sculptural serpent heads and the ornamental linear marks, is still highly three-dimensional and contains a number of absolute undercuts, a clear sign indicating the use of the lost-wax process. Here a comparison with the interior of the openwork base of the Fengxiang burner best reveals the difference in technical style between the lost-wax process (fig. 45b) and the section-mold process (fig. 40). Interlacing from the Pixian openwork cover clearly illustrates something that was invented by the lost-wax technique itself, something that could not have been inspired by two-dimensional openwork designs or wood-carved patterns. Instead, when the craftsmen first obtained the wax, it would have been easy to think of rolling it and making it into candlestick-like cylinders. Because the interior of the structure was not expected to be viewed, casters would have cared little about its appearance. The comparative regularity of the two-dimensional openwork might simply suggest the proficiency of the craftsmen.

In addition, small, laminated oval-shaped slices can be seen on the surface of each interior interlace of the bronze cover (fig. 44c). These slices are approximately 0.4 centimeters long by 0.2 centimeters wide, and are usually overlaid in two or three layers.[66] As discussed earlier, many other likely lost-wax castings, such as the aforementioned Metropolitan vessel (fig. 36), also have this feature (fig. 37b). The function of these oval-shaped slices is not completely clear. One possibility is that they were used to strengthen the wax model. As no mark of joinery (e.g., soldering, interlock casting) is visible, the entire cover was likely a single piece cast by the lost-wax process. Casting this bronze cover, though manageable when using the lost-wax process, must have been a daunting job, especially considering the great difficulty posed by the venting. Perhaps the application of the oblong layered slices could also have increased the space available for venting. This, of course, is only an unconfirmed guess.

The Diatrete Mirrors

Similar to the aforementioned diatrete vessels, certain Eastern Zhou double-tier mirrors also feature a combination of an openwork ornamentation with a plain, smooth inner shell or backing.[67] The double-tier feature of the mirror, in fact, did not introduce any major difficulty in their manufacture, because the reflective face and the decorative back could be cast individually and joined together by soldering, brazing, or other mechanical methods.[68] The openwork ornamentation itself, as Loveday asserted, also "does not require the use of lost-wax and can be easily made from a clay model."[69] This argument seems excessively general and simplified, because several double-tier mirrors with intricate openwork interlace do raise the possibility of lost-wax casting.

One such mirror, said to have been purchased from northwestern China, is now in the Shanghai Museum (fig. 46). The front and back tiers of the mirror were cast separately and fastened with each other, together weighing 150 grams. With a diameter of 10.2 centimeters, the front tier of the mirror has a plain reflective side.[70] The interior of the front tier is nearly undecorated, except for the rim, which has a series of continuously arrayed S-shaped patterns. Three notches equidistantly located along the interior rim of the front tier physically fit well with the three beast-shaped inserts from the back tier of the mirror.[71] With a total diameter of 8.0 centimeters, including a 0.8 centimeter-wide circular frame, the decorative back tier (fig. 46a) holds fifteen interlaced serpents curled up within a small circular area of 33 square centimeters. The serpents all have rounded, plump, sculptural bodies and visually convey a sense of undulation. Ten of the fifteen serpents are biting the mirror's frame, and the other five grip the openwork knob in the mirror's center.[72] It is generally believed that the mirror back alone was cast by the lost-wax process. For instance, as Chen Peifen commented on this particular mirror, "the openwork dragon design on the back of the mirror [...] was separately cast using the lost-wax method, after which it was laid into the framework of the mirror front."[73] Yet Chen did not specify any evidence supporting this assessment. In the case of another mirror, the proof that Chen provided as evidence of the lost-wax process is clearly unreliable,[74] a fact that greatly weakens the credibility of her judgment here.[75]

As X-ray photography indicates, the back-tier ornamentation of the mirror, including the sculptural openwork knob, seems to have been cast in a single piece.[76] This evidence does increase the chance that the mirror's back tier was cast by the lost-wax process, especially considering the intricacy and three-dimensionality of the entire openwork interlace. However, as Li Chaoyuan noted, technical experts in the Shanghai Museum, after a long-term discussion, have finally declared that the mirror back was cast entirely by the section-mold process. Their main evidence for this claim was the announced "mold marks" (fig. 47) found with the openwork ornamentation.[77] The characteristic of this section-mold process, as Li argued, lies in the modeling process in which long, narrow, and flexible clay pieces were coiled in an interlaced manner.[78] Li stressed this technical feature by comparing the mirror ornamentation with other section-mold cast interlaces with a predominant flatness on the interior of the openwork structures—as he explained, the models of the latter were achieved by "carving," but that for the current mirror ornamentation was rendered by *nitiao panzhu*, the clay-strip building technique.[79]

The technical report by Wu Laiming and Xu Huikang, included as an appendix to Li Chaoyuan's article, specifies the reasoning behind their conclusions. First, they detected a series of mold marks that can be seen from both the exterior (fig. 47a) and the interior (fig. 47b) of the mirror ornamentation. As they reasoned, "The lost-wax process did not require the segmentation of the mold, so the 'mold marks' would not appear on the corresponding castings. [...] Apparently, the openwork mirror back was cast within a clay (sectional) mold."[80] Moreover, as Wu and Xu indicated, except for the three-dimensional mirror knob, the openwork ornamentation has hardly any significant undercuts, a fact that would not have presented any difficulty for the section-mold casting process. The protruding knob, with mold marks within all three openwork spaces (see fig. 47a), is believed to have required multiple mold pieces at the corresponding positions.[81] Another important clue for Wu and Xu is the large number of small, straight segments, typically one centimeter in length, that constitute the interior of the mirror back, respectively marked as A-A, B-B, and so on, in their illustration (fig. 48). According to Wu and Xu, if the mirror back had been modeled in wax, this piece-by-piece connection would not have been necessary, because wax strips could have been freely bent and intertwined. The clay, with less plasticity and tensile strength than wax, could not have been so easily manipulated. Thus, individual curved interlaces, if they were made in clay, had to have been formed by linking small clay segments in a piecemeal manner.[82] During my visit to Shanghai Museum, Wu Laiming informed me that his group had even successfully replicated the model for the mirror ornamentation using *zisha*, the superb purple clay first found around Jiangsu Yixing, from which mold pieces could have been separated with little difficulty.[83] This successful experiment has provided him with greater confidence that the mirror ornamentation was cast by the section-mold process.

Some technical experts at the Shanghai Museum, such as Tan Derui and Lian Haiping, still favor the view that the mirror ornamentation was created through the lost-wax process. According to Tan, mold marks (fig. 47) cannot be taken for granted as section-mold cast marks, because wax-shaping molds could have created similar traces; the piece-by-piece linkage of small straight strips (fig. 48) cannot be construed as a decisive piece of evidence, because wax pieces could have been associated in the same manner.[84] Lian expressed a similar opinion but with more explicit reasons. According to her, the mold marks (fig. 47) might have been created by wax-shaping molds—as she observed, the application of "unit" molds for shaping wax produced visible traces;[85] this will be described in detail later. Furthermore, Lian suspected, the use of an investment mold might have also led to the formation of such mold marks: assuming the mirror ornamentation to have indeed been modeled in wax, the model would likely have had the first half of the investment mold applied from one side, and then the second half from the other side. The combination of the two halves of the investment mold might not have been perfect due to differences in the properties of the clay (especially as one half of the mold might have already been dry before wet clay or slurry was applied to the other portion). Therefore, the mold marks that appear in the cast objects should not be uncritically understood as section-mold cast marks.[86]

As for the phenomenon noticed by Wu Laiming on the interior of the mirror ornamentation, Lian explained why the piece-by-piece connection might also have been required when the modeling material was wax. During our joint viewing of the mirror, she meticulously demonstrated to me that individual segments on the interior of the mirror ornamentation (fig. 48) did, in fact, correspond closely to pattern sections on its exterior.[87] Because the independent serpentine interlaces from the exterior are visually identical to each other, Lian argued that "unit" molds for shaping the wax might have been used to make these designs in segments. Wax pieces taken from these "unit" molds, used as pattern segments on the exterior of the mirror ornamentation, would easily be perceived as nothing more than short, straight strips on the other side.[88] To me, Lian's opinions are highly insightful and provocative, but the features of the mirror ornamentation on which she has commented are merely compatible with the lost-wax theory and cannot confirm it outright.

The section-mold versus lost-wax discussion elaborated here is difficult to assess because the mold marks of unknown origin could have been left by either process, and the mirror ornamentation could have been modeled in either clay or wax. Even if unit molds, as Lian believed, were really used, both wax and clay strips could have been shaped from them. Perhaps the only consensus is that the model itself was not carved with the openwork patterns but was formed by coiling long, narrow, and flexible strips in an interlaced manner. Though no absolute undercuts have been detected, the openwork of the current mirror still conveys a strong sense of three-dimensionality (fig. 46). This type of openwork is usually termed *toukong*, as distinguished from the carved *loukong* openwork presented by another mirror in the Shanghai Museum (fig. 49) and the aforementioned square base of the Fengxiang burner (figs. 39, 40). Though very much flattened, perhaps so that it might better join with the mirror front, the interior of the current mirror contains raised and concave areas on its surface, as well as rough textures everywhere (fig. 46b). Such three-dimensionality, although not as pronounced, is completely absent from the interior of the *loukong* openwork structures (e.g., fig. 40). The interior of the openwork structures, usually invisible to the mirror users, best reveals the features inherent to this technique. The *loukong* openwork in the Chinese context here was undoubtedly created using the section-mold process. But what of the *toukong* openwork manifested in the ornamentation of the Shanghai Museum mirror (figs. 46-48)? How was it invented? As no similar pre-existing design is known in Chinese artifacts of other media, this openwork design itself seems to have been invented by the lost-wax technique. In other words, only in adopting the lost-wax process could Chinese casters have been awakened to thc possibility of modeling something with long, narrow, solid, and cylindrical pieces into a intertwined design. The first mirror ornamentation with this design must have been cast by the lost-wax process.

A design invented by the lost-wax technique does not necessarily mean that objects with this design were all lost-wax castings. This is especially true for mirrors, whose comparatively thin and flat shapes were not difficult to replicate. Accordingly, a type of lost-wax cast mirror ornamentation might have been reproduced or imitated by the section-mold process. This hypothesis may be particularly applicable to the current case, because no absolute undercuts have been detected in the interlaces of the mirror ornamentation, raising the possibility that sectional molds could have been serially taken from just such an initially lost-wax cast design, subsequently yielding a series of identical castings of sections of mirror ornamentation. In fact, it is possible that the Shanghai mirror ornamentation might be one of such a series rendered by the section-mold process. Another possibility is that the mirror-makers could have packed clay onto the already-existing back-tier ornamentations of these bronze mirrors to produce molds for replicating *surmoulage* mirrors of a slightly smaller size (as both the mold and the bronze materials would have shrunk during the manufacturing process) and less sharply replicated decoration.[89] Furthermore, as Greek records indicate, lost-wax casting could have combined sectional molds to preserve the original model; molds taken from the original model might not have been used for casting in metal but for replicating the entire original model in wax.[90] This indirect lost-wax process, if it was in use, could also nicely explain the formation of the visible mold marks on the mirror ornamentation. These alternate possibilities are not groundless, because a series of other mirrors with quite similar designs has been discovered, all of which can be checked against each of the hypotheses proposed earlier.

The first mirror close to the Shanghai piece is described in the catalogue *Chugoku ōchō no sui* (fig. 50a). With a diameter of 11.2 centimeters, this mirror seems slightly larger than the Shanghai mirror, but its total weight is surprisingly listed as 265 grams, making it far heavier than the Shanghai mirror (at 150 grams).[91] The *Chugoku* mirror also has two separately cast tiers joined together. Like the Shanghai mirror, the back rim of the front tier is decorated with a circle of S-shaped patterns. More remarkably, the exterior of the back tier contains the same number (fifteen) of serpentine interlaces as the Shanghai mirror, and the design of the interlaces is the same in every detail (see figs. 50b, c). According to

Namba Junko (Uchida Junko), the editor of this catalogue, the back-tier ornamentation was cast by the lost-wax process. Her evidence for this conclusion can be summarized as follows: (1) the intricate shape of its openwork structure seems to be entirely cast and (2) no section-mold cast marks have been detected on the back-tier ornamentation.[92] However, Li Chaoyuan asserted, based on the photo in Namba's catalogue,[93] the section-mold cast marks with the *Chugoku* mirror are even more manifest here than on the Shanghai mirror. Still, the nature of what are arguably "mold marks" remains unclear.

Hua Lin, an amateur but enthusiastic researcher, has calibrated the corresponding images of the Shanghai and *Chugoku* mirrors. According to his analysis, apart from the protruding knobs, the back-tier ornamentation of these two mirrors perfectly matches (see fig. 50b, c).[94] In other words, Hua argued that the fifteen serpentine interlaces of one mirror are identical to those of the other. Based on this announced discovery, Hua has declared that the openwork ornamentation of the two mirrors was cast from the same molds: the *Chugoku* mirror with the alleged sharper decoration was cast first, and the Shanghai mirror was cast later.[95] The inconsistency of the mirror knobs, according to Hua's explanation, is due to the extensive damage to the mold sections around the knob during the casting of the *Chugoku* mirror; these sections were replaced before the casting of the Shanghai mirror.[96] It is not completely improbable that mirrors were cast from used molds, but if they had been, the clarity of their delicate decorative patterns would undoubtedly have been lost.[97] Judging by the high degree of sharpness of the Shanghai mirror ornamentation (fig. 46), it is not likely that it was produced from used casting molds.

However, if Hua Lin's observation on the "perfect matching" of the décor is reliable, the two identical fifteen-serpent openwork interlaces must have followed the same original model. Because the three-dimensional knob was cast together with the rest of the back-tier ornamentation, the inconsistent knob designs do not favor the hypothesis that one mirror ornamentation was the *surmoulage* of the other, or that both were *surmoulages* from another mirror ornamentation. The three following possibilities should be borne in mind. *Hypothesis 1*: sectional molds could have been serially taken from the model of the serpentine interlaced design; they could then have been incorporated with mold pieces that had been rendered from different component models for the three-dimensional mirror knobs. In this case, the back-tier ornamentations would be identical except for the mirror knob. *Hypothesis 2* is the aforementioned indirect lost-wax process: sectional molds taken from the original model of the serpentine interlaced design could have been used to serially produce wax models that incorporated variously shaped mirror knobs, also modeled in wax. The combined casting models in wax would have finally yielded a series of mirrors, including the two pieces under discussion here. *Hypothesis 3* is that the two "identical" fifteen-serpent openwork interlaces are not exactly uniform but rather are extremely similar. This is possible because Hua Lin's observation was conducted simply by comparing printed catalogue photographs rather than the objects themselves, which he did not have the chance to examine in person.[98] Although photography indicates that the two-dimensional images of the corresponding mirror ornamentation do match, it does not prove that they are identical in all three dimensions. The surprising consistency could be attained by rendering individual serpentine interlaces coherently (in wax or clay) with the aforementioned "unit" molds. The final models could have strictly followed the original design or a design plan using unknown devices that could guarantee uniformity in both the general layout and specific details.

Another mirror similar to the Shanghai piece is in the HUAM (fig. 51), as a bequest of Grenville L. Winthrop, who purchased it from Yamanaka & Company in 1934.[99] This circular mirror also has double tiers, with its back-tier decoration consisting of intertwined serpentine openwork patterns, rendered in the same manner as those of the Shanghai and *Chugoku* mirrors. The openwork interlaces (fig. 51b) of the HUAM mirror also convey a strong impression of three-dimensionality, but careful observation does not detect any absolute undercuts. HUAM restrictions prevented close personal observation of the backside/interior of the openwork ornamentation. According to the HUAM official website, this circular mirror is 13.9 centimeters in diameter and weighs 595 grams,[100] making it significantly more substantial than the Shanghai mirror (10.2 centimeters and 150 grams, respectively) and the *Chugoku* mirror (11.2 centimeters and 265 grams, respectively).[101] According to the data provided by the HUAM website, the HUAM mirror is considerably larger than the Shanghai and *Chugoku* mirrors; presumably, its back-tier ornamentation is correspondingly larger, though the size of the ornamentation is not always specified. In any case, close analysis reveals visible inconsistencies between the HUAM mirror and its two counterparts. In handling both the HUAM mirror and the Shanghai mirror, I observed that the former (fig. 51b) had a rougher surface texture (compared with the latter in fig. 46a), probably owing to its more densely arranged surface patterns. The differences in the details of decoration and the size indicate that the openwork ornamentations of the HUAM and Shanghai mirrors did not follow the same model. The HUAM sample strongly favors the aforementioned *Hypothesis 3*, in which models of the back-tier ornamentation were probably constructed through "unit" molds and devices that could ensure the realization of the design or blueprint in a similar and neat manner.

Nevertheless, the great similarities between the HUAM, Shanghai, and *Chugoku* mirrors (figs 51, 46, and 50a), marked by the distinct interlaced openwork ornamentation on the exterior of the back tier, the continuously arrayed S-shaped patterns on the backside of the front tier, and the three or four beast-shaped inserts and corresponding notches that join the two tiers, are convincing evidence that they were all produced from the same or a few interconnected workshops. These features may also indicate that all the mirrors were produced using the same method. If so, the models of the back-tier ornamentation were probably all individually constructed with devices that could guarantee a high degree of uniformity, and the corresponding serpentine interlaces were manufactured component-by-component from a series of unit molds. Mirrors that can be attributed to the same group also include one in the Shōsō-in collection (Nara, Japan, fig. 52a), one owned by a private collector in China (fig. 52b), and a few others variously recorded in different sources.[102] Although none of these mirrors were scientifically excavated, many have long resided in the collections of renowned museums. Because mirror forgeries were already beginning to appear in the premodern period, however, it is still necessary to be cautious about authenticity. The possibility that mirrors could be reproduced should be kept in mind, although no evidence of *surmoulages* has yet been found. Regardless, the initial design of the back-tier ornamentation, as the artist's original, must have been executed in lost-wax, or at least so inspired. Because the lost-wax theory is thoroughly compatible with the known evidence, I take the openwork ornamentation of the samples here discussed as having been cast by the lost-wax process (although some uncertainty must be acknowledged).

Notes

1. Loveday, "Diversity in Eastern Zhou Bronze Casting," 102.
2. As Loveday reasonably claimed, "the date, the 'King's fifth year,' is believed to correspond to either 314 BC or 279 BC. These inscriptions, however, which were incised on the vessel after casting, may well post-date its actual date of manufacture and therefore provide us only with a firm *terminus ante quem*." "The fourth-century date is now generally considered the more probable. [...] An almost identical inscription, again engraved on the surface, appears on a larger inlaid *fang hu* in the University of Pennsylvania Museum. [...] Lawton (1982: 39) has argued for a fabrication date of between 350 and 300 BC at the latest for this vessel, with the possibility that it was made earlier in the fourth century" ("Diversity in Eastern Zhou Bronze Casting," 109, text and note 8). For detailed discussions, see Huang, "Xuyi Xinchu Tongqi, Jinqi ji xiangguang Wenti Kaobian"; Zhou, "Xuyi Suochu Zhongjin luo ling. Chen Zhang Yuanhu Dukao"; Li and Zhu, "Xuyi Hu Ming yu Qi Po Yan Niandai."
3. See Jiang, "Zhongguo Gudai Qingtongqi de Guibao," 57; Fan and Wan, "Chen Zhang Hu Zhizuo Jishu de Chubu Tantao," 1163.
4. Loveday, "Diversity in Eastern Zhou Bronze Casting," 109.
5. As Jiang observed, "The network around the vessel shoulder is made up of forty-eight serpents, with each serpent body twisting up and down three times. Two adjacent serpents, one's head facing the other's tail, form a group. Alternate serpents are connected with each other at turns, and at each conjunction a plum blossom is attached." "The network around the vessel belly (also has forty-eight serpents), with each serpent body twisting up and down nine times [...] alternate serpents are connected with each other at turns, and at each conjunction a plum blossom is also attached " ("Zhongguo Gudai Qingtongqi de Guibao," 57). For more details, see Loveday, "Diversity in Eastern Zhou Bronze Casting," 109; Wan, "Cong Chenzhang Hu kan Chunqiu Zhanguo de Yezhu Jishu," 37.
6. Jiang, "Zhongguo Gudai Qingtongqi de Guibao," 57; Fan and Wan, "Chen Zhang Hu Zhizuo Jishu de Chubu Tantao," 1167.
7. So far two theories of the section process, similar to the *Perspectives* 1 and 2 mentioned earlier in the discussion of the Zeng *zun-pan*, have been proposed for the casting of the three-layered network. These theories

have been convincingly and thoroughly criticized by Wan ("Cong Chenzhang Hu kan Chunqiu Zhanguo de Yezhu Jishu," 40).

8. Su Rongyu, oral communication, June 2015 (Location: Institute of Natural History, Chinese Academy of Sciences, Beijing).
9. Absolute undercuts, which completely separate layers of an openwork structure, obstruct the free flow of molten bronze and increase the difficulty of venting.
10. Wang Yong, oral communication, May 2015 (Location: Tongling, Anhui Province, China).
11. This is indicated by the extremely high concentric rate (99.4%) between the outside hoop, which tightens up the openwork ornamentation, and the inner shell of the vessel. See Fan and Wan, "Chen Zhang Hu Zhizuo Jishu de Chubu Tantao," 1167.
12. As Wan noted, "a carved line is present on the back of every serpent in the bronze network" ("Cong Chenzhang Hu kan Chunqiu Zhanguo de Yezhu Jishu," 40; see figs 5, 6, 15, and 16). These concave lines, by no means the section-mold cast marks (the latter being, without fail, convex) might form on wax because of the seam or interspace at the opening of the wax-shaping mold. However, Fan and Wan later found that such linear marks only occurred with bronze rods on the outside but not inside ("Chen Zhang Hu Zhizuo Jishu de Chubu Tantao," 1167). This phenomenon, truly bewildering, suggests that these linear marks may have been ornamental in function. This question awaits further study.
13. Fan and Wan, "Chen Zhang Hu Zhizuo Jishu de Chubu Tantao," 1165, 1167; 1166, fig. 5.
14. But such a device, if it was present, cannot be understood as a true "mold," because the highly three-dimensional and multi-layered structure could by no means be fashioned within any imaginable molds.
15. According to the careful observation of Wang Yong, the four ring-holder masks have exactly the same number of filaments, including a mistakenly carved one, in the feather-like surface ornament. Wang Yong, oral communication, May 2015 (Location: Tongling, Anhui Province, China).
16. According to Fan and Wan, "The (masks), if observed from the front, seem possibly to have been produced from molds. However, in the mouth of each serpents are twelve teeth, as well as a tongue. Considering that the serpent mouth is only 3–4 mm long, and the floating serpent tongue is 4–5 mm long, it is not an easy job to make the whole design within molds [...]" ("Chen Zhang Hu Zhizuo Jishu de Chubu Tantao," 1167). (The last sentence of the original text, "Within the serpent

mouth, a space of 3–4 mm long, to make a 4–5 mm long tongue by means of molds is a very difficult thing [...]" is not logical to me, and therefore I have revised it in a more understandable manner.) This statement, though bewildering, does offer some useful information.

17. In this case, the perhaps only way to solve the problem is to make the models of the two serpent tongues detachable. The detachment of the two projecting and suspended tongues from the original model would then allow the withdrawal of the mold in a feasible number of sections. But the procedure would be quite cumbersome and challenging as well, with the result that the two tongues would have to be precast, which means the ring-holder mask would no longer be a single-piece casting. For the use of the detachable model in a similar case, see Bagley, "Anyang Mold-making and the Decorated Model," 77.
18. This was probably also the case with the four movable rings held by the masks and the four beast-shaped ring handles, both of which seem to have been cast in a series after the same model.
19. Chase, *Ancient Chinese Bronze Art*, 72.
20. For details, see Chase, *Ancient Chinese Bronze Art*, 72; So, *Eastern Zhou Ritual Bronzes from the Arthur M. Sackler Collections*, 74, note 97; Loveday, "Diversity in Eastern Zhou Bronze Casting," 111–112. Indeed, compared with the tall, upright, and compelling bird on the *hu*'s vertex, the three legs are too slim and of insufficient length, conveying a top-heavy feeling for the whole vessel. This unbalanced feeling, however, could be made to vanish immediately either by replacing the legs with more robust ones or by replacing the top bird. It is unclear whether the problem lies in the legs or in the bird on top; perhaps it is even both. Alternatively, the object might have such an unsteady visual effect originally.
21. My study of the Metropolitan vessel (October 31, 2016) is supported by Jason Sun and Hwai-ling Yeh-Lewis.
22. Donna Strahan, "Debating the Use of Lost-Wax Casting in Ancient China," Public Lecture in the Freer Gallery Meyer Auditorium, Washington, D.C., January 15, 2015.
23. These linear marks are not consistent in form, direction of flow, or even occurrence, and they are very difficult to trace. According to Donna Strahan's proposal ("Debating the Use of Lost-Wax Casting in Ancient China"), there should be four "mold lines" in total to partition the whole openwork in four identical units. I have spent hours looking for them, but such "mold lines" are very often barely recognizable around the presumed boundary areas of the four ornamental sections proposed by

Strahan. My sense is that they are not real section-mold marks. Su Rongyu, one of the most celebrated experts on Chinese bronze technology, also agreed that these usually discontinuous and irregular linear marks are very difficult to detect, and they are very different from true section-mold marks (oral communication, Beijing, June 9, 2015).

24. As discussed earlier, such irregularly defined linear marks on the Xiasi *jin* were probably traces left from fusing wax pieces together. The "fusing" procedure, as Li Jinghua inferred, involves inserting a burning metal sheet into the miniscule interspace between two wax pieces; with the melting of wax, the two pieces could be quickly fused together after the extraction of the metal sheet. During this procedure, a small amount of liquefied wax might be pushed out of the interspace, forming a series of "fusing marks" that, after solidifying and being cast into bronze, might be mistaken as section-mold marks. For details, see Li, "Xichuan Chunqiu Chumu Tongjin Shila Zhuzaofa de Gongyi Tantao," 41, 44.
25. Strahan, "Debating the Use of Lost-Wax Casting in Ancient China." Strahan also stressed this point during our personal communication (December 17, 2015).
26. Although such inconsistencies could also occur when using the reconstructed section-mold procedure, the procedure itself was impossible to carry out.
27. Shown by Strahan, "Debating the Use of Lost-Wax Casting in Ancient China."
28. Strahan ("Debating the Use of Lost-Wax Casting in Ancient China") also noticed this phenomenon on these vessels, but she did not take it as evidence of their having been cast by the lost-wax method.
29. My study of these detached pieces from the Zeng *zun-pan* (May 21, 2015) was accompanied by Yung-ti Li and facilitated by Zhang Changping.
30. Chase, *Ancient Chinese Bronze Art*, 72.
31. The simple, symmetrical, and static design of the large bird, without any undercuts in shape, introduced little challenge for section-mold casting. However, the large bird seems to have been cast together with the upper hemisphere openwork, and if this were the case, then it would certainly have been cast by the same method as the openwork, which has been verified as lost-wax cast. As the upper hemisphere was fitted closely to the inner shell, without being allowed access to its interior, I could not reach a conclusive determination on this point during the time I had to observe it.

32. This was also suggested by Strahan in her public lecture ("Debating the Use of Lost-Wax Casting in Ancient China").
33. For image, see Peng, "The Lost-wax Casting in Bronze Age China," 223, fig. 5.14.
34. See Peng, "The Lost-wax Casting in Bronze Age China," 220, fig. 5.8.
35. Chase, *Ancient Chinese Bronze Art*, 72.
36. Loveday, "Diversity in Eastern Zhou Bronze Casting," 111–112; Jing and Wang, "Fengxiang Faxian Zhanguo Fengniao Xianhuan Tong Xunlu," 57.
37. Jing and Wang, "Fengxiang Faxian Zhanguo Fengniao Xianhuan Tong Xunlu," 57.
38. Yang, "Handai Tong Xunlu Yanjiu," 88. According to Yang, ceramic incense burners, probably initially used to get rid of mosquitoes through the burning of a plant (generally believed to be the *Lysimachia foenum-graecum* Hance), appeared in the hot and humid Chu cultural region in the south of China ("Handai Tong Xunlu Yanjiu," 88). Later such vessels were also used to release aromatics.
39. Jing and Wang, "Fengxiang Faxian Zhanguo Fengniao Xianhuan Tong Xunlu," 57; Loveday, "Diversity in Eastern Zhou Bronze Casting," 111.
40. Loveday, "Diversity in Eastern Zhou Bronze Casting," 111.
41. Ibid.
42. Ibid.
43. According to the measurement by Jing and Wang, the interlace of the Fengxiang vessel is 0.3 centimeters in thickness ("Fengxiang Faxian Zhanguo Fengniao Xianhuan Tong Xunlu," 57).
44. Jing and Wang, "Fengxiang Faxian Zhanguo Fengniao Xianhuan Tong Xunlu," 57.
45. Ibid.
46. Ibid.
47. The original Chinese term for this is *loukong gao fudiao*, used by Jing and Wang ("Fengxiang Faxian Zhanguo Fengniao Xianhuan Tong Xunlu," 57).
48. Here the comparison with the interior of the openwork sphere of the Metropolitan vessel (as that of the Fengxiang burner is not accessible) offers an important clue. As described earlier, the interior of the Metropolitan vessel shows a similarly coarse appearance, but also a greater diversity in minute shapes and a higher degree of three-dimensionality (fig. 37b)—an unintentional result of applying the wax model for that case.

49. This vessel is variously referred to as *fou* (So, *Eastern Zhou Ritual Bronzes from the Arthur M. Sackler Collections*, 28, fig. 23; Loveday, "Diversity in Eastern Zhou Bronze Casting," 110) or *pou* (=*bu*瓿) (see the official website of the Harvard University Art Museums).
50. So, *Eastern Zhou Ritual Bronzes from the Arthur M. Sackler Collections*, 27.
51. See the official website of the Harvard University Art Museums (Object Number: 1943.52.70).
52. Loveday, "Diversity in Eastern Zhou Bronze Casting," 110.
53. Ibid.
54. Ibid.
55. As Loveday explained, "The turned heads and long protruding tongues of the zoomorphic handles are a prominent feature of wood and bronze sculpture from the Chu cultural region, known from the mid-sixth century at least. [...] Large feline-shaped handles with curled tongues were adopted in North China on a series of *fang hu* and *hu*, which continued to be cast over a long period, throughout the sixth and fifth centuries with turned heads and long protruding tongues. In the light of this, the dating of fifth to fourth century has been retained here as the most appropriate" ("Diversity in Eastern Zhou Bronze Casting," 110–111). Loveday's main concern here was dating, and she correctly rectified So's opinion of viewing objects with the double-shelled diatrete structure as manufactured around the sixth century BCE (So, *Eastern Zhou Ritual Bronzes from the Arthur M. Sackler Collections*, 26–28).
56. Jingmen, *Jingmen Zilinggang*, 154–155.
57. As the report states, artifacts like this *lei* vessel in such a particular form and with the turquoise inlays have not been discovered in any contemporary Han tombs. Besides, in cemeteries dated to the Qin and Han periods, many bronzes in addition to this *lei* vessel could also be dated to the Warring States or the even earlier Spring and Autumn periods. For details, see Jingmen, *Jingmen Zilinggang*, 168.
58. Jingmen, *Jingmen Zilinggang*, 155.
59. The Jingmen zoomorphic designs have a curled protruding tongue, a slender neck, a lean, curvilinear body, and a long, curled tail. The archaeological report (Jingmen, *Jingmen Zilinggang*, 154–155) interprets these creatures as *baihu* ("White Tigers,") a well-known symbol of the Chinese constellations representing the west, which probably originated during the Eastern Zhou period. However, this symbolic meaning should not be associated with the zoomorphic designs of the Jingmen *lei* solely based on their appearance.

60. See Jingmen, *Jingmen Zilinggang*, 154.
61. These bronzes include two *fang hu*, two *fu* (one with eight inscriptions), one *yi*, one *ding*, an openwork *bu*, an openwork square vessel cover, as well as a few miscellaneous artifacts. Nanjing, "1959 nian dong Xuzhou diqu kaogu diaocha 1959," 26–27.
62. Ibid.
63. Wan, "Nanjing Bowuyuan cang Zhanguo Toukong Panlongwen Qingtongqi," 73.
64. For detailed descriptions, see Wang, "Tan Zeng Hou Yi Zun Pan de Zhuzao Gongyi," 71–72; Wan, "Nanjing Bowuyuan cang Zhanguo Toukong Panlongwen Qingtongqi," 72–73.
65. Wan, "Nanjing Bowuyuan cang Zhanguo Toukong Panlongwen Qingtongqi," 72.
66. Ibid., 75.
67. Loveday, "Diversity in Eastern Zhou Bronze Casting," 119; Li, "Xinhuo Zhanguo Toukong Fuhejing Yanjiu," 40–53.
68. Chen, "Early Chinese Bronze Mirrors," 264.
69. Loveday, "Diversity in Eastern Zhou Bronze Casting," 119.
70. For images of the front tier of the mirror, see Li, "Xinhuo Zhanguo Toukong Fuhejing Yanjiu," 42, Figs. 16.1, and 16.2.
71. Ibid.
72. Ibid., 42, 53.
73. Chen, "Early Chinese Bronze Mirrors," 264.
74. As Chen argued, "From the background patterns of some mirrors of the Warring States period, such as the *Siyewen* 四葉紋 mirror (Plate 6 of the book), we could see apparent defects caused by the shrinkage of the wax model" ("Luelun Chu Jing de Zhuzao Jishu," 31). The so-called lost-wax defects mentioned here, such as several short linear marks left on the example mirror (see Shanghai Bowuguan, *Lianxing Shenye, Yingzhi Lianggong*, Plate 6), are not at all convincing and, worse yet, somewhat bewildering.
75. Besides, Chen believed that the lost-wax process was "widespread" in use for the casting of bronze mirrors during the Warring States period; this estimation does seem too optimistic. For details, see Chen, "Early Chinese Bronze Mirrors," 263.
76. See Li, "Xinhuo Zhanguo Toukong Fuhejing Yanjiu," 43, fig. 17.
77. Ibid., 43.
78. Ibid. The English abstract of Li's article translates the word *mo* ("model") as "mold" (53), which is incorrect.

79. Ibid.
80. Wu and Xu, "Shanghai Bowuguan Wenwu Baohu yu Kaogu Kexue Shiyanshi Guance yu Jiance Baogao," 50.
81. Ibid.
82. Ibid.
83. Wu Laiming, oral communication, Shanghai Museum, April 28, 2015.
84. Tan Derui, oral communication, Shanghai, May 6, 2015.
85. Lian Haiping, oral communications, Shanghai Museum, October 21–22 and November 24, 2015.
86. This statement is based on my understanding of Lian Haiping's opinion, voiced during our viewing of the bronze mirror in the storage room of Shanghai Museum, November 24, 2015.
87. Lian Haiping, oral communication, Shanghai Museum, November 24, 2015.
88. This statement is based on my understanding of Lian Haiping's opinion during our viewing of the bronze mirror in the storage room of Shanghai Museum, November 24, 2015. Lian's proposal of applying the wax-shaping "unit" molds is indeed very thought-provoking. It goes a long way to explain not only why the flexible wax had to be rendered into short straight strips, but also why individual interlaces of the mirror ornamentation are close to each other in almost every detail. Additionally, the use of wax-shaping "molds" also raises one possibility with regard to the formation of the aforementioned "mold marks."
89. For the discussion of making *surmoulages* from existing bronze mirrors, see Li, "Mirrors from 500–200 BC Middle Yangzi Region," 260–262.
90. This indirect lost-wax process was not an unrealistic design in this case, because molten bronze would do a great deal of damage to the casting molds, which could not then be used repeatedly to cast serial products. As Li clearly explained, molds with "delicate decorative patterns would have been damaged due to the strong force of the molten bronze when pouring it into the mold, the solidification of the bronze, and the removal of the mirror. It is not impossible that mirrors were produced from used molds, but if they were, the sharpness and clarity of their decorative patterns would surely be lost" ("Mirrors from 500–200 BC Middle Yangzi Region," 262).
91. Namba, *Chugoku ōchō no sui*, No. 41. Quoted in Li, "Xinhuo Zhanguo Toukong Fuhejing Yanjiu," 49, No. 9. These measurements have nevertheless been questioned by Hua Lin. See Hua, "'Toukong Jiaolongwen Jing' yanjiu," report 5.

92. Namba, *Chugoku ōchō no sui,* No. 41. Quoted in Li, "Xinhuo Zhanguo Toukong Fuhejing Yanjiu," 43.
93. See Namba, *Chugoku ōchō no sui,* No. 41; Li, "Xinhuo Zhanguo Toukong Fuhejing Yanjiu," 43, Fig. 18.
94. Hua, "'Toukong Jiaolongwen Jing' yanjiu," reports 2 and 3.
95. Ibid., report 3.
96. Ibid., report 3.
97. Li, "Mirrors from 500–200 BC Middle Yangzi Region," 262.
98. Hua, "'Toukong Jiaolongwen Jing' yanjiu," reports 2 and 3.
99. See official website of Harvard University Art Museums (acc. 1943.52.150).
100. Ibid.
101. The data on the HUAM official website are inconsistent with the data provided by Li Chaoyuan on the same mirror (13.9 versus 10.9 centimeters in diameter). See Li, "Xinhuo Zhanguo Toukong Fuhejing Yanjiu," 49. Something must be wrong with the measurements.
102. For details, see Li, "Xinhuo Zhanguo Toukong Fuhejing Yanjiu," 43, 49; Hua, "Tongjing Zonghe Yanjiu," unpaginated.

Chapter 6

Further Confirmed Lost-wax Castings III

Objects with "Interpenetrating" Openwork Appendages

Beyond the Xiasi *jin* and the diatrete artifacts, a series of artifacts with "interpenetrating"[1] openwork appendages embellishing the key components can also serve as useful demonstrations of the lost-wax process.[2] The first example is a *zhan* now displayed in the Metropolitan Museum of Art, New York (fig. 53; called a *dui* in the museum's official credit line). The inscription indicates that the Metropolitan *zhan* was a royal vessel of Chu, likely for King Gong (r. 590–560 BCE).[3]

The Metropolitan *Zhan*

With its provenance unclear, the Metropolitan *zhan* was first made known to the public in June 1986 at Christie's, New York;[4] it was later bestowed as a gift by Charlotte C. and John C. Weber on the Metropolitan Museum of Art in 1992.[5] This slightly squat bowl, together with its domed lid, is 20 centimeters high overall, with a diameter of 24 centimeters.[6]

The disk-shaped crown situated on top of the vessel lid is the most compelling and delicate part of the vessel (fig. 53b). With a height of only two centimeters and a diameter of 11.8 centimeters, this openwork crown can be understood as three concentric rings tied together by a number of intricately shaped and spaced bronze rods.[7] The three rings are labelled R1, R2, and R3, respectively,[8] in the diagram figure 53c, which also depicts the shape and distribution of distinct features, including different types of animal heads and serpent tails, within the crown. The diagram in figure 54 depicts how the different ornamental elements are assembled.[9]

According to Hua Jueming, the innermost two rings, R1 and R2, with diameters of 1.4 centimeters and four centimeters, respectively, are linked by eight bronze rods (fig. 55).[10] Four of these rods, as a group, are decorated with two buffalo heads ("H1a" in the diagram) and two rams' heads ("H1b") in symmetrical alternation (fig. 55a).[11] Each of these four animal heads, overhanging the innermost ring R1, is correspondingly supported by a curved bronze rod underneath. The other four rods ("T1"), when viewed from above, form the shape of a serpent's tail curled in a clockwise direction, at the same level as the rings (see fig. 55a, b).[12]

Twelve serpent-like bronze rods, from the second ring R2 (fig. 55), extend outwards horizontally in a tortuous manner to the point where they join the outer ring R3.[13] The other twelve bronze rods, even more tangled in construction, can be divided into two groups, each containing six rods. The first group, with six serpent heads ("H2") suspended over the ring R2, collectively stretch toward the ring R3 (see fig. 53c). The other group, with six beast heads ("H3a" and "H3b") that overhang above, also spread collectively outward until they touch R3 (see fig. 53c).[14] These twelve rods in fact support the weight of the entire crown (fig. 56b),[15] even though the area they occupy, with a diameter of 5.5 centimeters,[16] is comparatively small (see fig. 56a). Surrounding this supporting structure are two types of serpent tails (not shown in the diagram), each numbering six, all occupying a circle that is 6.4 centimeters in diameter.[17] Rods with these serpent tails branch out from rods with the beast head H3a and those with the serpent head H2 (see fig. 53c).[18] Such bifurcations of bronze rods further increase the intricacy of the openwork crown.

The outer ring R3 is composed of two layers (see figs. 54, 56). The upper layer is made up of multiple serpent-like bronze strands interlaced together. Each strand, at 0.3 centimeter in width, has a long, narrow groove on its flattened round back (fig. 54).[19] With twenty-four "knots" forming a full circle,[20] this interlaced structure is even more tangled than a regularly knotted braid (fig. 56). Twenty-four curled "flower petals" below, formed by bronze rods extending from the inside, correspond to the twenty-four "knots" in the upper level.[21]

In total, thirty-two animal heads of eight types, along with sixteen serpent tails of three types, can be identified within the crown (fig. 53c).[22] The entire structure is extremely elaborate, yet at the same time arranged in remarkable order.[23] This openwork crown, as the visual evidence suggests, was by no means a result of free or experimental creation and must have been designed in meticulous detail before fabrication. It also must have been cast in its entirety as a single piece, because no traces of soldering, brazing, or interlock casting have been discovered, and careful examination excludes the possibility that it was hammered into shape.[24] With its tremendously tangled bronze rods and especially its many deep and densely spaced undercuts within a space a mere two centimeters high, this entirely cast openwork must have been manufactured by the lost-wax process.[25] Ornamental elements of the same type, such as the aforementioned animal heads and serpent tails, are almost identical to each other in size and shape (see fig. 53c), so it seems likely that the corresponding wax pieces were shaped within molds.[26] The smoothly winding, intricately entwined rods and strands, however, could not have been *completely* mold-shaped; instead, they must have been twisted and intertwined by hand.[27] Considering that these rods or strands have similarly shaped profiles, they might also have been first formed within molds,[28] but this possibility has not been verified.

As Hua Jueming reasonably suggested, the design of this openwork crown was probably inspired by woven works in rattan, bamboo, or wood.[29] Wax, with its great plasticity, could be fashioned into nearly any form imaginable. Metalworkers could first shape wax into long, flexible rods or straps and then "weave" them into their final desired form. The final model would, as a result, have been based on woven materials. This result would be barely imaginable were the method in use the section-mold process, and so such a woven-like openwork structure can be rationally regarded as a product of the lost-wax technique.

Similar to the disk-shaped crown, two ornate appendages attached to the *zhan* body also have openwork interlaces of great intricacy, comprising crisscrossed bronze rods spaced three-dimensionally (fig. 57a). For reasons similar to those applied to the disk-shaped crown, one can deduce that these two intricately shaped appendages, lacking mold marks on their surfaces, must have also been cast by the lost-wax process.[30] Similarly, the two handle-shaped ornaments attached to the vessel lid,[31] though greatly damaged, were probably also cast by the lost-wax process, judging by its three-dimensional structure with interlaced rods.[32] The remaining parts of the *zhan* seem to have been predominantly cast by the section-mold process, despite the three cabriole legs (fig. 57b), which are still controversial.

In view of their intricate structures, the three *zhan* legs (fig. 57b) might also have been cast by the lost-wax process. Although such an openwork structure with three-dimensional openwork interlaces on the sides would not have been impossible to cast by the section-mold process, doing so would have been fairly cumbersome and difficult.[33] Because the lost-wax process had already been used to cast other openwork structures of the same vessel, there would have been no special reason to cast the legs singly in a particularly laborious way. Yet surprisingly, raised marks appear, bisecting the surface of each leg (fig. 57b),[34] a feature usually associated with the section-mold process. Using this observation as evidence, Zhou Weirong and Huang Wei argued that the section-mold process was used here, along with the aforementioned process proposed by Wang Jingchao.[35] The method put forward by Wang has, however, already been disproven in the discussion in chapter 3. In fact, such "mold marks" would have come into being if the wax model of each vessel leg had been pre-shaped into two halves in a bivalve mold before being further embellished.[36] This process not only permitted an intricate shape to be cast but also ensured the standardization of castings with the same shape and size so they could be functionally equivalent parts of the same vessel.[37] Nevertheless, the possibility that these openwork legs were cast by the section-mold process, a method capable of achieving the final effect, still cannot be entirely ruled out; the chance of that case, in fact, is even not too trivial, considering the sometimes extraordinary influence of tradition.

The Xiasi *Zhan* Vessels

The dating of the Metropolitan *zhan* to the reign of the King Gong of Chu (590–560 BCE) is supported by the finding of a similar *zhan* with a slightly less magnificent interlace from the mid sixth century BCE tomb (M1) at Xichuan Xiasi, Henan province (fig. 58).[38] Found broken into dozens of pieces, the lidded Xiasi *zhan* has been restored to its original shape, with an overall height of 19 centimeters and a diameter of 23.5 centimeters at its greatest dimension at the belly.[39] Like its Metropolitan counterpart, the Xiasi *zhan* also has a thickset silhouette, with more or less the same dimensions and a domed lid with a disk-shaped crown. Comparable details between the two *zhan* also include the gently concave contour just below the rim, a slightly bulging belly, and a broad, round bottom.[40]

Of course, the most imposing similarity is the exceptionally delicate crown composed of openwork serpentine interlaces on the top of the lid. The three-dimensional mesh of the Xiasi *zhan* (fig. 59), less compelling than its counterpart on the Metropolitan *zhan* (fig. 53b), is slightly more simplified in structure and reduced in detail. The disk-shaped openwork is composed of three-dimensional serpentine interlaces in eight symmetrically spaced groups.[41] The upward or outward stretching heads of the serpents form the outer part of the disk, and the serpent tails form the lower part of its stem, with the winding and mingled serpent bodies occupying the area in between.[42] Careful observation yields no traces of soldering, hammering, or section-mold marks on the crown,[43] verifying that this crown was cast entirely by the lost-wax process. These intricate braided and interlaced patterns were undoubtedly shaped and constructed in wax, and later cast in bronze by the lost-wax process. Apparent "repair marks" on the interlaced rods, observed by Tang Wenxing, can accordingly be understood as having been left by retouching on the handmade wax model of the crown.[44] The inconsistencies in form and dimensions of the individual wax interlaces caused by freehand shaping were probably finally trimmed by tools.[45]

In addition to the crown portion of the lid, the lost-wax process also cast a pair of ornate openwork handles attached to the vessel body and three equally opulent vessel legs. The openwork handles are arranged evenly around the vessel belly in alternation with two lug handles. The two simple lug handles, like the four similar handles around the lid, would not have presented any challenges for section-mold casting, and it follows logically that these ring handles were indeed all precast pieces manufactured by the section-mold process. The difficulties posed by the two openwork handles, however, would have been tremendous. With the overall shape of a dragonhead, the openwork handle is actually composed of two pairs of serpentine interlaces interwoven in an exceedingly complex manner (see fig. 58).[46] As no section-mold marks, soldering, or hammering traces have been discovered on the surface,[47] the entire openwork structure must have been cast in a single piece through the lost-wax process.[48] For the same reasons, each of the openwork vessel legs (see fig. 58), in a more complex mesh that comprises multiple pairs of serpentine interlaces,[49] was also likely entirely cast by the lost-wax method.[50] These lost-wax cast interlaces, inconsistent in shape and thickness, were probably formed directly in wax and mainly by hand.[51] Retouching traces left by tools have also been detected on their surface.[52]

The three legs were likely interlocked with the vessel later through casting. The two openwork handles were, in contrast, soldered to the separately cast vessel with mortise-and-tenon devices, the tenons having been predesigned along with the section-mold cast vessel body, with the mortises dug into the clay core of the handles.[53]

The fabrication of the Xiasi *zhan*, which was generally similar to that of the Metropolitan *zhan*, also used both lost-wax casting and section-mold casting. The *zhan* lid was cast by the section-mold process with precast appendages joined by interlock casting. The precast pieces include the lost-wax cast crown and four section-mold cast ring handles. The body was also section-mold cast with precast parts joined by casting or soldering; its precast pieces include two ornate handles and three openwork legs finished by the lost-wax process, as well as two section mold-cast ring-shaped lug handles. Although interlaces of the lost-wax cast parts, largely inconsistent in shape and thickness, were formed mostly by hand with the assistance of essential tools (e.g., for carving, wax-fusing, and retouching), the possibility of using molds to shape wax cannot be excluded, especially considering the comparatively standardized components of the most embellished crown.

Another *zhan* vessel, seriously damaged and in dozens of pieces, has been unearthed from the tomb M3 in the same cemetery.[54] With its handles, legs, and nearly all other components detached from its body, this fragmentary vessel has not undergone any restoration,[55] yet archaeologists could still tell that it was identical in form to the Xiasi *zhan* from the tomb M1, discussed earlier. According to the archaeological report, some handle-like appendages of this vessel were cast by the lost-wax process.[56] This judgment seems reasonable in the light of the fully three-dimensional shapes and the curled intertwined bronze rods within the structures of these appendages.[57]

In addition, a single openwork *zhan* handle (fig. 60), similar to the ornate handles of the *zhan* vessel from Xiasi M1, has been discovered on its own in another mid-sixth-century BCE tomb, M2, of the same cemetery.[58]

With a length of 6.2 centimeters and a width of 4.5 centimeters,[59] this *zhan* handle in profile resembles a crawling beast with a perplexing openwork "headdress," an inverted U-curved hollow "body," and an upward curling "tail" (fig. 60a).[60] Partly broken, the openwork "headdress" still reveals a highly intricate three-layered structure (fig. 60b). The inner layer is composed of a dozen straight rods positioned vertically in relation to the handle surface (figs. 60b, and 60e); the middle layer comprises winding rods entwined in bewildering circles (fig. 60d); and the exterior consists of many ornamental elements similar to spiral water plants, drifting in multiple directions (fig. 60c).[61] Small, hollow raised circles, which may originally have contained inlay, occur at intervals throughout the décor elements on the exterior.[62] As no adjoining traces of soldering or interlock casting have been discovered,[63] this openwork handle, with its full three-dimensionality in space, must have been cast by the lost-wax process in a single pour.

THE YEXIAN *ZHAN* HANDLE

Another openwork *zhan* handle (fig. 61), virtually identical to the single piece from Xiasi M2, was unearthed in 2002 in Yexian, Henan Province. Yielding more than 280 bronzes in total, including one inscribed *ge* dagger axe, this tomb was probably for Duke Ning of Xu, who reigned from 592 BCE until his death in 547 BCE.[64] Xu, who was constantly harassed by the state of Zheng, moved its capital from Xuchang to Yexian in 576 BCE and thereafter became a subsidiary polity of the Chu state.[65] It is thus believed that this *zhan* with the openwork handle, accompanied with many bronzes made in the artistic style of Chu, was likely manufactured between 576 and 547 BCE.[66] This speculation is reasonable, but the possibility of a slightly earlier fabrication for the *zhan* cannot be completely dismissed.

With a length of 5.8 centimeters, a width of 4.8 centimeters, and thicknesses in the range of 0.2 to 0.3 centimeters,[67] this *zhan* handle has a multi-layered ornamental openwork linked by a dozen straight rods vertical to the handle's surface (fig. 61b). The stunning, complex ornamental openwork is composed of thirty pattern elements of sixteen types (fig. 62a), of which fourteen are symmetrical in distribution around a central axis.[68] Each pattern element is usually supported by one or two bronze rods. Such rods, interspersed at staggered intervals, greatly enhance the complexity of the overall visual effect of the openwork structure.[69] Computerized tomography has verified that the entire handle was cast in shape without any metallurgical or mechanical adjoining traces.[70] Hence, this openwork handle, with its full three-dimensionality in space, can be viewed as a single piece of lost-wax casting.

Wang Jinchao, the scholar who proposed *Perspective 2* on the making of the rim appendage of the Zeng *zun* (see chapter 3), has also argued that the openwork *zhan* handle was cast entirely by the atypical section-mold technique.[71] However, a series of details in the replica of the *zhan* handle that Wang rendered (fig. 63b) has proven his method vis-à-vis the manufacture of this particular object historically incorrect.[72] In addition, a series of details on the openwork handle has further supported the use of the lost-wax process. For instance, the bronze rods within the openwork structure are of different thicknesses, with a variety of bumps on their surfaces but at the same time remain free from any section-mold marks, indicating that they were shaped primarily by hand.[73] Recognizable marks of burning and fluid flow[74] suggest that various parts of the openwork handle were rendered in wax and joined together by heating the corresponding surfaces to the point of melting.[75] The irregularly defined and usually discontinuous linear traces along two components, easily misunderstood as soldering marks for joining bronze pieces, were actually such wax-joining marks.[76] The pattern elements with comparatively smooth surfaces and fluent contours were probably shaped within molds or templates, which could handily explain some of the seam-like linear marks on them.[77]

Chen et al. replicated the openwork handle by using the "traditional" Chinese lost-wax casting they reconstructed from received texts and ethnographical materials.[78] The individual pattern elements they produced in series from the wooden template (fig. 62b) usually carried similar seam-like linear marks.[79] The hand-shaped wax rods, rendered through rubbing, kneading, pinching, and drawing, have coarse and uneven surfaces and are of various thicknesses.[80] This feature also occurs on the bronze rods within the openwork *zhan* handle. The welding of different wax components, whose minimum interspace is only 0.1–0.2 centimeters, had to be facilitated by needle-like tools.[81] The final replica successfully cast by the lost-wax process (fig. 63a) is very close to the original object (fig. 61a), especially in comparison to its counterpart cast by Wang Jinchao's atypical section-mold process (fig. 63b; look closely at the stiff bronze rods of the openwork structure). Although it is not clear how the lost-wax process was applied in detail, the method reconstructed by Chen et al. offers at least one possibility and some useful inspiration for testing such hypotheses.

THE YUEYANG *ZHAN*

Another *zhan* with appendages that were likely cast by the lost-wax process comes from Yueyang Fengxingzuishan M1, Hunan Province (fig. 64), excavated together with other bronze items, including two *ding*, one *fu*, one *he*, one *pan*, one *yi*, one *bi* spoon, and two *ge* weapons.[82] Because the bronze *he*, *pan*, and *yi* and the *zhan* from the Fengxingzuishan M1 are all very close to their counterparts from Xichuan Xiasi M1 in form and style, the archaeological report asserts that the two tombs were close in date.[83] With an overall height of 17.4 centimeters and a greatest diameter of 20.4 centimeters,[84] the Yueyang *zhan* is usually titled after its maker, "Yun Er," whose name was mentioned in its inscriptions.[85] This vessel has a bulging domed lid, a squat round body, two openwork lug handles, three openwork legs, and a series of fixed ring handles. But its most embellished and compelling part is the disk-shaped crown on the apex of the lid (fig. 65).

Like its counterparts on the Metropolitan and Xichuan vessels, the crown is also a fully three-dimensional network composed of serpentine interlaces. Its structure, in contrast with these previously discussed examples, is less complex. Seen from overhead, the crown seems to comprise eight neatly connected ornamental units with a moderately intricate form.[86] Yet a non-vertical perspective reveals that the mesh still has a complex shape and spacing (see fig 65). At least two levels of ornament with space between them are present in every unit of interlace. As discussed earlier, this feature of absolute undercuts is a reliable sign of lost-wax casting. The eight ornamental units can be divided into two forms, marked by two types of creature heads (fig. 65).[87] According to my own measurements, the large heads are 1.8, 1.6, 1.8, and 1.9 centimeters in width, respectively, roughly one third broader than the small heads, which are 1.4, 1.5, 1.3, and 1.3 centimeters wide, respectively. These slightly uneven dimensions indicate that all the creature heads were probably modeled individually. Indeed, none of the heads are completely identical in the details of form and decoration. This in and of itself does not entirely rule out the use of molds or templates in forming individual wax components, because everything made in wax could later be altered in shape. Linear marks in decoration are found on every interlaced rod of the openwork (see fig. 65a). It is similarly unclear whether they were directly shaped in wax from molds/templates or further carved on individual wax pieces.

Tan Derui concluded that beyond the crown on the lid, the lost-wax process was also used to cast the three openwork vessel legs.[88] Hua Jueming regarded the opulent handles attached to the vessel body as also having been lost-wax cast.[89] Though Tan and Hua did not specify the reasons for these assertions, their judgments seem to be generally correct. Vessel legs from the current piece (see fig. 64) are rendered in the three-dimensional form stylistically like the crown (fig. 65). When casters took up the lost-wax process, they surely tried to take advantage of its expanded possibilities to make something special that no section-mold casters had tried before. The twisted-spaghetti openwork represented by the three vessel legs seems to be such a design inspired by this technique. The openwork structure, which was not functional in that it could not contain beverages, would have been added with an inner shell inside it in the process of manufacturing the vessel body, in the same manner exemplified by the diatrete vessels (see figs. 35–43). More often, this design would have been chosen to make other parts of the vessel, such as the ornamental crown, the legs, and the handles. The two openwork handles on the Yueyang *zhan* (see fig. 64), with similarly noodle-like interlaces wrapped around their upper parts, were also likely lost-wax cast appendages. Although it would not have been impossible to fabricate the shapes of the two openwork handles and the three openwork legs by the section-mold process, the twisted-spaghetti design on all seems to have been invented by the lost-wax technique. As no mold mark is visible on any surface,[90] all these openwork parts seem to have been entirely cast by the lost-wax method.

OTHER EXAMPLES OF *ZHAN*

An additional *zhan* with likely lost-wax cast appendages is from M38 of the Mingmenhuafu construction area, Nanyang, Henan Province (fig. 66), excavated together with some other bronze items.[91] Because many of these bronze vessels are stylistically very close to their counterparts from the tombs M2 and M3 of Xichuan Xiasi, Henan, the archaeological report dates this tomb to the early phase of the late Spring and Autumn period.[92] "Peng Zi She/Ye," or "She/Ye"[93] for short, a name mentioned by almost all the bronze inscriptions from M38, probably refers to the host of the tomb.[94] Inscriptions from a covered *ding* specifically refer to She/Ye as "Shen Gong zhi Sun" ("the grandson of the Duke of Shen"). The tomb host was apparently a descendant of a preeminent aristocrat of Shen, a region around Nanyang administered by Chu during the Eastern Zhou period.[95] With an overall height of 20.9 centimeters and a belly diameter of 23.1 centimeters,[96] this Nanyang *zhan* has a domed lid, a stocky body, two openwork lug handles, three openwork legs, and several fixed ring handles. The most captivating part of the vessel is the disk-shaped crown on the lid vertex. Like its counterpart on the Yueyang *zhan*, the crown is garland-shaped and surrounds a central void (fig. 66b). Although even less involved in form than the Yueyang *zhan*, the crown is similarly rendered in a fully three-dimensional network composed of multiple layers of serpentine interlaces. Also present in the crown of the Nanyang *zhan* are the absolute undercuts (see fig. 66b)[97] that are such a reliable sign of the lost-wax method. The archaeological report also identifies the two openwork lug handles (see fig. 66a) as having been lost-wax cast.[98] It is possible that the three stylistically similar but less complicated openwork legs (see fig. 66a) were cast by the lost-wax process as well.

Two more bronze *zhan* vessels with lost-wax cast appendages are now held in a private overseas collection.[99] Inscriptions indicate that these two *zhan*, as well as other bronze(s) in the same group, were made for the "Huang Zi," Viscount(s) of Huang.[100] One of these two *zhan* (fig. 67) is extremely close to the Metropolitan *zhan* (fig. 53) in shape, style, and decoration. This is especially true in that the disk-shaped crown on top of the vessel lid is decorated with overhanging heads and tortuous bodies of many intricate, tightly spaced serpent-like creatures in the central area; the same number of ornamental "knots" (twenty-four in total) constructed in a braid-like manner also form the full circle of the outer ring (fig. 67b), just as on the Metropolitan *zhan*. The ornamental crown, with a highly intricate structure and many apparent absolute undercuts (see fig. 67b),[101] must have been cast by the lost-wax process. For the same reason, the two twisted-spaghetti lug handles (see fig. 67a) were probably also lost-wax cast. The three openwork legs (see fig. 67a),[102] stylistically a bit different, might likewise have been cast by the lost-wax process, although this is not easy to determine.

A second "Huang Zi" *zhan* (specifically, "Huang Zi Lou" *zhan*, fig. 68) in the same collection presents more moderate openwork designs when compared with the first. For instance, the circular crown on the vessel lid (fig. 68b) in this case is much less compelling in form, while still rendered in the same "technological style."[103] This style is clearly displayed by the rounded, string-like rods staggered three-dimensionally. Careful observation reveals several absolute undercuts in the openwork structure,[104] firmly demonstrating the use of lost-wax casting. The same process might have been applied to the casting of its three openwork legs as well (see fig. 68a).

Because the Huang state was destroyed by Chu in 648 BCE, it is usually presumed that bronzes inscribed with "Huang Zi" were manufactured earlier than this date. If so, these two "Huang Zi" *zhan* would be the earliest verified lost-wax samples in the Central Plains region of China. Even if the authenticity of the received records is granted, it is still hard to demonstrate whether the "Huang" in this case really refers to the state of Huang that was vanquished by Chu in 648 BCE,[105] and in any case, it was not uncommon historically in China for a conquered headman and his successors to retain their original title. No matter what, the two *zhan*, especially the first, are stylistically very close to the Metropolitan *zhan*, and their dates must not be far apart. It is thus more reasonable to date these two "Huang Zi" *zhan* to the neighborhood of the early sixth century BCE.

THE WENFENGTA *PAN*

Beyond the *zhan* vessels, verified lost-wax cast appendages have also been found with other types of objects, but with lesser frequency. For instance, one vessel demonstrating lost-wax use is the bronze *pan* unearthed from the tomb M33 at Suizhou Wenfengta, Hubei Province (fig. 69).[106] This *pan* has a shallow, circular basin, three vertical hoof-shaped legs supporting the basin, two appended beast-shaped figurines, and two U-shaped handles topped by a square openwork appendage (fig. 69b). As the most eye-catching parts of this handsome vessel, the two openwork appendages both display a three-dimensional network constructed by many freely spreading and twisting bronze rods. Their twisted-spaghetti-like structures[107] are reminiscent of many of castings rendered in the same technological style, such as the openwork crown of the *hu* cover from Xichuan Xiasi (fig. 73) and especially the handle-shaped rim appendages of the *pan* from the set of Zeng *zun-pan* (fig. 15). Most significantly, wrinkle marks (fig. 70a), usually believed to be caused by the flow of melting wax during the modeling process, have been found on many bronze rods within the Wenfengta *pan*. As the archaeological report announces,

> (The Wenfengta cemetery) for the first time provides direct evidence of lost-wax casting that was used in the manufacture of the bronzes of the state of Zeng during the middle or late Spring and Autumn period. The bronze *pan* from the tomb M33 reveals extensive flow marks of the melting wax. This discovery offers new evidence for the long-term debate concerned with lost-wax casting.[108]

The Chinese Newspaper of Cultural Relics shows an even more affirmative attitude: "this newly discovered definite evidence provides a conclusion to the endless lost-wax debate."[109]

In opposition to these statements, Zhou Weirong and Huang Wei argued,

> the openwork appendages of a bronze *pan* [...] were identified as evidence of lost-wax casting based on the presence of a wrinkled surface on some of the metal crosslinks, which were considered to be flow marks from the liquid wax. This can now be seen to be incorrect, and the wrinkled appearance should be attributed to piece-mold casting because the liquid metal cannot always maintain a continuous profuse flow over the long and bent cavity of the inner clay core, which finally results in the wrinkled structure after cooling.[110]

As Zhou and Huang pointed out, similar wrinkle marks are very common on ancient silver ingots, which were cast not by the lost-wax process but within an iron mold.[111] Thus, they remarked, the wrinkle mark cannot be viewed as proof of lost-wax casting. In addition, Zhou and Huang stated that such "flow marks of the flowing wax" were not first discovered on the Wenfengta *pan*, because Zhang Changping had previously noticed a similar feature with fragments from the handle-shaped appendages of the *pan* from the Zeng *zun-pan* set (fig. 71). Zhang similarly regarded these bronze rods as having been modeled in wax, then heated, and then melted, finally creating the wrinkle marks.[112] Apparently unconvinced by the lost-wax theory, Zhou and Huang instead used the process proposed by Wang Jinchao (the aforementioned *Perspective 2* objecting to the lost-wax theory on the Zeng *zun-pan*; see chapter 3) to explain why the cavities for casting the openwork appendages of the Wenfengta *pan* and the Zeng *pan* (of the *zun-pan* set) were "long and bent," which led to the poor flow of bronze fluid and therefore the formation of the wrinkle marks on the corresponding casting surfaces.

It is indeed a somewhat hasty conclusion to regard the wrinkle marks on the casting as representing an idiosyncrasy of the lost-wax process alone. Zhou and Huang were prudent to consider other possibilities, such as the poor flow of the molten metal, which, according to them, could have been caused by the "long and bent" casting cavities. However, their argument, which relies on the example of "wrinkled" silver ingots, does not seem reasonable to me. For one thing, the silver ingots are not bronze; for another, they were certainly not cast in "long and bent cavities." Zhou and Huang gave no evidence that "long and bent cavities" have the effect they attributed to them. Moreover, my detailed discussion has proven the process proposed by Wang Jinchao, borrowed by Zhou and Huang to serve their argument, to be historically incorrect. It therefore has greatly weakened the credibility of Zhou and Huang's contention. In fact, it is very likely that some of the wrinkle marks were indeed caused by the flow of melting wax. Specifically, a single wax model for casting might have comprised multiple wax pieces that were heated, partially melted, and finally adhered together.[113]

Close observation has detected new clues about the creation of the "wrinkle marks." These traces on the Wenfengta *pan* are a series of small round, oval, or irregularly shaped slices laminated together on the surface of individual bronze rods within the openwork structure (fig. 70b). Huang Fengchun, the excavator of the Wenfengta *pan*, gave vivid analogical examples, such as "round pancakes" and "flat-topped terraces," to describe the forms of these small, variously shaped layered slices.[114] The usually arc-shaped edges of these staggered slices form a set of ripples, which might easily be misunderstood as the flow of a viscous fluid. As these slices were deliberately set in layers—and folded, bended, and adhered at leisure—they had an apparently strong plasticity and great viscosity. Some of these slices even seem to have been melted to a thinner, more liquid state, because their edges show an irregular shape of diffusion. All these phenomena indicate that these slices were rendered in heated, glutinous wax. This assumption has been confirmed by the extremely intricate shape of the openwork structure, its twisted-spaghetti-like technological style, the numerous absolute undercuts, and the absence of any section-mold cast marks, all of which point to lost-wax casting as the process in use.

Moreover, the distinct feature of laminated slices also appears within the openwork structures of several other likely lost-wax castings, such as the Metropolitan tripod (fig. 37b), a square openwork vessel lid in the Nanjing Museum (fig. 44c), an ornamental finial collected by Paul Singer,[115] and, most notably, the Zeng *pan* (of the *zun-pan* set, figs. 71 and 72d). To better understand this feature, I have conducted a detailed visual and microscopic analysis of the Zeng *pan* fragments in the storage room of the Hubei Provincial Museum.[116] Among the different linear marks found on the fragments, the most noteworthy is the ripple-like type (fig. 71), which used to be understood as indicative of the flow of melting wax. However, this kind of trace, seen almost universally on every section or unit of the bronze rods, was apparently not randomly formed by the flow of liquefied wax but was instead consciously rendered by craftsmen. At first glance, these seemingly regular marks even seem to present the visual effect of ornamental patterns. Yet their presence on the interior side of bronze rods negates the possibility that they were made for the sake of decoration. Close observation reveals that these ripple-like marks likewise took form from a series of small staggered slices with arc-shaped edges. These small slices were probably set in a laminated manner on purpose, judging by two fingerprints that have been left on one slice, surely when it was pressed and attached to the layers below (fig. 72d). Perhaps this was done to strengthen individual wax rods for the sake of stabilizing the whole structure before applying the investment mold.

Additionally, a tenuous, raised mark consistently appears along the middle of the other side of each bronze rod (fig. 72a; for a magnified image, see fig. 72b). At first, this linear trace might be misunderstood as the section-mold cast mark. Yet it is easy to tell that this mark never occurs on the parting surfaces[117] but always along the freely meandering and twisting bronze rods. Another possibility is that the mold may have cracked along these lines during casting, because they look like cracks in the mold. This linear trace, though usually discontinuous in appearance, is routinely present on almost every bronze rod. Was this mark left by the fusing of wax? Perhaps the candlestick-like rods were usually combined with two semi-cylindrical halves fused together. This is only a bold guess, but my hunch is that sometimes one half of the rod might not have been well aligned with the other half; it also seems easier to shape wax from a semi-cylindrical mold than with a fully cylindrical one. Occasionally bronze rods have a more obviously "folded" shape (fig. 72c), and this phenomenon might be explained by the poor flow of bronze fluid or the melting of wax. Although many issues still await further investigation, the detailed technical discussion in chapter 3 makes it clear that all these traces were left by the lost-wax process and probably occurred on the wax model.

FURTHER EXAMPLES

After a comprehensive examination of the Wenfengta *pan*, a vessel with considerable useful clues for the identification of the lost-wax casting process, I now move to another example. In the aforementioned

Xichuan Xiasi M2, archaeologists have found six fragments of a square ornamental crown (fig. 73), probably from the lid of a *hu* vessel.[118] Among the fragments, two pieces belong to the bottom rim of the *hu* crown (figs. 73a and 73b). Both pieces have retained a long and a short edge, of thicknesses around 0.6–0.7 centimeters, connected with an angular turn in the middle.[119] Finer bronze rods, of thicknesses around 0.2–0.25 centimeters, extend from the two edges and are coiled into multiple layers of a three-dimensional network.[120] The networks, though seriously damaged, with most of the bronze rods broken, still look highly intricate and tangled. The other preserved fragments (see figs. 73c–73f), with a more complex, involuted structure and freely stretched rods, belong to the walls of the *hu* crown. The rods often have irregular contours, inconsistent surface patterns, and varying thicknesses that range from 0.2 to 0.4 centimeters.[121] Two such rods, too close to each other, have been adhered together like fried dough twists.[122] In addition, no soldering or section-mold cast mark has been detected on any of the fragments.[123] Together, these phenomena suggest that the whole crown, as an appendage to the *hu* cover, was cast entirely by the lost-wax process.[124]

As the lost-wax process did not necessarily leave any telltale traces, to verify that an object was manufactured by this technology, it is usually necessary to rely on the shape of the design—only those objects with a high degree of dimensionality, intricacy, and observable absolute undercuts can be reliably identified as having been fabricated by the lost-wax method. This standard might seem a little too strict at times, and it is worth considering whether some comparatively simple designs might have been lost-wax cast as well. Specifically, some objects have shapes that did not require the lost-wax casting method, but they are stylistically related to the aforementioned verified bronzes that were definitely cast by this method, thereby suggesting that they were made by casters who were familiar with the lost-wax process. For instance, in Xichuan Xiasi M2, a tomb dated to around 552 BCE, archaeologists have discovered a set of seven *ding* with the same shape and decoration but of different sizes. Each of these *ding* has a flat lid with a pair of inverted

U-shaped handles; a shallow, inward-slanted, and flat-bottomed body; three highly decorated cabriole legs; and six appended climbing-beast figurines. Among them, the largest *ding*, labeled M2: 38 (fig. 74a), is 67.4 centimeters high and 68 centimeters in diameter around the belly and weighs 110.4 kilograms after restoration;[125] the smallest *ding*, labeled M2: 28, is 61.3 centimeters high and 59 centimeters in diameter around the belly and weighs 82.8 kilograms after restoration.[126] By analyzing fragments of the most damaged *ding*, M2: 34, which has broken into more than one hundred pieces, archaeologists have recognized that many *ding* components, such as the U-shaped handles, the cabriole legs, and the beast-shaped figurines, were all separately manufactured and joined to the section-mold cast vessel body by means of casting or soldering.[127] It has been clearly observed that each of the six beast-shaped figurines (similar to fig. 74b) comprises an individually cast beast body, a tail, an antler-like structure, and a so-called "waist-decoration."[128] These four parts were soldered together with the assistance of mortise-and-tenon joints.[129] Some openwork decorations of the beast-shaped figurine, especially the antler-like structure, display a high three-dimensionality in space and a great intricacy in structure and were probably cast by the lost-wax process.[130] This inference is further supported by the phenomenon that in terms of technical style, the beast-shaped figurines of these *ding* tripods were very close to the verified lost-wax castings, such as the highly decorative woven-like crown on the cover of many *zhan* vessels and especially the beast-shaped figurines appended to the Xiasi *jin* altar table (fig. 26b).

From Xichuan Xiasi M1, a tomb also dated to the mid-sixth century BCE, archaeologists have discovered two *ding* of different dimensions (similar to fig. 74) but both in the same form as the aforementioned seven *ding* from Xiasi M2. Tang Wenxing has specifically examined the casting process of the larger *ding*. According to Tang, the *ding* body and most vessel parts, such as the U-shaped handles and the cabriole legs, were all separately cast by the section-mold process and joined together by casting or soldering, very often with the reinforcement of mortise-and-tenon joints.[131] Yet Tang, through repeated careful observation, could not find any section-mold marks on the surfaces of the six beast-shaped figurines appended to the *ding* body.[132] Very likely, the intricately shaped parts of these beast-shaped figurines, with freely twisting three-dimensional designs and serious undercuts, were cast by the lost-wax process.[133]

The same type of *ding*, usually called *sheng ding*, with similarly intricate, three-dimensional appended climbing beasts, has also been seen by the author in the Poly Art Museum, Beijing (two examples[134]), the Shanghai Museum (one example[135]), and the Minneapolis Institute of Art (one example[136]). Moreover, in the aforementioned tomb of Duke Ning of Xu at Yexian, Henan, archaeologists have found three more examples in the same form.[137] In fact, before this discovery at Yexian, four other *sheng ding* had already been looted from the same tomb. It is said that at least one *sheng ding* at the Poly Art Museum was originally from the tomb of Duke Ning of Xu.[138] Having been constantly invaded and harassed by the state of Zheng to the north, the state of Xu, during the reign of Duke Ning, moved its capital to Ye (present-day Yexian) and was thereafter geographically and politically closer to the state of Chu to its south.[139] Still unable to prevent the invasion of Zheng, Duke Ning later went to Chu to request military assistance but unfortunately died there and was eventually buried by the Chu people.[140] It is therefore possible that the seven *sheng ding* buried with him, if not cast locally by the state of Xu, might have been provided by the state of Chu. It seems that all these *sheng ding* with more or less the same design and style were manufactured in either a single workshop or a handful of interconnected workshops around the Chu cultural region. Likewise, the elaborately decorated openwork parts of the beast-shaped figurines on these *sheng ding*, with their many curves, turns, and serious undercuts, seem to have been cast by the lost-wax process in every case.

Stylistic association with the verified lost-wax castings indicates that some other objects, or their three-dimensional appendages, were also likely cast by the lost-wax process. Such objects, beyond the aforementioned series of *sheng ding*, also include other vessels like the "Huang Zi Lou" *he* from a private overseas collection (fig. 75),[141] chariot fittings like one in the Nanjing Museum (fig. 76),[142] ornamental finials similar to the one collected by Paul Singer,[143] bronze weapons like the *ge* in the Guyuege collection, Taipei,[144] and perhaps artifacts elsewhere in other categories. Of course, judgment based on stylistic association alone is not clear-cut, because it is necessary to consider the possibility that a caster who did not know the lost-wax technique saw a lost-wax cast artifact and adapted the design to a piece-mold process. In other words, something invented by a lost-wax caster might have been "simplified" or "revised" for section-mold casting. For this reason, it is important to maintain an open mind with regard to these objects and tolerate a healthy degree of uncertainty.

It is possible to set too conservative or "rigorous" a standard in the judgment of lost-wax castings and accept only those with an extremely intricate shape as belonging to the lost-wax process (as it is impossible to think of another reasonable way to make them). Even when stylistic association is used as another basis for judgment, one may still be inclined to view castings with relatively intricate shapes but ignore those with simpler forms. It would not be terribly surprising if the number of lost-wax castings done in Bronze Age China greatly exceeded present expectations. It is usually said that the *zun-pan* (fig. 15) is the *only* casting with lost-wax use in the tomb of Marquis Yi of Zeng, but this might not be true. Consider the pair of giant *hu* in the tomb (fig. 77): their handles have "antlers" at both top and bottom. Might the caster have used the lost-wax process for the handles or at least for the antlers? Furthermore, might the lost-wax process have been involved in even less complicated designs? The complexities of such investigations should caution against assuming that objects with more ordinary shapes were necessarily made by non-lost-wax processes. Much of the time, the fabrication methods of a given object are not as clear as we hope; a decision should not be forced when the evidence is not decisive.

Notes

1. "Interpenetrating," defined by Notis and Wang as a remarkable feature of the lost-wax cast structures, refers to "a true three dimensional composite openwork structure with separate interpenetrating parts" ("Ancient Chinese Bronze Casting Methods," 1746–1747).
2. In fact, the Zeng *zun-pan* discussed in chapter 3, with interpenetrating openwork rim appendages verified as lost-wax cast, also belongs to this group.
3. The inscription of the vessel is 楚王酓審之盞 ("The *zhan* of Yin Shen, the King of Chu"). As Colin Mackenzie reasonably explained, "Of the two characters which name the ruler, the first can be identified without difficulty as 酓. This character occurs in a number of royal Chu inscriptions and was presumably in antiquity approximately homophonous with *xiong* 熊, the character which occurs as the first half of the names of many Chu kings as recorded in the historical texts. [...] The crucial second character must be *shen* 審, because it differs from examples of that character given in Gao Ming 1980 only in the addition of an extra horizontal stroke across the graph at the bottom. Of the sixth-century Chu kings, the only ones whose names bear any phonetic resemblance to this character are Shen 審 [Gong Wang] and Zhen 珍 [Zhao Wang], of which the former is clearly closer." ("Chu Bronze Work: a Unilinear Tradition or a Synthesis of Diverse Sources?" 155, note 71). The book "Gao Ming 1980" that Mackenzie mentioned is Gao Ming, *Gu Wenzi Lei Bian* (Beijing: Zhonghua Shuju, 1980).
4. Li, "Chu Bronzes and Chu Culture," 11.
5. Hua, *Zhongguo Gudai Jinshu Jishu*, 177.
6. Li, "Chu Bronzes and Chu Culture," 11.
7. Hua, "Niuyue Daduhui Bowuguan Suocang Shangzhou Qingtongqi de Jishu Kaocha," 61; Hua and Xiao, "Chu Wang Yu," 81.
8. "R," standing for "ring," is used by Hua Jueming to illustrate the three rings in the diagram (fig. 55). In the main text, it is written as "B," probably a typo. See Hua and Xiao, "Chu Wang Yu," 81–84.
9. Hua and Xiao, "Chu Wang Yu," 84.
10. Hua, "Niuyue Daduhui Bowuguan Suocang Shangzhou Qingtongqi de Jishu Kaocha," 61; Hua and Xiao, "Chu Wang Yu," 81.
11. Ibid.

12. Ibid.
13. Ibid.
14. Ibid.
15. As Hua Jueming observed, some of these bronze rods are even visible from the interior of the vessel lid. See Hua, "Niuyue Daduhui Bowuguan Suocang Shangzhou Qingtongqi de Jishu Kaocha," 62; Hua and Xiao, "Chu Wang Yu," 81.
16. Hua and Xiao, "Chu Wang Yu," 81.
17. Hua, "Niuyue Daduhui Bowuguan Suocang Shangzhou Qingtongqi de Jishu Kaocha," 62; Hua and Xiao, "Chu Wang Yu," 81, 84.
18. Ibid., 84.
19. Ibid.
20. Ibid.
21. Ibid.
22. Hua, *Zhongguo Gudai Jinshu Jishu*, 177; Hua and Xiao, "Chu Wang Yu," 84.
23. Hua, *Zhongguo Gudai Jinshu Jishu*, 178.
24. Hua, "Niuyue Daduhui Bowuguan Suocang Shangzhou Qingtongqi de Jishu Kaocha," 62; Hua and Xiao, "Chu Wang Yu," 85.
25. Hua and Xiao, "Chu Wang Yu," 85.
26. Ibid.
27. Ibid., 85–86.
28. Ibid., 86.
29. Ibid., 80.
30. Ibid., 84–85.
31. For image, see Peng, "The Lost-wax Casting in Bronze Age China," 253, fig. 6.10-3.
32. Hua and Xiao, "Chu Wang Yu," 81, 85.
33. Ibid., 85.
34. Ibid.
35. Zhou and Huang, "Lost-wax Casting in Ancient China," 1635.
36. At the first glance, such raised marks are reminiscent of the aforementioned traces on the body of the bronze *jin* from Xichuan Xiasi (fig. 26). It seems plausible, therefore, that the same explanation might be relevant here to the modeling of the *jin*—that the fusing of wax pieces might have caused such raised marks. Because the raised marks on the *zhan* vessel appear to be more continuous, Hua Jueming's explanation may be more applicable here, that is, the wax model of each vessel leg was pre-shaped into two halves in a bivalve mold. As Hua mentioned, this fabrication

method might also have been applied to the legs of a similar *zhan* from Xichuan Xiasi (fig. 58). See Hua and Xiao, "Chu Wang Yu," 85.

37. Ibid.
38. So, *Eastern Zhou Ritual Bronzes from the Arthur M. Sackler Collections*, 120.
39. Henan et al., *Xichuan Xiasi Chunqiu Chumu*, 75.
40. Ibid.,; Hua and Xiao, "Chu Wang Yu," 81.
41. Henan et al., *Xichuan Xiasi Chunqiu Chumu*, 75; Su et al., *Zhongguo Shanggu Jinshu Jishu*, 310.
42. Ibid.
43. Tang, "Xichuan Xiasi Yihao Mu Qingtongqi de Zhuzao Jishu," 176.
44. Ibid.
45. Ibid.
46. Henan et al., *Xichuan Xiasi Chunqiu Chumu*, 75; Su et al., *Zhongguo Shanggu Jinshu Jishu*, 310.
47. Tang, "Xichuan Xiasi Yihao Mu Qingtongqi de Zhuzao Jishu," 176.
48. Su et al., *Zhongguo Shanggu Jinshu Jishu*, 310.
49. Henan et al., *Xichuan Xiasi Chunqiu Chumu*, 75.
50. Su et al., *Zhongguo Shanggu Jinshu Jishu*, 310.
51. Hua, *Zhongguo Yezhu shi Lunji*, 227; Tan, *Canlan de Zhongguo Gudai Shila Zhuzao*, 39.
52. Tang, "Xichuan Xiasi Yihao Mu Qingtongqi de Zhuzao Jishu," 176; Su et al., *Zhongguo Shanggu Jinshu Jishu*, 310.
53. Tang, "Xichuan Xiasi Yihao Mu Qingtongqi de Zhuzao Jishu," 175; Tan, *Canlan de Zhongguo Gudai Shila Zhuzao*, 40.
54. Henan et al., *Xichuan Xiasi Chunqiu Chumu*, 230.
55. Ibid.
56. Ibid.
57. For image, see Henan et al., *Xichuan Xiasi Chunqiu Chumu*, Plate 85.3.
58. Henan et al., *Xichuan Xiasi Chunqiu Chumu*, 137.
59. Ibid.
60. Zhao, "Chunqiu Shiqi Shilafa Zhuzao Gongyi Wenti Tantao," 86.
61. For details, see Ibid.
62. Ibid.
63. Ibid.
64. Li et al., "Xu Gong Ning Toukong Panhui Wen Qingtong Shijian," 96.
65. Zhao, "Chunqiu Shiqi Shilafa Zhuzao Gongyi Wenti Tantao," 87; Li et al., "Xu Gong Ning Toukong Panhui Wen Qingtong Shijian," 96.
66. Li et al., "Xu Gong Ning Toukong Panhui Wen Qingtong Shijian," 96.

67. Ibid., 97.
68. Li et al., "Zhongguo Zaoqi Shila Zhujian," 192.
69. Ibid., 193.
70. For details, see Li et al., "Xu Gong Ning Toukong Panhui Wen Qingtong Shijian," 99–100, figs. 10–12.
71. For details, see Wang, "Nifanfa Fuyuan Xugong Ning Loukong Panhuiwen Qingtong Shijian Yanjiu," 35–36.
72. In particular, it can be imagined that the bronze rods cast from the previous "runners" that have been tunnel-drilled by Wang within the inner mold (or "core-mold" in Wang's own term) must be stiff and awkward (see fig. 63b), and so different from the freely bent and spaced rods of the original artifact. For details, see Wang, "Nifanfa Fuyuan Xugong Ning Loukong Panhuiwen Qingtong Shijian Yanjiu," 34–36.
73. Li et al., "Zhongguo Zaoqi Shila Zhujian," 193.
74. See Chen et al., "Xu Gong Ning Panhuiwen Toukong Shijian Fuzhi Yanjiu," 48, figs. 9, 10.
75. Li et al., "Zhongguo Zaoqi Shila Zhujian," 193; Chen et al., "Xu Gong Ning Panhuiwen Toukong Shijian Fuzhi Yanjiu," 48.
76. Li et al., "Zhongguo Zaoqi Shila Zhujian," 194; Chen et al., "Xu Gong Ning Panhuiwen Toukong Shijian Fuzhi Yanjiu," 48.
77. Ibid.
78. For details, see Chen et al., "Xu Gong Ning Panhuiwen Toukong Shijian Fuzhi Yanjiu," 49–50; Chen and Xu, "Xu Gong Ning Panhuiwen Toukong Shijian Fuzhi Yanjiu," 301–305.
79. Chen et al., "Xu Gong Ning Panhuiwen Toukong Shijian Fuzhi Yanjiu," 49; Chen and Xu, "Xu Gong Ning Panhuiwen Toukong Shijian Fuzhi Yanjiu, 301.
80. Ibid.
81. Ibid.
82. For details, see Yueyang, "Hunan sheng Yueyang xian Fengxingzuishan yi hao mu Fajue Jianbao," 2–7.
83. Ibid., 8.
84. The overall height of the *zhan* presented by the report, 18 centimeters, is slightly inaccurate (Yueyang, "Hunan sheng Yueyang xian Fengxingzuishan yi hao mu Fajue Jianbao," 4). According to my own measurements, the vessel is 11.5 centimeters high, the lid is 5.9 centimeters high, and the overall height of the lidded *zhan* is 17.4 centimeters.

85. The vessel is cast with eight characters, "Yun Er Zi Zuo Zhu Qi ZhanYu 愠兒自作鑄其盞盂" ("Yun Er himself made and cast this *zhan-yu*"). See Liu, *Chutu Jianbo Wenzi Congkao*, 235–38. .
86. See Peng, "The Lost-wax Casting in Bronze Age China," 264, fig. 6.31.
87. For more images, see Peng, "The Lost-wax Casting in Bronze Age China," 266, fig. 6.33.
88. Tan, "Zhongguo Zaoqi Shila Zhuzao Wenti de Guancha yu Sikao," 37.
89. Hua, *Zhongguo Gudai Jinshu Jishu*, 179.
90. For more clear images of the openwork legs and handles, see Peng, "The Lost-wax Casting in Bronze Age China," 267, figs. 6.34, 6.35.
91. Nanyang, "Henan Nanyang Chunqiu Chu Pengshe Mu Fajue Jianbao," 5–15.
92. Ibid., 29–30.
93. In the English translation of the original report (Nanyang, "Henan Nanyang Chunqiu Chu Pengshe Mu Fajue Jianbao," 2011), Paul Nicholas Vogt reads the name 射 as "She" (see Nanyang Municipal Institute of Cultural Relics and Archaeology, "Peng She's Tomb of the Chu State of the Spring-and-Autumn Period in Nanyang, Henan," 2012). However, "Ye" (and less commonly, "Yi") are two other possible readings for this character.
94. Nanyang, "Henan Nanyang Chunqiu Chu Pengshe Mu Fajue Jianbao," 5–23, 29.
95. Ibid., 5, 7; Qiao, "Zuixin Kaogu Faxian," 147, 150–151.
96. Nanyang, "Henan Nanyang Chunqiu Chu Pengshe Mu Fajue Jianbao," 12.
97. For more clear image of the "absolute undercuts," see Peng, "The Lost-wax Casting in Bronze Age China," 269, fig. 6.37-2.
98. Nanyang, "Henan Nanyang Chunqiu Chu Pengshe Mu Fajue Jianbao," 9.
99. This collection's name/location shall not be mentioned in this book according to its request.
100. Personal communications, Su Rongyu, November 14, 2016. For information of another bronze vessel (*he*) in the same group, see Wu, *Shang Zhou Qingtongqi Mingwen ji Tuxiang Jicheng Xubian*, vol. 3, 337 (object No.: 0973, "Huang Zi Lou *he*").
101. For more clear image of the "absolute undercuts," see Peng, "The Lost-wax Casting in Bronze Age China," 271, fig. 6.40.
102. For more clear images of the *zhan* legs, see Peng, "The Lost-wax Casting in Bronze Age China," 272, fig. 6.42.

103. According to Hughes, "Technological style can be defined as the technical characteristics that give a machine, process, device, or system a distinctive quality. Out of local conditions comes a technology with a distinctive style" (*Networks of Power*, 405).
104. See Peng, "The Lost-wax Casting in Bronze Age China," 273, fig. 6.44.
105. For instance, scholars like Shi Quan and He Hao suspect that more than one Huang state existed during the Eastern Zhou period. For details, see He, "Chunqiu shi Chu Mieguo Xintan," 59–60.
106. The archaeological report, though without specifying the dimensions of the *pan*, does provide the transcription of its inscription, which may be translated including the sentence, "*Bu* selects auspicious metal and makes [this] washing *pan* for herself." Apparently, *Bu*, a female elite, was the maker of the vessel, and probably also the figure buried in the tomb. See Huang et al., "Hubei Suizhou shi Wenfengta Dongzhou Mudi," 29.
107. For more images, see Peng, "The Lost-wax Casting in Bronze Age China," 274-75, figs. 6.46, 6.47.
108. Huang et al., "Hubei Suizhou shi Wenfengta Dongzhou Mudi," 33.
109. See *Zhongguo Wenwubao*, April 11, 2014, p. 6.
110. Zhou and Huang, "Lost-wax Casting in Ancient China," 1635.
111. Zhou and Huang, "Shi Lun Qingtong Shidai Toukong Qingtongqi de Gongyi Tese," 154.
112. Zhang, "Guanyu Zeng Hou Yi Zun Pan Shifou Caiyong Shilafa Zhuzao de Shuping," 90.
113. Taking a step back, even if Zhou and Huang's interpretation of the formation of the wrinkle marks is accepted and the marks are regarded as a result of the imperfect flow of the molten bronze within "long and bent cavities," the lost-wax theory is not in contradiction with this case. Likewise, the lost-wax application could create the "long and bent" casting cavities for the openwork appendages of the Zeng *zun-pan* and Wenfengta *pan*. Thus, following Zhou and Huang's theory, it could also result in the imperfect flow of the molten bronze and give rise to the wrinkle marks. But in such a case the wrinkles appearing on the casting surface would cease to be evidence exclusively in favor of the lost-wax process. Whether the wrinkled marks were formed by the melting of the wax model or the poor flow of molten bronze, the lost-wax theory either fits it well or at least does not contradict it.
114. Huang Fengchun, public presentation, Qiulin Hall at The School of International Studies, Peking University, April 10, 2014.

115. Shown by Strahan, "Debating the Use of Lost-Wax Casting in Ancient China."
116. This viewing was arranged by Zhang Changping and guided by Yung-ti Li on May 21, 2014.
117. "Parting surface" refers to the contact surface of any two mold pieces.
118. Zhao, "Chunqiu Shiqi Shilafa Zhuzao Gongyi Wenti Tantao," 87.
119. Ibid., 87–88.
120. Henan et al., *Xichuan Xiasi Chunqiu Chumu*, 136; Zhao, "Chunqiu Shiqi Shilafa Zhuzao Gongyi Wenti Tantao," 88.
121. Zhao, "Chunqiu Shiqi Shilafa Zhuzao Gongyi Wenti Tantao," 88.
122. Ibid., 89.
123. Ibid., 88–90.
124. Ibid., 90.
125. Henan et al., *Xichuan Xiasi Chunqiu Chumu*, 114.
126. Ibid., 118.
127. Henan et al., *Xichuan Xiasi Chunqiu Chumu*, 384–385; Su et al., *Zhongguo Shanggu Jinshu Jishu*, 323.
128. Henan et al., *Xichuan Xiasi Chunqiu Chumu*, 384; Su et al., *Zhongguo Shanggu Jinshu Jishu*, 323.
129. Ibid.
130. Henan et al., *Xichuan Xiasi Chunqiu Chumu*, 385–386; Tan, *Canlan de Zhongguo Gudai Shila Zhuzao*, 40.
131. Tang, "Xichuan Xiasi Yihao Mu Qingtongqi de Zhuzao Jishu," 174–176.
132. Ibid., 176.
133. Ibid.; Henan et al., *Xichuan Xiasi Chunqiu Chumu*, 385.
134. For images, see Peng, "The Lost-wax Casting in Bronze Age China," 285–286, figs. 6.63 and 6.64.
135. For images, see Peng, "The Lost-wax Casting in Bronze Age China," 287, fig. 6.66.
136. See the official website of the Minneapolis Institute of Art (acc. 2005.92).
137. For images, see Peng, "The Lost-wax Casting in Bronze Age China," 288, fig. 6.65.
138. One such *sheng ding* looted from the tomb of Duke Ning of Xu was acquired by He Yonglin and illegally sold to the businessman Li Zhijia. After hearing of He's arrest, Li "donated" this artifact to the "China Foundation for the Development of Social Culture" (CFDSC) in Hong Kong.
139. *Zuo zhuan*, "Year Fifteen of Duke Cheng of Lu."

140. *Zuo zhuan*, "Year Twenty-six of Duke Xiang of Lu."
141. The "Huang Zi Lou" *he* is in the same group along with the two aforementioned "Huang Zi" *zhan* (figs. 67, 68). For images manifesting the three-dimensional openwork appendages to the "Huang Zi Lou" *he*, see Peng, "The Lost-wax Casting in Bronze Age China," 290, fig. 6.68. For discussion of the vessel's inscriptions, see Wu, *Shang Zhou Qingtongqi Mingwen ji Tuxiang Jicheng Xubian*, vol. 3, 337 (object No.: 0973, "Huang Zi Lou *he*").
142. See Wang, "Zhongqiu Hua Yue Ye, Jiajing Gong Shang Shi," 26–29.
143. For image, see Chase, *Ancient Chinese Bronze Art*, 71, no. 33. For discussion, see Singer, "The 'Unique' Object in Chinese Art."
144. See Li, "Guyuege suo cang Qingtong Bingqi Xuancui," 23–24, figs. 7, 8; Li, *Sihai Xunzhen*, 128.

Chapter 7

Design as the Driving Force

It is not surprising that many of the verified Eastern Zhou lost-wax castings have an openwork interlace structure. Openwork interlace emerged in China long before the noticeable use of the lost-wax process. Jenny So proposed that the use of lost-wax casting "was prompted by requirements of style."[1] But how did this "style" come into being?

According to Jessica Rawson, to render an interlace pattern, two or more motifs had to be interwoven (one motif passes in front of and then behind another), creating a third dimension.[2] Rawson also considered interlace and openwork, accompanied by small sculptural animal figures, to be "three features crucial to the vitality of Eastern Zhou design" inherited from Late Western Zhou.[3] Although admitting the significant role of southeastern China,[4] Rawson concluded that Inner Asia, specifically the Iranian Plateau, was the source of interlace, as evidenced by artifacts such as the bronze harness ring from the Luristan Plateau (c.1000–700 BCE) (see fig. 78, similar to Rawson's Luristan example, with the Western Zhou chariot fitting and *hu* in figs. 79, 80).[5] As Rawson asserted, "interlace was a common feature of the material culture of Inner Asia, where intertwined ropes and leather thongs must have been essentials of daily life."[6]

Rawson's argument, though not very convincing, finds supporting evidence in the handle of a bronze knife from Ningcheng Nanshangen of Inner Mongolia (fig. 81), an unprovenanced piece cast by a two-piece mold, probably from a similar context,[7] and in several daggers recovered from Beijing Yanqing Jundushan in the early first millennium BCE.[8] Similar examples illustrating interlace openwork in the northern and northwestern borderlands include a series of bronze scabbard fittings discovered in several early Western Zhou sites, such as Gansu Lingtai Baicaopo,[9] Shaanxi Baoji Zhuyuangou,[10] and Beijing Fangshan Liulihe (fig. 82a).[11] Especially notable is an unprovenanced scabbard fitting of this type currently in the Freer Gallery of Art, Washington, D.C.(fig. 82b), highly similar to the excavated Liulihe piece (fig. 82a), but with identifiable Europoid figures.[12] Similar styles of openwork interlace structure also gradually permeated the bronze assemblages of Zhou and its subordinate states, represented by the Ke *bo* bell of Late Western Zhou from Shaanxi Fufeng Renjiacun (fig. 83) and the box framework of Early Eastern Zhou from Shaanxi Hancheng Liangdaicun.[13]

"Intertwined ropes and leather thongs" are hardly persuasive as the original source of inspiration for the interlace of Chinese bronzes. The models for interlace had long been around in the form of baskets, mats, rattan, bamboo, and so on. Basketry patterns in interlace form have been imitated in pottery from Neolithic China (e.g., the Songze Culture in the Lake Tai area during the fourth millennium BCE) and the aforementioned openwork scabbard fittings (e.g., fig. 82) clearly imitate objects made by basketry or similar interweaving techniques.[14] It is therefore unnecessary to trace Western Zhou interlace in bronzes or jades to foreign sources, though artistic exchange between the Western Zhou and far western and northern regions must have existed. Furthermore, interlace and openwork are usually manifested individually in Western Zhou bronzes[15] and both could be revised in form, recombined with other ornaments, or even reinvented elsewhere.

According to Rawson, Eastern Zhou interlace benefited from several different traditions.[16] Probably because of its connection with the metropolitan centers, especially after the beginning of the Middle Western Zhou, southeastern China was another early center where interlace was adopted and developed; this is reflected in an inscribed Middle Western Zhou *you* from Anhui Tunxi (fig. 84) and some related jades.[17] Unlike bronze decorations from the metropolitan areas, which were thoroughly altered after the hypothetical Late Western Zhou "Ritual Revolution/Reform,"[18] southeast bronzes preserved some archaic features (e.g., the *taotie*, plumed birds, and a few vessel shapes, especially *you*, trumpet-shaped *zun*, and *gui* on arched bases)[19] and localized the metropolitan zoomorphic handles into filigree openwork flanges, such as those on a *gui* from the Harvard University Art Museums (fig. 85).[20] The openwork flanges present on southeastern castings might be linked to the life-like animal appendages of the early sixth-century BCE Xinzheng *lian he fang hu*[21] and those shown as a net of openwork belonging to the even later Houma *hu* (fig. 86). The most exotic and elaborate products of this development, as Rawson suggested, are the Xichuan Xiasi bronzes (especially the *jin*, fig. 26), thereafter the Zeng *zun-pan* (fig. 15), and finally the Xuyi *hu* (fig. 35).[22] In other words, Rawson argued that the direct source of the three-dimensional interlace openwork, a design often related to the lost-wax process, was the southeastern Chinese Central Plains region, where bronzes had been decorated with filigree openwork before the Eastern Zhou period.[23] This feature was absorbed by what she considered the "central southern tradition of Chu," which in turn influenced Xinzheng and Houma.[24] Rawson believed the strong two-dimensional interlace openwork found, for example, on the lid and foot ring of the aforementioned Houma *hu* (fig. 86) to have been derived from bronze castings in the north and west.[25]

In an argument similar to Rawson's, So also took the intertwined serpentine openwork as a characteristic of "Chu decoration."[26] Yet unlike Rawson, who posited two direct sources for the Eastern Zhou openwork interlaces,[27] So considered such structures to represent a single stylistic tradition widely followed in the central Henan (e.g., the Xinzheng *pan long fang hu*, fig. 87), Shanxi (e.g., the Xinjiang Liuquan *ding*, fig. 88), and the Chu-influenced regions (e.g., the Xiasi and Suizhou castings, figs. 26, 15).[28] So described the Xinzheng casting, reflected in the latest artifacts from the Xinzheng cache (e.g., the monumental *fang hu*, Fig. 87), as a stylistic leader: "the Xinzheng Lijialou cache remains unchallenged in size, monumentality, and artistic creativity among known middle Spring and Autumn finds."[29] Although confirming the openwork interlace as a Late Western Zhou invention, So considered it to be one of the two "most dramatic and distinctive traits of the Xinzheng bronzes."[30] The openwork interlace had become slightly three-dimensional at Xinzheng, most clearly shown on the *pan long fang hu* (fig. 87b), where "the intertwined serpents seem to raise their heads above the surface, creating an especially lively effect."[31]

According to So, the openwork interlace of Xinzheng castings not only developed further in the Chu cultural region, such as in Xichuan Xiasi,[32] but also benefited the Houma castings, as reflected on both artifacts and mold fragments. Though it had a broad impact, the "monumental and exuberantly ornamental" Xinzheng style was still treated by So as associated with the distinctive tradition of the spreading Chu,[33] whose castings might have felt the influence of wood carving.[34] So suggested that the three-dimensional effect of the intricate interlace openwork, as an example of "technical bravura," was likely inspired by the comparatively easy-to-make wood carving, suggested by the correspondence between the bronze drum stand from the tomb of Marquis Yi of Zeng and the carved wooden box from Hubei Jiangling Yutaishan (fig. 89).[35] This suggestion has been criticized by Colin Mackenzie who argued that such correspondence "bears witness to the wood carvers' emulation of rather than inspiration to the casters" because the most elaborate fabrications of the interlace openwork were expressed in bronzes.[36] However, considering that artists may have used the more valuable material to imitate other media in pursuit of a more compelling visual effect and aesthetic appeal, So's proposal might still be correct. Based on the current knowledge, such three-dimensional interlace openwork seems to be the most frequent carrier for lost-wax casting.

By examining a group of diatrete vessels with the double-tier openwork structure, Helen Loveday has found a technical distinction behind their outward similarity: two of these vessels with strongly two-dimensional interlace openwork, the aforementioned Liuquan *ding* (fig. 88) and an unprovenanced *he** cup in the MEG (Musée d'ethnographie de Genève), Switzerland,[37] can be reasonably considered to be castings executed solely by the section-mold process, probably from workshops related to Houma. The rest of those with three-dimensional filigree-like interlace, namely the aforementioned Xuyi *hu* (fig. 35), the Fengxiang incense burner (fig. 38), the unprovenanced Winthrop jar (fig. 41), and the Metropolitan tripod (fig. 36), are thought to be castings combining the lost-wax and section-mold processes.[38] At first glance, Loveday's argument seems to resemble Rawson's dichotomy of interlaces. However, Loveday did not differentiate the openwork interlaces into those that are southeastern and those that are northern and western in origin. In fact, So's proposal for the two derivations of the same Xinzheng prototype of interlace openwork, one related to Houma workshops and the other from the supposed Chu cultural region,[39] is compatible with Loveday's suggestion. Because Loveday recognized that the four castings in her analysis that seem to have been cast with the lost-wax process (figs. 35, 36, 38, and 41) were both stylistically and geographically distant from each other, she argued that lost-wax was not the prerogative of the Chu cultural region and that the north-south/Jin-Chu dichotomy is too simple.[40] Indeed, the four diatrete vessels that Loveday believed to have used the lost-wax process are stylistically different from their Xiasi and Suizhou counterparts (e.g., figs. 58 and 15), where lost-wax casting was used frequently on the appendages. Considering that the dates of the four vessels appear to be considerably later than the first lost-wax castings of China, it is likely that the lost-wax process was initially mastered solely in the Chu cultural region and only later became geographically widespread.

Wan Li viewed the openwork interlace in a unilineal tradition and divided its development into four stages. The openwork box from Shaanxi Hancheng Liangdaicun[41] in the early Spring and Autumn period and the aforementioned Liuquan *ding* (fig. 88) from Shanxi Xinjiang in the middle Spring and Autumn period represent Stage One, marked by two-dimensional openwork and carved lines on the interlace.[42] The aforementioned Metropolitan *zhan* (fig. 53), Yueyang *zhan* (fig. 64), and Yexian *zhan* appendage (fig. 61), believed to come from the first quarter of the sixth century BCE, are examples of Wan's Stage Two, featuring multi-layered three-dimensional openwork used only on the appendage.[43] The Xiasi *jin* (fig. 26) (no later than c. 552 BCE) and Zeng *zun-pan* (fig. 15) (no later than c. 433 BCE) stand for Stage Three, illustrating the possibilities of using three-dimensional interlace openwork as the main decoration of the vessel; the aforementioned Pixian *bu* (fig. 43) and the stylistically similar Fengxiang incense burner (fig. 38) are also examples of Wan's Stage Three.[44] The most intricate and elaborate gold-and-silver inlaid Xuyi *hu* (fig. 35), combining various techniques, is viewed as representative of Stage Four.[45] By roughly dating and mapping these unearthed artifacts, Wan proposed that the two-dimensional interlace openwork was transmitted from the northwestern Zhou states (e.g., Shaanxi Hancheng and Shanxi Xinjiang) to the Chu cultural region (e.g., Henan Yexian, Xichuan, and Hubei Suizhou) and developed into three-dimensional décor, finally spreading to other regions with further elaboration.[46] Wan inferred that the gold-and-silver inlaid Xuyi *hu* manufactured by the state of Yan, which was stylistically distinct from its counterparts in the Chu cultural region, was likely influenced by the nearby non-Chinese state, Zhongshan.[47]

Among these opinions, the single evolutionary scheme proposed by Wan Li is the most likely candidate for criticism. Clearly Wan believed that the design of the openwork interlace developed for several centuries and across different regions, but only along a single path. Yet it bears asking: do all the openwork interlaces that Wan listed belong to the same strain? Wan seems to have understood the making of openwork interlace as a secret recipe or craft that was passed down along a single line. This view is unrealistic. Accepting it means assuming that designs spread slowly and gradually and that states within China had distinctive traditions over long periods because they were largely isolated from each other. However, judging by verifiable levels of interstate trade,[48] it is entirely likely that many states were in constant contact with each other, exchanging works of art, artists, and artistic ideas.

In contrast to Wan, Rawson established a dichotomy of Eastern Zhou openwork interlaces in two separate traditions (northern and western castings versus southeastern castings), both of which she held to have originated in Inner Asia. Interlace was indeed widely used in Inner Asia in the form of intertwined ropes and leather thongs, as Rawson argued, but her argument is predicated on the idea that these materials constituted a unique source. Did the Chinese not have these same things? The Chinese certainly had baskets, which are, the world over, the most ubiquitous source for interlaced designs.[49] Moreover, the strong two-dimensional openwork interlace represented by the lid and foot ring of the Houma *hu* (fig. 86), believed by Rawson to have been derived from regions north and west,[50] has been widely found on bronzes in Chu-influenced regions; the evidence for this includes such items as the censer cup from Hubei Jiangling Yutaishan (fig. 90) and even the climbing felines appended to the Zeng *zun-pan* (see fig. 15). Yet Rawsom considered the three-dimensional openwork interlace on the very same *zun-pan* set to have been gradually derived from the Western Zhou castings of the southeastern Chinese Central Plains. In the absence of credible evidence, this dichotomy remains unconvincing.

Furthermore, Rawson's argument implies that evolution towards a fully three-dimensional form of openwork interlace had already occurred in the Western Zhou period. In fact, three-dimensional openwork began to take form only during the Eastern Zhou, and its source of inspiration was likely contemporary. As So has speculated, the desire to imitate the decorative effect achieved by wood carving might have stimulated the emergence of three-dimensional openwork interlace in bronze. So's opinion is insightful but suffers from only seeing one aspect of the event. The design of some verified lost-wax cast openwork interlaces, most notably that of the comparatively simple interlace design on the *fang hu* cover from Jiangsu Pixian Liulin (figs. 44, 45), did not originate from wood carving but instead illustrates something invented by the technique of the lost-wax process. When craftsmen obtained the wax, it would have been natural to think of rolling it and forming it into strips and then using them to fabricate something like the current design (e.g., see fig. 45).

To add a layer of complexity to this issue, because the interlace itself is conceptually involved with the third dimension, the physically two-dimensional interlace could not fully manifest this pattern until it became tangibly three-dimensional. Accordingly, it is not completely unhistorical to view Eastern Zhou three-dimensional interlace as a realization of the previous two-dimensional interlace. It is quite possible that the lost-wax method helped craftsmen realize this transformation or that the motivation to achieve this realization facilitated the use of the lost-wax process. For instance, as So indicated, compared with the likely lost-wax precast openwork interlaced accents on the *zhan* from Xichuan Xiasi M1 (fig. 59), the crown of an unprovenanced Sackler *zhan* (fig. 91b), although far simpler was also precast, and "such consistency in casting technology suggests that the same technical traditions may have guided the workshop or workshops that produced the two vessels."[51] Meanwhile, So also proposed that "significant strides in both design and casting technology must have occurred after the manufacture of the Sackler vessel,"[52] based on the far more complex, multi-layered openwork interlace of the Xiasi *zhan* (fig. 59).[53] Do "the same technical traditions" that produced the Sackler *zhan* and the Xiasi *zhan* include the use of the lost-wax process? Or alternatively, was the lost-wax process used only to make the more formidable openwork structure, which the Sackler *zhan* (fig. 91) had not obtained yet? Further investigation of the Sackler *zhan* and more or less contemporary artifacts with similar openwork appendages, such as the famous *pan long fang hu* (fig. 87b)[54] and the three large *ding* from Henan Xinzheng Lijialou,[55] might offer some new insight on this issue.

A *ding* with a visually multi-layered openwork crown from Hubei Dangyang Zhaojiahu (fig. 92) should be of special interest in this regard. At 35.7 centimeters high and 32.9 centimeters in diameter at the belly, this *ding* cauldron was excavated from the tomb ZHM8 of the cemetery. Similar to the aforementioned Yueyang *zhan* (figs. 64, 65) and Nanyang *zhan* (fig. 66), the Dangyang *ding* has an exquisite three-dimensional circular crown in a garland-like form with a central circular void. Although it exhibits a less prominent three-dimensionality and greater regularity in its spatial disposition, the crown is still rendered in a multi-layered network composed of massed serpentine interlaces. The technical analysis section in the archaeological report does not address the lost-wax issue.[56] The official website of Yichang Municipal Museum, however, announces that the crown ornament of this *ding* lid was cast by the lost-wax method, without providing any specific reasons to support this claim.[57] Without being able to view this object in person or obtain a high-resolution photograph, I remain less than confident that its openwork crown was cast by the lost-wax method. Apart from the still questionable crown, no other parts of this *ding* seem to have been cast by the lost-wax process. With a slightly domed lid, a pair of angular inverted U-shaped handles issuing immediately below the rim, a bulbous body, and three cabriole legs, this vessel seems to have been cast during the middle Spring and Autumn period (c. 670–570 BCE). Archaeological materials from the same tomb verify this date.[58] Were the crown of this *ding* to be confirmed as lost-wax cast, it would be among the earliest known lost wax samples from the Central Plains of China.

The question remains: which came first in China, the design of the three-dimensional interlace or the technology of the lost-wax process? More precisely, did the motivation for the use of this process to achieve the three-dimensionality of the interlace or to imitate a contemporary design in another medium (e.g., woodcutting, basketry, rattan, or bamboo weaving), precede the application of lost-wax technology, or vice versa? This causality dilemma is nearly impossible to resolve. In the light of the recognizable diversity present within the three-dimensional interlaces verified to have been made by the lost-wax method, perhaps the question could be re-posed as a dilemma about design and technology. That is, more than one of the possibilities discussed earlier might have happened in the history of explorations by Chinese craftsmen. Regardless of the conclusion, design cannot be ignored as a driving force behind the rise of lost-wax castings in China. As I will discuss in chapter 8, Chinese casters are likely to have known the basic principles of lost-wax technology long before the Eastern Zhou period. But why would the lost-wax method, which Chinese casters had ignored for so long, have suddenly emerged in metalworking in the Eastern Zhou? The reason was probably closely related to the stimulation of design. Once the lost-wax process began to be adopted, craftsmen might have experimented with the wax and rendered additional designs, such as the interlaced pattern of the Pixian *fang hu* cover (e.g., fig. 45).

Notes

1. So, "Hu Vessels from Xinzheng," 26. Strictly speaking, So's formulation here makes something called "style" into an agent, something that actively causes historical change. That is simply fiction, and there is no such agent. For a systematic critic of "styles" that are misused in art history, see Bagley, *Gombrich among the Egyptians*, 23–47.
2. For details, see Rawson, *Western Zhou Ritual Bronzes from the Arthur M. Sackler Collections*, 114.
3. Ibid.
4. For details, see Rawson, "Western Zhou Sources of Interlaced Motifs."
5. Rawson, "Carnelian Beads, Animal Figures and Exotic Vessels," 2–3.
6. Ibid., 16–17.
7. See So and Bunker, *Traders and Raiders on China's Northern Frontier*, 122–123, no. 40.
8. For details, see Rawson, "Carnelian Beads, Animal Figures and Exotic Vessels," 16–17.
9. See *Zhongguo qingtongqi quanji*, vol. 6, 191.
10. See Rawson, "Carnelian Beads, Animal Figures and Exotic Vessels," 19, fig. 26.h.
11. For details, see So and Bunker, *Traders and Raiders on China's Northern Frontier*, 123–125; Rawson, "Carnelian Beads, Animal Figures and Exotic Vessels," 19, fig. 26.
12. This scabbard fitting was given to the Freer Gallery of Art from the "Therese and Erwin Harris Collection, formerly John D. Rockefeller 3d Collection." See So and Bunker, *Traders and Raiders on China's Northern Frontier*, 48, Plate 9; 124.
13. For details, see Rawson, "Western Zhou Sources of Interlaced Motifs"; Rawson, "Carnelian Beads, Animal Figures and Exotic Vessels," 18, fig. 24.b.
14. Robert Bagley, personal communication, July 19, 2017.
15. For details, see Rawson, *Western Zhou Ritual Bronzes from the Arthur M. Sackler Collections*, 111–123.
16. For details, see Rawson, "Western Zhou Sources of Interlaced Motifs."
17. Rawson, *Chinese Bronzes*, 50; Rawson, "Western Zhou Sources of Interlaced Motifs"; Rawson, *Western Zhou Ritual Bronzes from the Arthur M. Sackler Collections*, 116.

18. For details, see Rawson, "A Bronze-Casting Revolution in the Western Zhou and Its Impact on Provincial Industries"; Falkenhausen, *Chinese Society in the Age of Confucius*, 52, note 36. In fact, it is not certain whether the "Ritual Revolution/Reform" is historical fact. For instance, one of Rawson's arguments for it is the "sudden" appearance of bells around 900 BCE. But the Zhou was very likely using imported southern bells long before that. Robert Bagley, personal communication, August 2014.
19. Rawson, *Chinese Bronzes*, 50.
20. See Rawson, "Western Zhou Sources of Interlaced Motifs."
21. See *Zhongguo qingtongqi quanji*, 7, Plate 22.
22. For details, see Rawson, "Western Zhou Sources of Interlaced Motifs."
23. Ibid.
24. See Rawson, *Chinese Bronzes*, 52; Rawson, "Western Zhou Sources of Interlaced Motifs."
25. Rawson, *Chinese Bronzes*, 56.
26. See So, "Hu Vessels from Xinzheng"; So, *Eastern Zhou Ritual Bronzes from the Arthur M. Sackler Collections*, 26. Although very few castings (e.g., the inscribed Chu bell from the tomb of Marquis Yi of Zeng) can be considered to be unquestionable "Chu bronzes," most scholars, including So ("Hu Vessels from Xinzheng"), still take bronzes with eccentric, flamboyant, and sometimes-monumental features found at the central southern China to represent the "unrecognized Chu decorative style."
27. For details, see Rawson, "Western Zhou Sources of Interlaced Motifs."
28. So, *Eastern Zhou Ritual Bronzes from the Arthur M. Sackler Collections*, 26–28, 35.
29. Ibid., 21–29.
30. Ibid., 26, 35. The other such trait is the "ornamental animal appendages" (So, *Eastern Zhou Ritual Bronzes from the Arthur M. Sackler Collections*, 35), also frequently accompanied by many lost-wax castings such as the Xiasi *jin* (fig. 26) and Zeng *zun-pan* (fig. 15).
31. So, *Eastern Zhou Ritual Bronzes from the Arthur M. Sackler Collections*, 26.
32. Ibid., 27–29, 35. In both Xinzheng and the Chu cultural region (e.g., Henan Xichuan and Hubei Suizhou), the openwork interlace was mainly used as secondary decorations on the circular crowns of vessel lids, as well as rim and animal appendages. It seems the supposed Chu bronzes were stylistic successors of the Xinzheng castings. For details, see So, *Eastern Zhou Ritual Bronzes from the Arthur M. Sackler Collections*, 26–27, 30–31, 53–54.

33. Ibid., 35.
34. For details, see So, "Hu Vessels from Xinzheng." The woodcutting might have already been exerting an influence on the Xinzheng castings. For instance, as Mackenzie argued, the miniature bird and feline figures of the Xinzheng *lian he fang hu* (see *Zhongguo qingtongqi quanji* [7], Plate 22), plausibly "represent a borrowing from an already established tradition of southern wood sculpture" ("The Chu Tradition of Wood Carving").
35. For details, see So, "Hu Vessels from Xinzheng."
36. For details, see Mackenzie, "The Chu Tradition of Wood Carving." Apparently, Mackenzie has not considered the great majority of the lacquers that have been lost and remain unknown to us.
37. For images of this *he** cup, see Loveday, "Diversity in Eastern Zhou Bronze Casting," 105, fig. 3; 115, fig. 8.
38. Ibid., 119–133.
39. See So, *Eastern Zhou Ritual Bronzes from the Arthur M. Sackler Collections*, 27–29, 35.
40. Loveday, "Diversity in Eastern Zhou Bronze Casting," 132.
41. See Rawson, "Carnelian Beads, Animal Figures and Exotic Vessels," 18, fig. 24.b.
42. Wan, "Shixi Zhongguo Xianqin Shiqi Toukong Panlongwen Qingtongqi Zhuzao Jishu de Fazhan yu Chuanbo," 110–111.
43. Ibid., 111.
44. Ibid., 111–112.
45. Ibid., 112–113.
46. Ibid., 113–114.
47. Ibid., 114.
48. Beyond evidence from received texts, interstate trade during the Eastern Zhou period has also been corroborated by archaeological records. For instance, the "Jin Jiang" *ding* excavated from Shaanxi Hancheng and the "Rong Sheng" bell set in the Poly Museum, Beijing, both explicitly described a significant trading event in which the Jin state exchanged metals for salt. For details, see Li, "Rongsheng Bianzhong Lunshi," 79–82.
49. Robert Bagley, personal communication, August 2014.
50. But during the Erligang period there was already a bronze *gu* vessel from Hubei Panlongcheng with an openwork foot, whose textile pattern apparently copies baskets or woven cloth.
51. So, *Eastern Zhou Ritual Bronzes from the Arthur M. Sackler Collections*, 120, note 11.

52. Ibid., 120.
53. An even more impressive case is the openwork interlace of the Metropolitan *zhan* in fig. 53.
54. Zhang proposed that the slightly three-dimensional interlace crown of this *fang hu* was cast by the "lost-lead" method, a variation of the lost-wax process ("Zhongguo Zuizao 'Shilafa'"). His argument, while unconvincing, still offers one possibility. For descriptions of this *fang hu* vessel, also see So, "Hu Vessels from Xinzheng"; So, *Eastern Zhou Ritual Bronzes from the Arthur M. Sackler Collections*, 22–26.
55. E.g., see Peng, "The Lost-wax Casting in Bronze Age China," 306, fig. 7.22. For discussions, see Sun, *Xinzheng Yiqi*, 47–51; So, *Eastern Zhou Ritual Bronzes from the Arthur M. Sackler Collections*, 119–120; Henan and Taibei, *Xinzheng Zhenggong Damu Qingtongqi*, 79–86.
56. See Hubei and Beijing, *Dangyang Zhaojiahu Chumu*, 257–265.
57. For details, see: http://www.ycbwg.com/view.asp?id=788 (accessed March 6, 2018).
58. For more details, see Hubei and Beijing, *Dangyang Zhaojiahu Chumu*, 118.

Chapter 8

The Origin of the Lost-wax Technique in Bronze Age China

As I have argued, the evidence for the existence of the lost-wax technique in Bronze Age China, based on unquestionable lost-wax castings from the Eastern Zhou period, is incontrovertible. But was the lost-wax process developed independently in China, or was it introduced from or stimulated by outside sources?[1] If the latter is true, when, where, why, and how was lost-wax casting transmitted? To answer these questions, three primary hypotheses are circulating today: (1) the lost-wax process was developed locally in China from the earlier "burn-out" casting technique; (2) the lost-wax process was introduced from the southwest; and (3) the lost-wax process was introduced from the north. In the following pages, I will test these hypotheses with known archaeological materials and examine new clues.[2]

INDEPENDENT INVENTION IN CHINA?

The Eastern Zhou lost-wax castings clearly demonstrate that early Chinese foundries still relied on section molds even when they were already using the lost-wax process with a high degree of sophistication. This already advanced level of technology and craft indicates that Chinese casters may have mastered the lost-wax or similar processes considerably earlier than the Eastern Zhou. Noel Barnard has reported finding fiber-like imprints on the swing handles of some early bronzes in the Lingnan region (Guangdong, Guangxi, and Hainan provinces), but no mold marks are visible, showing that the model in this case might have been made of a rope that could be incinerated and then cleared away in the process.[3] According to Barnard, such a "lost-cord" process emerged in the Lingnan region no later than the beginning of the Spring and Autumn period.[4] Tan has also observed a similar phenomenon on even earlier bronzes.[5] For instance, the swing handle of a double-owl *you* in the Shanghai Museum attributed to the Anyang period of the late second millennium BCE (figs. 93a, b) is designed to look like an intertwined rope; neither mold marks nor erased traces have been found on the main part of it, but a mold mark and the imprint of a string bundle appear on each end of the handle (see fig. 93c).[6] According to Tan, the manufacturing procedure for this swing handle was as follows: the rope model with the string-tied loop at its ends was coated with the investment mold, which was made sectional only around the "loop" portion; thereafter, the rope-model was burnt away through the baking process and cleared off the "loop" areas; after the small "section-molds" around the loops were reassembled, the molten bronze was poured in, and the handle was finally cast.[7] As Tan noted, a swing handle with the same features has also been found on another Anyang-style vessel in the Changsha museum in Hunan.[8] According to Tan's observations, a rope-like swing handle lacking any mold mark appeared no later than the Erligang-Anyang Transition Period (c. 1300 BCE), as seen in a *hu* from Beijing Pinggu Liujiahe.[9] Later examples include three *yan* from Anhui Tongling in the Spring and Autumn period (e.g., see fig. 94).[10] These artifacts, according to Tan, were all made by

Barnard's lost-cord process or by the similar but more inclusive "burn-out" process described by Tan himself.[11]

Tan's burn-out examples include not only the supposed lost-cord castings mentioned earlier but also other bronzes in the following categories. (1) Artifacts with narrow and deep sunken décor over the entire body but cast from a limited number of section molds, such as the tiger from Jiangxi Xin'gan (c. 1300 BCE) (fig. 95a) and a *nao* bell from Zhejiang Changxing (fig. 95b).[12] According to Tan, when a mold section was removed in a certain direction from the model, some of the protruding ridges, which correspond to the groove-shaped ornaments of the model, would have been unavoidably broken off (see diagram, fig. 95c).[13] (2) Artifacts with sunken ornament with an external opening narrower than the interior, such as the *gui* container from Zhejiang Changxing (figs. 96a, b; for the corresponding diagram, see fig. 96c).[14] (3) Artifacts with hooked, undercutting ornament, such as one of the *nao* bells from Jiangxi Xin'gan.[15] Tan claimed that for all of these artifacts, the mold decoration could not have been kept intact when the mold section was removed from the model. As a result, the "burn-out" process was necessarily used in their manufacture.[16] As Tan argued, the model for such items had to be at least partly made of a material that was as easy to carve on as to burn (e.g., wood, plant fiber, starch). Furthermore, to thoroughly dispose of the resultant ash, any investment mold for these artifacts would have to be wholly or partially fashioned in sections, which accounts for the mold marks on the objects.[17]

Although Tan was among the first to designate bronzes in categories (1) to (3) as burn-out castings, his arguments about the difficulties of withdrawing molds from models are not new. Discussions centered on the transfer of décor between positive and negative have a long history in the literature.[18] Most other scholars, such as Zhou Weirong et al. and Lucas Nickel, have chosen to clear up this "problem" by suggesting that the undercutting decorations were created in the molds, not the models.[19] However, undercuts as intricate as those with openwork ornament (see fig. 12 for illustration) were still achievable through transfer by Houma foundries during the mid-first millennium BCE, as demonstrated by baked-clay pattern blocks with identical decorations (fig. 13).[20] None of Tan's examples has undercuts as deep or intricate as the Houma piece. Although Tan's diagrams (figs. 95c, 96c) seem to justify his technical logic, just as Nickel's comment on a similar diagram did,[21] "diagrams and logic are very poor guides to the capabilities of a craftsman with years of practice and centuries of knowhow behind him."[22]

Barnard and Tan's lost-cord examples are thought-provoking. Some of their examples, however, such as the handles of the Tongling *yan* (fig. 94a), as Zhou et al. noticed, seem to have been cast from section molds, judging by the presence of a long, indistinct mark (figs. 94c, d).[23] Additionally, a piece from the Fogg Art Museum (fig. 97) with a design almost identical to that of the double-owl *you* in the Shanghai Museum, unmentioned by Tan, also possesses the same type of swing handle with a "string bundle" on the ends. Yet this swing handle was clearly cast by the section-mold process, a conclusion supported by the presence of a pronounced mold mark running through it. Was this section-mold cast handle an imitation of the supposed lost-cord cast piece of the Shanghai Museum *you*? Or was the Shanghai *you* also a section-mold casting whose mold mark was perfectly polished away? Is it possible, as Zhou et al. have argued,[24] that all the supposed lost-cord castings listed by Tan were in fact section-mold castings?

According to Zhou et al., the hypothesized burn-out/lost-cord process is not tenable, because (1) it was unnecessary to burn out the cord, even if it was really used as the model (mold sections could have easily been made from such a cord-model); (2) the molds, after baking, would not typically be reopened before casting, and therefore the burnt ashes would have been difficult to clear. The latter is an especially convoluted explanation for objects that lack any mold marks—and with ashes inside the mold, a casting cannot be successful.[25] Such objections, though not thoroughly persuasive, provide useful points of departure for further inquiry. Indeed, if one does not open the mold after burn-out, the ashes are difficult to clear; if one does open the mold, why bother to burn the cord at all?[26]

Was the supposed lost-cord/burn-out process, if it in fact existed, truly the source for the lost-wax process of early China? Scholars like Tan and Hua have believed so.[27] As Zhang Changping has indicated, however, the burn-out process "used on the bronzes of Late Shang or the Shang-Zhou Transition periods" stood apart from the Eastern Zhou lost-wax process both in time and in technology and, as a result, cannot be considered the source for the latter.[28] Although several assumed lost-cord castings may date to the Spring and Autumn period[29] (i.e., not as early as Zhang supposed) and although Zhang's argument requires some clarification and support,[30] it is indeed doubtful that the lost-wax process derived from this technically distinct burn-out process. The burn-out process might have been a step on the way to the invention of lost-wax casting, but it does not really prove anything; in particular, it does not prove independent invention of lost-wax casting in China. (Consider, for example, the possibility that the burn-out process was instead inspired by lost-wax casting.) In addition, the jump from rope to wax is not an obvious one.[31] Burn-out casting is an attempt to find a local origin for lost-wax casting. The attempt is not convincing, but this does not mean that lost-wax casting was *not* locally invented; it says nothing one way or the other.

Diffusion from the Southwest?

Ever since the discovery of realistic and highly three-dimensional bronzes that are supposed to be lost-wax cast at the necropolis of Jinning Shizhaishan in the Dian area of Yunnan Province in the 1950s,[32] many scholars have believed that the Chinese lost-wax process was introduced from the southwest region of modern China or locations even farther away (e.g., Southeast Asia or South Asia). For instance, Barnard has suggested that investment casting, including several variations such as the lost-wax, lost-cord, and "lost-lead" methods, was introduced to the

Chu region via Yunnan and perhaps also the Lingnan region, coming ultimately from Southeast Asia (e.g., Thailand).[33] However, lost-wax technology in Yunnan is probably not nearly as early as Barnard has supposed (c. 7th–6th centuries BCE).[34] According to recent research, incipient copper-based metallurgy in Yunnan (c. 13th–8th centuries BCE), represented by the hammering and open/bivalve-mold casting activities at the Haimenkou site, might have been introduced from steppe-related cultures in the north during the late second millennium BCE.[35] Around the seventh century BCE, a group of copper and bronze drums, like those discovered at Chuxiong Wanjiaba, appeared in west-central Yunnan.[36] Unlike Barnard's date for the appearance of lost-wax in Yunnan, this is a date I find supportable by tangible evidence. It should be noted, however, that the date of the seventh century BCE here is for section molds, not lost-wax casting.

The first lost-wax castings emerged in Yunnan probably no earlier than the middle to late Warring States period (c. 350 BCE). One example is a three-dimensional openwork plaque from Chenggong Tianzimiao M41 (fig. 98a);[37] slightly later lost-wax examples include several similar plaques, such as the hunting-scene plaque from Jiangchuan Lijiashan M13, dated to around the third century BCE (fig. 98b).[38] The Shizhaishan assemblage, although it includes many lost-wax bronzes, falls generally within the range of the Western Han dynasty (202 BCE–8 CE).[39] The reason Barnard[40] and other scholars[41] have dated the Yunnan lost-wax casting too early is by and large related to a single and problematic radiocarbon dating of Jiangchuan Lijiashan M21, which yielded lost-wax artifacts shown to belong to the range 625±105 BCE (or 550±105 BCE, if taking the radiocarbon half-life as 5570 rather than 5730 years).[42] This date has been questioned by scholars such as Wang Ningsheng[43] and Pirazzoli[44] and has since been abandoned by academia.[45] As no local technical tradition is known, the lost-wax process seems to have been imported, probably from the steppes of Eurasia, along with other metallurgical techniques like tinning, gilding, chasing, and inlay.[46] Some newly excavated designs and artifacts, such those with animal-combat and horseback-riding motifs, trident-shaped iron swords with bronze hilts, and animal-shaped openwork plaques, also support the hypothesis of this "steppe" influence.[47] Chu cultural impact, although limited, also began to appear, perhaps partly related to the arrival of the Chu army led by Zhuang Qiao in the early third century BCE, according to the received texts.[48] At present, Barnard's suggestion that the investment-casting technique was transmitted from Yunnan and the southern regions of China to the north[49] seems to be incorrect and may even be the diametric opposite of the real events.

The Yunnan bronze industry was, in fact, the earliest large-scale lost-wax production in all of southwest China and Southeast Asia, although small objects such as figurines and bracelets might have been made by this process during the early Bronze Age of Indochina.[50] For instance, bangles (bracelets, anklets, and rings) believed to have been made by the lost-wax process seem to be characteristic of northeast Thailand:[51] at the Ban Na Di site (lasting from c. 1200 BCE to 200 CE?), insect wax[52] was said to have been found between the central clay core and the corroded metal of one bracelet from Mortuary Phase 1b, Level 7 (c. 900–500 BCE?) and on two other bracelets there.[53] Yet it is doubtful that wax could survive the casting process. Bangles assumed to be made by the lost-wax process have also been found at other northeast Thai sites, such as Non Nok Tha and Ban Chiang.[54] Technically parallel to northeast Thailand, Vietnam might have also have known investment casting no later than the Go Mun phase (which has been taken to represent c. 1000–500 BCE[55]), judging by certain items that have been claimed to have been produced by the lost wax process,[56] though no convincing evidence has been given to support this idea.[57] The chronology of Bronze Age Southeast Asia, especially in the region of northeastern Thailand, is a long-debated issue. Joyce White has insisted that bronze metallurgy appeared in northeast Thailand around 2000 BCE, represented by the early metallurgy at Ban Chiang (c. 2000–1700 BCE).[58] Charles Higham, in contrast, believed that bronze technologies emerged at Ban Chiang and elsewhere in Southeast Asia no earlier than 1500 BCE.[59] Recently, following Higham's new chronological framework based on seventy-five radiocarbon dates from the well-stratified site Ban Non Wat[60] (close to the site of Ban Chiang), more scholars[61] have been willing to believe that metalworking started in Southeast Asia just shortly before 1000 BCE. If this new chronology is accepted, the chronology for lost-wax casting there would also have to be postdated accordingly. Now it seems that lost-wax casting in Southeast Asia was practiced not much earlier, and in fact possibly later, than in China. In truth, there is no conclusive evidence that the lost-wax process was ever practiced in Bronze Age Southeast Asia, because castings as simple as the bangles mentioned earlier could easily have been cast by the piece-mold method and any visible mold marks polished away. The announcement that wax was discovered within one casting, as mentioned earlier, seems irrelevant.

Compared with Yunnan and Southeast Asia, Sichuan is closer to the Central Plains of China (centered around the Middle Yellow River Valley) and the Middle Yangzi region in bronze technology, as indicated by the section-mold casting practiced at Sichuan Sanxingdui.[62] Recently Christopher Davey has suggested the use of the lost-wax process based on his inspection of a mask from Sanxingdui Pit 2, a burial unit dated to the late second millennium BCE (fig. 99a).[63] According to Davey, parts of an "investment mold" remain on the rear attached to the nose cavity (fig. 99b). Davey also identified a scratch close to an opening as possibly a "false cut" originally made in the wax model (fig. 99c). Based only on this evidence, Davey has suggested, "The wax model of the mask would have been made from sheets of wax. The openings would have been cut with a knife...."[64] By speculating that the 4,600 cowrie shells came from the Indian Ocean, Davey linked Sanxingdui with South Asia, one center where the lost-wax process was practiced early and argued that the technique could have been transmitted by the same path as the later "south-western silk road."[65] What Davey characterized as "investment mold" remains could just as well have been left by mold sections, and the scratch could also have been a false cut made on the clay model. Therefore, Davey's argument has proved far from persuasive. The proposal of a connection between Southwest China and South Asia based on the cowrie shells is thought-provoking, but cowries are easily transported valuables that can travel great distances and pass simply from hand to hand. The exchange mechanism that moved cowries such a distance would not necessarily have been used to transmit metallurgical expertise, which represents a body of knowledge that must have been carried by the metalworkers themselves.[66] Nor have the shells from Sanxingdui been verified as originating from the Indian Ocean; furthermore, similar discoveries have been widely distributed throughout China, such as at Anyang in the Chinese Central Plains and Dadianzi in Inner Mongolia. Does this mean that Anyang and Dadianzi also necessarily got their casting technique from India? This is clearly not the case. Perhaps the only useful point to be made with regard to these cowries is that they indicate that trade was probably taking place everywhere and over vast distances.

Diffusion from the North?

If the diffusion hypothesis is accepted, the lost-wax technique is more likely to have come from the north than from the southwest, via the region variously termed the Northern Corridor, the Northern Frontier, and/or the Northern Zone.[67] This region, extending from Xinjiang and Gansu in the west to Liaoning and Jilin in the east,[68] includes two geographically and culturally distinct zones: the Xibei (the northwest) and the Dongbei (the northeast), divided by the Taihang Mountains, roughly following the boundary between modern Shanxi and Hebei Provinces.[69] As Robert Bagley has argued,

> whatever the date of its arrival, the lost-wax process is likely to have come to China by way of the northern steppes, and highly sophisticated technological transfer along this route is attested as early as the first century of the Anyang period by the horse-drawn chariot, a West Asian invention which figures in Anyang burials.[70]

As horse-drawn chariotry was a technological complex involving construction, use, and maintenance,[71] it seems that Anyang around 1200 BCE had learned chariotry, horse breeding, and management, as well as other necessary knowledge and skill, through some sort of exchange.[72] This exchange is also supported by the mutual borrowings between Anyang and the Northern Zone, such as northern artifacts (e.g., bow-shaped objects, mirrors, and knives) found at Anyang and bronze vessels found in the Northern Zone.[73] It is noteworthy that some Anyang-period knives with intricately formed animal-head terminals unearthed in northern Hebei and Shaanxi were likely lost-wax castings (fig. 100).[74] Although such artifacts spread both westward and eastward to the Taihang Mountains, they seem to have been left by the same or culturally similar migratory communities[75] and to have a common connection with the far north (e.g., southern Siberia and Mongolia), probably the Karasuk culture of the Minusinsk Basin during the late second millennium BCE,[76] which belonged to a much larger metallurgical sphere of "Eurasian steppe belt cultures."[77] A hair ornament from the formerly Arthur M. Sackler Foundation collection (fig. 101), with no visible mold marks and presumably lost-wax cast, together with two superb cast-gold ornaments of the same Karasuk style from central Mongolia[78] (c. 13th – 11th centuries BCE), clearly shows the mastery of the lost-wax process in the region of Karasuk culture.[79] Given the proven technical exchange between the Eurasian steppes and China, as well as the likely use of the lost-wax process in the Northern Zone as early as the Anyang period, Bagley suggested that "Chinese metalworkers may well have known of the technique long before they chose to make any appreciable use of it."[80] In other words, Anyang casters might not have been completely unaware of the lost-wax process, but they simply chose not to use it. Simplified in design, the lost-wax lookalike cast northern knives could have been imitated by Anyang casters using the bivalve casting process.[81]

During the Western Zhou period (c. 1050–771 BCE), the openwork interlace structure (e.g., fig. 83), as the basic form for many verified Eastern Zhou lost-wax castings, appears to have emerged in China.

Compared with the Anyang period, this interaction between China and Inner Asia seems to have been more frequent during the early first millennium BCE, to judge by the influx and borrowing of various steppe artifacts, designs, and techniques.[82] Although closely linked with Inner Asia, neither the Zhou aristocrats in the Wei River Valley (who were driven to Luoyang after 770 BCE) nor their active and usually hostile semi-nomadic neighbors in the Xibei region made much, if any, use of the lost-wax casting process during the early first millennium BCE. But the case seems to have been altogether different in Dongbei. In contrast to the scarce and isolated caches dating to the late second millennium BCE, many early first-millennium BCE Dongbei finds occur in burials, signifying the existence of more stable settlements there.[83] The sudden proliferation of both Zhou and northern artifacts in the Dongbei region seems to have been caused by the eastward expansion of Zhou rule and the colonization of China's Northern Frontier.[84] According to Emma Bunker, the Dongbei people had close contact with Mongolia and southern Siberia via the "Fur Road,"[85] roughly corresponding to the trans-Siberian railway and bypassing the Xibei regions.[86] This contact might well explain why horseback riding emerged earlier in Dongbei than in Xibei, and why certain motifs from the far north appeared in Dongbei vocabularies but never in those of Xibei.[87] Compared with the stylized Xibei zoomorphic motifs, Dongbei designs of animals, either copulating or moving along the axis of the knife hilt, seem far more naturalistic, indicating a strong representational tradition connected to Mongolia and southern Siberia.

Bunker believed that lost-wax casting occurred in Dongbei earlier than in the Chinese heartland.[88] The evidence she listed includes a number of jingles or ornaments (e.g., fig. 102) manufactured in the Dongbei region during the seventh to sixth centuries BCE.[89] As Bunker asserted, Dongbei casters probably knew both direct and indirect lost-wax processes during this period.[90] However, many of what are assumed to be lost-wax castings would not have been difficult to achieve with the section-mold process.[91] The assertion that the indirect lost-wax process was used is also dubious, as Bunker has never specified her reason for it beyond the presence of visible mold marks on the casting surfaces. As a result, one can suspect that this judgment might be used only to explain away such mold marks. Although Bunker's reasoning is in itself less than persuasive, her opinion may happen to be correct, to judge by the close contact of Dongbei with Mongolia and southern Siberia, where lost-wax was probably in use no later than the late second millennium BCE. This may especially have been the case with many Dongbei castings, which were iconographically connected to the far north and so may have been technically stimulated by the latter. In other words, the shapes of these Dongbei objects did not require the lost-wax process, but some of them were stylistically related to the lost-wax metalworking tradition in the far north, making it likely that they were made by casters who were familiar with lost-wax.

Northern Hebei, which belongs geographically to the Dongbei region but whose artifacts lack certain Dongbei features,[92] requires particular attention here. With access to the "Fur Road" via the area corresponding to the modern city of Zhangjiakou, northern Hebei could also connect to the far north while bypassing the Xibei regions, a fact that could easily explain why some Hebei motifs traceable to the tradition of the far north did not appear in Xibei until considerably later.[93] A series of intricate zoomorphic finials from areas in the center of northern Hebei (fig. 103) are commonly taken to be lost-wax castings from around the seventh and sixth centuries BCE.[94] As such zoomorphic finials are more common in southern Siberia,[95] where canopies and funerary carts were ritually buried with the elites,[96] it is quite likely that lost-wax casting was transmitted from the far north. Long-distance trade, historically moving through Zhangjiakou and Ulaan Baatar, is a likely vehicle for the interaction between these two regions. As So and Bunker alleged, the existence of lost-wax casting in northern Hebei can also be attested by a few gold pectorals, such as the crouching feline plaque from Beijing Yanqing Jundushan (fig. 104a).[97] Similar bronze pectorals also exist in northern Hebei, such as the crouching feline plaque in the Levy and White collection (fig. 104b) and two others of a very similar type in the Arthur M. Sackler collection.[98] According to Bunker's research, both pectorals were made from a wax model formed in a two-piece mold.[99] This proposal, while not completely convincing, deserves consideration.

As mentioned, it is quite likely that Anyang casters became aware of the lost-wax technique through communication with northerners during the late second millennium BCE.[100] A technique, however, usually has to be maintained through continuous practice. Because of the dominance of section-mold casting, the lost-wax process, which may have been known by Chinese metalworkers even as early as the late second millennium BCE, could still have been forgotten over time. Following the same lost-model principle, but with objects displaying comparatively "primitive" and crude features, the aforementioned burn-out/lost-cord castings, if they were in use, might provide a clue that the lost-wax process was unknown or came to be forgotten in the corresponding space and time. It is otherwise possible that Chinese metalworkers, even if they already knew the lost-wax process, still developed the burn-out/lost-cord process, as the fiber-like texture of the cord might have been a desirable visual effect that casters wished to achieve. Whatever the case, lost-wax manufacture in northern Hebei and perhaps also in the Dongbei region during the early-to-middle first millennium BCE (but no later than the 7th–6th centuries BCE) would have been either a direct source for or at least a "reminder" of the lost-wax technique used in the Chinese heartland not long thereafter. The "reminder" here can be understood in a technical sense, but as technology and art are two sides of the same coin, it could also act as a reminder of artistic possibilities.[101]

In addition to issues related to proper timing, the Dongbei and northern Hebei regions were geographically adjacent to the main Zhou states of Yan in the north and Qi in the northeast, as well as other minor Zhou states nearby. Northern Hebei, especially, was nearly within the sphere of Zhou; even if the border people there were not the Shanrong of received texts, they were undoubtedly interacting with the Zhou states.[102]

Cultural intrusion from the Dongbei region and northern Hebei, for instance, is reflected in certain gold articles found at Shandong Yishui

Liujiadianzi M1, a tomb dated to the middle or late seventh century BCE.[103] Gold was used in the Northern Zone as early as the late second millennium BCE and was introduced to the Zhou states no later than the eighth century BCE (e.g., the states of Rui at Shaanxi Hancheng and Guo at Henan Sanmenxia).[104] The recently discovered gold *huang* from the Western Zhou cemetery of Shanxi Yicheng Dahekou[105] most clearly attests to the use of gold in early China. However, according to So, a gilt ladle from Liujiadianzi M1 is the earliest Eastern Zhou example of sheet gold used on vessel exteriors.[106] The triangular and cowrie-shaped pieces of sheet gold[107] are probably related to the gold foils covering bronze cowrie shells from the Yuhuangmiao cemetery, Beijing Yanqing Jundushan.[108] The use of gold sheet to decorate lacquered vessels was also a feature in the Dongbei and northern Hebei regions.[109] A northern-style gold-hilted dagger also from Liujiadianzi M1[110] further reinforces the conclusion that there were contacts between the Zhou world and the northerners by the seventh century BCE.[111] In light of such verified contact, it is quite possible that the lost-wax process also "sneaked" into the Zhou states from the Dongbei region and/or northern Hebei during this period.

Large numbers of ornamental gold sheets with repoussé designs of interlaced serpentine patterns from the first half of the sixth century BCE, discovered at Henan Xinzheng Lijialou,[112] show further exploitation of sheet gold and related steppe techniques later in the Chinese Central Plains. Although stimulus from the northwest cannot be ruled out, such gold sheets, presumably used to decorate bronze surfaces,[113] are more likely to be the result of influence from the north or northeast. It is also notable that turquoise inlays, used earlier in the ornamentation of bronze, though seldom during the Western Zhou, were finally revived at Xinzheng in the early sixth century BCE, indicating an increased contact with the nomadic world.[114] The "nomadic world" here can probably be more precisely linked to the Dongbei and/or especially the northern Hebei regions, because such a practice did not occur in the Zhou states in the northwest until considerably later. With artifacts stylistically succeeding Xinzheng, the Chu tombs at Henan Xichuan Xiasi (c. the second quarter of the sixth century BCE) not only revealed ornaments of sheet gold and inlays of semiprecious stone[115] but also confirmed the use of the lost-wax process, as I described at length earlier in this book.

Moreover, the cast-copper inlay and gold inlay practiced at Xiasi seem to have been novel techniques.[116] The origin of the unconventional "cast-inlay" process is still unclear.[117] This technique was likely invented in early China, inspired by the use of spacers in casting,[118] but the idea might have been motivated by the adornment of northern textiles, woodwork, or metalwork, as suggested by Jessica Rawson.[119] Gold inlay, with the more conventional hammering technique,[120] was certainly influenced by the north, because both the material (gold) and the technique (hammering) seem to have been features of the Eurasian Steppes. Considering the stylistic succession of bronzes from Xinzheng to Xiasi, both the cast-copper and the gold inlays at Xiasi might have been partially inspired by the turquoise inlay first revived in the northern-influenced Xinzheng. Considering the comparatively late emergence of these techniques in Jin and the other nonwestern Zhou states, it seems highly probable that the foreign source of motivation for the novel techniques seen in objects at Xiasi, such as the cast-copper and gold inlays, as well as the lost-wax process, came from first-millennium BCE Dongbei and/or especially the northern Hebei region.

NOTES

1. As Robert Murowchick pointed out (personal communication, September 2018), these two choices are not necessarily mutually exclusive. All technologies might be "invented" multiple times in many places, and stimulation or influences from "outside" can occur at any point in the developmental process. This point will be further explored in the discussion to come.
2. In fact, like most attempts dealing with "origins," confirming the source of the lost-wax process in early China might be ultimately unachievable. The only course is to lay out possibilities and rank them based on the evidence available. It is especially important to be fully aware of the limitations of the archaeological record, the often random and certainly uneven nature of the preservation and discovery of that record, the wide range of possible scenarios around the development and spread of technologies, and the creativity of humans faced with a problem to solve. Robert Murowchick, personal communication, September 2018.
3. See Barnard, "The Entry of Cire-Perdue Investment Casting," 75; Hua, *Zhongguo Gudai Jinshu Jishu*, 179.
4. Hua, *Zhongguo Gudai Jinshu Jishu*, 179.
5. See Tan, "Zhongguo Gudai Shila Zhuzao Qiyuan Wenti de Sikao."
6. Ibid., 38.
7. Tan, "Zhongguo Zaoqi Shila Zhuzao Wenti de Guancha yu Sikao," 38–39; Hua, *Zhongguo Gudai Jinshu Jishu*, 179–180.
8. Tan, "Zhongguo Zaoqi Shila Zhuzao Wenti de Guancha yu Sikao," 38; Hua, *Zhongguo Gudai Jinshu Jishu*, 180.
9. For image, see Tan, "Shang Zhongqi de Tiliang Hu," 492, Color Plate. For discussion, see Tan, "Zhongguo Zaoqi Shila Zhuzao Wenti de Guancha yu Sikao," 38; Tan, "Shang Zhongqi de Tiliang Hu," 491.
10. Tan, "Zhongguo Zaoqi Shila Zhuzao Wenti de Guancha yu Sikao," 38; Hua, *Zhongguo Gudai Jinshu Jishu*, 180.
11. Ibid.
12. Tan, "Zhongguo Zaoqi Shila Zhuzao Wenti de Guancha yu Sikao," 37.
13. Ibid.
14. Ibid., 38.

15. For images, see Steinke, "Erligang and the Southern Bronze Industries," figs. 28, 29a. For discussion, see Tan, "Zhongguo Zaoqi Shila Zhuzao Wenti de Guancha yu Sikao," 38.
16. Tan, "Zhongguo Zaoqi Shila Zhuzao Wenti de Guancha yu Sikao," 37–38.
17. Ibid., 38.
18. Bagley, "Anyang Mold-making and the Decorated Model," 49.
19. See Zhou et al., "Zai Lun 'Shila Gongyi Bushi Zhongguo Qingtong Shidai de Xuanze,'" 47; Nickel, "Imperfect Symmetry," 11–16, 33.
20. Bagley, "Anyang Mold-making and the Decorated Model," 50–51.
21. Nickel, "Imperfect Symmetry," 24, fig. 13; 25, fig. 19.
22. Bagley, "Anyang Mold-making and the Decorated Model," 51.
23. Zhou et al., "Zai Lun 'Shila Gongyi Bushi Zhongguo Qingtong Shidai de Xuanze,'" 48.
24. Ibid., 46–48.
25. Ibid., 47.
26. One possibility, as pointed out by Robert Murowchick (personal communication, September 2018), is that burning out the organic material might result in less damage to the finer details of the design compared to lifting out the cord model (it would be easy to test how "sticky" the mold clay is when pressed against a fiber cord; the fine striations of the cord might indeed make it difficult to pull off).
27. See Tan, "Zhongguo Gudai Shila Zhuzao Qiyuan Wenti de Sikao"; Tan, "Zhongguo Zaoqi Shila Zhuzao Wenti de Guancha yu Sikao"; Hua, *Zhongguo Gudai Jinshu Jishu*, 179–180.
28. For details, see Zhang, "Guanyu Zeng Hou Yi Zun Pan Shifou Caiyong Shilafa Zhuzao de Shuping."
29. See Tan, "Zhongguo Gudai Shila Zhuzao Qiyuan Wenti de Sikao"; Tan, "Zhongguo Zaoqi Shila Zhuzao Wenti de Guancha yu Sikao"; Hua, *Zhongguo Gudai Jinshu Jishu*, 179.
30. Standing apart in time is true of the spread of all influences from one place to another; and standing apart in the details of the technology is also to be expected, as changes and adaptations take place accidentally or on purpose. Neither time nor technology rules out the possibility that the burn-out process could have been the source, or a source, of lost-wax developments. Certainly, this does not mean that the burn-out method was the source of the lost-wax process, only that a deeper explanation and clarification is needed to support Zhang's claim that it was *not* the source. Robert Murowchick, personal communication, September 2018.

31. Robert Bagley, personal communication, July 19, 2017. Of course, it is still possible to imagine some possible scenarios for the jump from rope to wax, such as the one proposed by Robert Murowchick (personal communication, September 2018): a caster realizes that a model made of a "removable" material works well. "Gee, this wax melts away—that's useful. But I don't have ample supplies of wax, so what else could I use? I can grow an endless supply of fibers with which to make rope, and the rope burns out pretty easily..."
32. See a series of archaeological reports: Yunnan, "Yunnan Jinning Shizhaishan Guyizhi ji Muzang"; Yunnan, *Yunnan Jinning Shizhaishan Gu Muqun Fajue Baogao*; Yunnan, "Yunnan Jinning Shizhaishan di san ci Fajue Jianbao"; Yunnan, "Yunnan Jinning Shizhaishan Gumu di si ci Fajue Jianbao"; Yunnan, "Yunnan Jinning Shizhaishan Gumuqun Chutu Tongtieqi Buyi."
33. Barnard, "The Entry of Cire-Perdue Investment Casting," 75–76.
34. Ibid., 76.
35. For details, see Chiou-Peng, "Incipient Metallurgy in Yunnan"; Chiou-Peng, "The Technical History of Early Asian Kettledrums."
36. Chiou-Peng, "The Technical History of Early Asian Kettledrums," 20–21.
37. Li and Han, *Gu Dianguo Jinshu Jishu Yanjiu*, 64. For burial context, see Kunming, "Chenggong Tianzimiao Dian Mu."
38. Li and Han, *Gu Dianguo Jinshu Jishu Yanjiu*, 64; Meyers, "Casting Technology in Cambodia and Related Southeast Asian Civilizations," 34–35. For burial context, see Yunnan, "Yunnan Jiangchuan Lijiashan Gumuqun Fajue Baogao."
39. Li and Han, *Gu Dianguo Jinshu Jishu Yanjiu*, 5.
40. Barnard, "The Entry of Cire-Perdue Investment Casting," 76. Barnard wanted to believe that the Chinese did not know about lost-wax casting, because for him, that was proof of the independent origin of Chinese civilization. To Barnard, the Chinese used section molds only because they did not know the lost-wax process.
41. E.g., Hua, *Zhongguo Gudai Jinshu Jishu*, 179; Tan, *Canlan de Zhongguo Gudai Shila Zhuzao*, 42–43.
42. See the Radiocarbon-dating Report 4 (*Kaogu* 3[1977], 200-204), p. 203.
43. Wang, *Yunnan Kaogu*, 67.
44. Pirazzoli, "Dian Wenhua de Niandai Wenti," 78.
45. Li and Han, *Gu Dianguo Jinshu Jishu Yanjiu*, 13.
46. Ibid., 152–155.

47. See Qiu, "Yunnan Qingtong Wenhua de Qima Wenyang," 239–240; Han and Li, "The Bronze-using Cultures in the Northern Frontier of Ancient China and the Metallurgies of Ancient Dian area in Yunnan Province," 178–179; Li and Han, *Gu Dianguo Jinshu Jishu Yanjiu*, 152–155.
48. See *Shiji*, vol. 116; *Hou Hanshu*, vol. 86.
49. Barnard, "The Entry of Cire-Perdue Investment Casting," 75–76.
50. Meyers, "Casting Technology in Cambodia and Related Southeast Asian Civilizations," 34, 39.
51. White, "Early East Asian Metallurgy," 178.
52. Insect wax refers to the wax or waxy substance secreted by certain scale insects (e.g., *Ceroplastes ceriferus* and the related *Ericerus pela*).
53. Higham and Kijingam, *Prehistoric Investigations in Northeastern Thailand*, 81, 124; White, "Early East Asian Metallurgy," 178.
54. For details, see Agrawal, *Ancient Metal Technology and Archaeology of South Asia*.
55. For details, see Murowchick, "The Development of Early Bronze Metallurgy in Vietnam and Kampuchea."
56. For details, see Higham, "Prehistoric Metallurgy in Southeast Asia."
57. The "chronology" of Vietnam of the developmental "stages" from Neolithic to the era of state formation is oversimplified and has been heavily influenced by a longstanding political need to interpret the archaeological data in a way that supports an entirely indigenous cultural development, free of any influences from China or other regions in these early times. Robert Murowchick, personal communication, September 2018.
58. E.g., White, "Early East Asian Metallurgy"; White, "Banqing yu Taiguo Dongbeibu Qingtong Wenhua de Biannian Wenti"; and White, "Dating Early Bronze at Ban Chiang, Thailand."
59. E.g., Higham, "Prehistoric Metallurgy in Southeast Asia"; Higham, *The Bronze Age of Southeast Asia*, 240.
60. For details, see Higham and Higham, "A New Chronological Framework for Prehistoric Southeast Asia."
61. E.g., Bunker and Latchford, *Khmer Bronzes*, 1–3.
62. Xu, "Bronze at Sanxingdui," 60.
63. Davey, "The Early History of Lost-wax Casting," 151.
64. Ibid.
65. Ibid., 152.
66. Robert Bagley, personal communication, July 19, 2017.

67. Bunker et al., *Ancient Bronzes of the Eastern Eurasian Steppes from the Arthur M. Sackler Collections*, 18.
68. Bagley, "Shang Archaeology," 221.
69. So and Bunker, *Traders and Raiders on China's Northern Frontier*, 34.
70. Bagley, *Shang Ritual Bronzes in the Arthur M. Sackler Collections*, 63–64, note 245. Strictly speaking, "Northern Steppe" and "Northern Zone" could be treated as separate sub-regions, with the former referring to the Gobi and steppe of Mongolia, to the further north of the latter. See Cao, "The Loess Highland in a Trading Network," 87, note 44.
71. Piggott, *Wagon, Chariot, and Carriage*, 45–48; Bagley, "Shang Archaeology," 207.
72. Bunker et al., *Ancient Bronzes of the Eastern Eurasian Steppes from the Arthur M. Sackler Collections*, 31.
73. Lin, "A Reexamination of the Relationship Between Bronzes of the Shang Culture and of the Northern Zone," 238–241; Bagley, "Shang Archaeology," 222.
74. Bagley, *Shang Ritual Bronzes in the Arthur M. Sackler Collections*, 44.
75. According to Jenny So, most artifacts east of the Taihang Mountains during the second millennium BCE, including the probably lost-wax cast animal-pommeled knives, come mainly from caches, not from burials, and are far fewer in number compared with artifacts west of the Taihang Mountains. As So inferred, "the scattered finds of non-Chinese artifacts east of the Taihang Mountains probably mark temporary settlements or seasonal migrations of peoples from the west who came for trade, pasture, or hunting" (So and Bunker, *Traders and Raiders on China's Northern Frontier*, 37–38).
76. Loehr, "Weapons and Tools from Anyang, and Siberian Analogies," 135; Bagley, "Shang Archaeology," 224.
77. For details, see Chernykh, *Ancient Metallurgy in the USSR*, 264–295; Chernykh, "Ancient Metallurgy in the Eurasian Steppes and China," 3–5.
78. For images, see Bunker et al., *Ancient Bronzes of the Eastern Eurasian Steppes from the Arthur M. Sackler Collections*, 142, figs. 32.1, 32.2.
79. For details, see Bunker et al., *Ancient Bronzes of the Eastern Eurasian Steppes from the Arthur M. Sackler Collections*, 142–143.
80. Bagley, *Shang Ritual Bronzes in the Arthur M. Sackler Collections*, 44.
81. Ibid., 63, note 244.
82. In her controversial article, Rawson asserted that a harness ring similar to fig. 78 from the Luristan Plateau offered a source not only for the interlace design commonly seen in the Western Zhou bronzes (e.g.,

fig. 80), but also for the intricate chariot fitting from Shaanxi Chang'an Zhangjiapo (fig. 79), suggesting the possibility of contact in the early first millennium BCE between Zhou-ruled China and the Iranian Plateau ("Carnelian Beads, Animal Figures and Exotic Vessels"). This argument does not seem tenable, but Rawson may be right that Zhou aristocrats themselves might have been more closely linked with Inner Asia than their Anyang predecessors. According to Rawson, an influx of steppe influences arrived in China during the dynastic Zhou period, probably with immigrants from the borderlands. The evidence she listed includes carnelian beads, realistic steppe animal designs (e.g., felines and horses), vessels supported by tigers or human figures, various exotic vessel shapes, stone-lined tombs, face coverings, and the considerable use of gold and iron. Although many points in this article are problematic (e.g., the argument that the horse emblem on a bronze *gui* from Shanxi Lingshi means that the vessel was a local product of nomads), some of Rawson's ideas deserve consideration.

83. For details, see Sun, "Colonizing China's Northern Frontier."
84. So and Bunker, *Traders and Raiders on China's Northern Frontier*, 41, 44. The founding of Zhou's subordinate state of Yan in northern Hebei undoubtedly produced a tremor in the local political and cultural landscapes. For details, see Sun, "Colonizing China's Northern Frontier."
85. The basis for this name is the historical fur trade linking these areas. It is unknown whether such fur trade had occurred during the Bronze Age.
86. So and Bunker, *Traders and Raiders on China's Northern Frontier*, 65; Bunker et al., *Ancient Bronzes of the Eastern Eurasian Steppes from the Arthur M. Sackler Collections*, 154.
87. Bunker et al., *Ancient Bronzes of the Eastern Eurasian Steppes from the Arthur M. Sackler Collections*, 154.
88. Bunker, "First Millennium BCE Beifang Artifacts as Historical Documents," 279.
89. Bunker, *Nomadic Art of the Eastern Eurasian Steppes*, 60–64. With archaeological remnants of Dongbei mining and casting dating to the early first millennium BCE, these pieces of metalwork were probably local Dongbei products. For details, see Bunker et al., *Ancient Bronzes of the Eastern Eurasian Steppes from the Arthur M. Sackler Collections*, 72.
90. Bunker, *Nomadic Art of the Eastern Eurasian Steppes*, 60–64.
91. For instance, for the purpose of casting jingles like those in fig. 102a, the section-mold process was more than sufficient—casters only needed to bury the bronze pellet inside the clay core of the outer shell of each

jingle. A similar process could also be applied to cast the finial with water bird shown in fig.103f (notice the pellet within the openwork body of the bird).

92. Bunker, "First Millennium BCE Beifang Artifacts as Historical Documents," 279.
93. Ibid., 281.
94. Bunker et al., *Ancient Bronzes of the Eastern Eurasian Steppes from the Arthur M. Sackler Collections*, 177; Bunker, *Nomadic Art of the Eastern Eurasian Steppes*, 58–60.
95. Bunker et al., *Ancient Bronzes of the Eastern Eurasian Steppes from the Arthur M. Sackler Collections*, 177.
96. Ibid.; Bunker, *Nomadic Art of the Eastern Eurasian Steppes*, 58.
97. So and Bunker, *Traders and Raiders on China's Northern Frontier*, 50. Because most Hebei pectorals are still made of bronze and are piece-mold cast, Bunker believed that the technological sophistication of the lost-wax process and a deep understanding of metal properties are represented by these gold pectorals ("First Millennium BCE Beifang Artifacts as Historical Documents," 280). Her reason was that gold cast by the lost wax method tends to result in a better-quality artifact when compared with those made by section-mold casting, which is, simply, baseless and unfounded. The way in which the mold was made has no relation whatsoever to the quality of the gold casting.
98. See Bunker et al., *Ancient Bronzes of the Eastern Eurasian Steppes from the Arthur M. Sackler Collections*, 188–189, Nos. 111, 112.
99. So and Bunker, *Traders and Raiders on China's Northern Frontier*, 112–113; Bunker et al., *Ancient Bronzes of the Eastern Eurasian Steppes from the Arthur M. Sackler Collections*, 189.
100. For details, see Bagley, *Shang Ritual Bronzes in the Arthur M. Sackler Collections*, 44.
101. Robert Bagley, personal communication, July 19, 2017.
102. Individuals buried in the northern Hebei tombs during the 7th–6th centuries BCE are usually associated with the Shanrong from references in received texts. The Shanrong moved eastwards from northern Shanxi in 714 BCE and thereafter attacked the Zhou states Qi in 706 BCE and Yan in 664 BCE. With only limited evidence, Bunker claimed that the northern Hebei inhabitants might have been traders of exotic furs and leather working with nearby Chinese neighbors. Bunker also argued that the presence of daggers in their graves indicates their warlike character. Yet daggers were in fact not very important in warfare. For

details, see Bunker et al., *Ancient Bronzes of the Eastern Eurasian Steppes from the Arthur M. Sackler Collections*, 175; So and Bunker, *Traders and Raiders on China's Northern Frontier*, 50; Bunker, "First Millennium BCE Beifang Artifacts as Historical Documents," 282.

103. See So, *Eastern Zhou Ritual Bronzes from the Arthur M. Sackler Collections*, 19.
104. For details, see Bai, "Xia Shang Xizhou Jinqi Yanjiu"; Rawson, "Carnelian Beads, Animal Figures and Exotic Vessels."
105. For image, see: http://www.sxkaogu.net/webshow/aritcleDetail.shtml?articleId=2121. Accessed March 6, 2018.
106. So, *Eastern Zhou Ritual Bronzes from the Arthur M. Sackler Collections*, 19–20.
107. For image, see Luo, "Shandong Yishui Liujiadianzi Chunqiu Mu Fajue Jianbao," 6, fig. 10. For discussion, see So, *Eastern Zhou Ritual Bronzes from the Arthur M. Sackler Collections*, 19.
108. See Bunker et al., *Ancient Bronzes of the Eastern Eurasian Steppes from the Arthur M. Sackler Collections*, 64.
109. See Bai, "Xia Shang Xizhou Jinqi Yanjiu," 45. This practice could also be found in nearby areas, such as Beijing Fangshan Liulihe, where a lacquered *gu* vessel decorated with gold sheets was excavated from M1043, a tomb usually associated with the state of Yan of Western Zhou.
110. For image, see Luo, "Shandong Yishui Liujiadianzi Chunqiu Mu Fajue Jianbao," Plate 2.4.
111. See So, *Eastern Zhou Ritual Bronzes from the Arthur M. Sackler Collections*, 19–20. As Jenny So pointed out, "the Qi state, under the leadership of its greatest ruler, Duke Huan (685–643 B.C.), was the first of five hegemons to come to power in the Spring and Autumn period" ("New Departures in Eastern Zhou Bronze Designs: The Spring and Autumn Period," 254). As a superpower with many allied states, even the Zhou court (as reflected in a dowry bronze *yu* basin found at Luoyang, see So, "New Departures in Eastern Zhou Bronze Designs: The Spring and Autumn Period," 253), Qi (in present-day Shandong) probably interacted with other Zhou states frequently during the seventh and sixth centuries BCE. Especially with regard to preventing the spread of Chu, Qi founded alliances with many Zhou states, including some petty states in the Hanhuai region close to Chu. Zhou's closeness with many Hanhuai states was, in fact, long-lasting. For instance, the marriage alliance between Qi and Chen in southeastern Henan is suggested by a

set of eleven eighth-century BCE bronzes found in Shandong, among which was "a pair of wine vessels with inscriptions indicating that they were made to accompany the daughter of a marquis of Chen in marriage" (So, "New Departures in Eastern Zhou Bronze Designs: The Spring and Autumn Period," 253). It is possible that the lost-wax process was introduced from the Dongbei/northern Hebei region via Shandong, Henan, and the Hanhuai region, finally flourishing in the Chu region.

112. See So, *Eastern Zhou Ritual Bronzes from the Arthur M. Sackler Collections*, 24.
113. Ibid.
114. So and Bunker, *Traders and Raiders on China's Northern Frontier*, 101–102; So, *Eastern Zhou Ritual Bronzes from the Arthur M. Sackler Collections*, 24, 73–74: note 88.
115. So, *Eastern Zhou Ritual Bronzes from the Arthur M. Sackler Collections*, 34–35.
116. Ibid.
117. Ibid., 34.
118. For details, see Zhang, "Guoli Gugong Bowuyuan suocang Dongzhou Xiangqianqi Yanjiu."
119. Rawson, *Chinese Bronzes*, 52–53.
120. So, *Eastern Zhou Ritual Bronzes from the Arthur M. Sackler Collections*, 35.

Chapter 9

Coda

Additional Issues and Comments

This coda is not a conclusion: many questions remain unanswered, and some have yet to be asked. Thus far, I have analyzed how the lost-wax process was used in each confirmed casting of its type from Bronze Age China. Many of the lost-wax examples discussed, especially the earliest, incorporate three-dimensional openwork interlace and are concentrated in or stylistically connected to the region of southern Henan and northern Hubei within the presumed "Chu sphere." It is precisely the openwork interlace that attracts scholarly attention to the issue of lost-wax casting, but it does not necessarily follow that all the lost-wax castings from Bronze Age China are exclusively of this design. Because the lost-wax process may not leave any telltale traces, it remains unclear whether it was used in other types of castings. For this reason, providing a statistically based argument on this issue is impossible. I cannot say, for example, that *most* lost-wax castings in Bronze Age China have openwork interlace. In fact, some castings that used the lost-wax process were probably not rendered in three-dimensional interlaced form. The most convincing example of this is the forty-six realistic life-size bronze waterfowl (e.g., see two broken pieces in figs. 105, 106) unearthed from the mausoleum of the First Emperor of Qin (r. 247–210 BCE, mausoleum constructed from 247 to 208 BCE).

OTHER TRADITIONS OF LOST-WAX?

A comprehensive study by Shao Anding indicates that lost-wax technology was used to manufacture the waterfowl from the mausoleum of the First Emperor of Qin.[1] Shao's evidence includes the absence of section-mold casting marks from even the "dead angles,"[2] intricate configurations that do not permit the withdrawal of mold sections (e.g., see fig. 105b),[3] and the "innovative" use of chaplets and armatures, comparable to the lost-wax methods used in the Hellenistic world.[4] The inner surface of the body of the No. 10 swan in particular reveals obvious brush marks (fig. 106b), a feature usually seen on life-sized statues in the Mediterranean world, indicating that the indirect lost-wax process was probably used to make the body of the waterfowl.[5] Shao mentioned that the brush marks are relevant to lost-wax casting in general, but did not specifically stress the indirect lost-wax process here.[6] In fact, as the Greek cases show, lost-wax castings manufactured by the direct process are not likely to have this feature. Nevertheless, Shao noticed other features within the body cavities of the waterfowl that are associated with the indirect lost-wax process: craftsmen probably used brushes to coat the interior of the permanent mold with semiliquid wax to make the model, filled it with core material, attached to the wax model the layers of feathers that had probably been directly shaped in wax, and finally cast the body in a single piece.[7]

The casting of the long-curved necks of these waterfowl, usually composed of several metallurgically joined portions,[8] is yet another issue, one that has not been thoroughly understood.[9] Like the naturalistic, life-sized sculpture-making represented by the Qin terracotta warriors,[10] these Qin waterfowl stand alone in the history of early China, apart from what is understood to have been the main track of development in art and technology at that time. The lost-wax process seen in the fabrication of these waterfowl, with its parallels in the casting of the Mediterranean world, is clearly distinct from the methods used earlier in the Chinese context. It can be argued that this new type of lost-wax casting filtered into China after the Asian campaigns of Alexander the Great in the second half of the fourth century BCE.[11]

The case of the Qin waterfowl is a reminder that lost-wax technology could certainly have been reintroduced into China after its earliest application in Chinese workshops: not all lost-wax casting processes in use in early China necessarily represent the same technological tradition. For instance, lost-wax casting was probably already practiced in the northern borderlands of China in the late second millennium BCE. During the early-to-middle first millennium BCE, lost-wax manufacturing appeared in the northern Hebei and perhaps also the Dongbei regions, which might have provided a technical source or a reminder of the artistic possibilities of this process, which was later used in the Central Plains region of China (for a detailed discussion, see chapter 8). Despite the Qin waterfowl, which seem to represent an outlying tradition, the aforementioned lost-wax fabrication process, albeit with different stylistic variations, fell more or less within this same technological tradition.

It has sometimes been suggested that many belt ornaments and fasteners made in China and dated from the fifth to third centuries BCE, such as the belt hooks represented by figure 107 and the belt-buckle plaques represented by figure 108, were also cast by the lost-wax process.[12] Unfortunately, the evidence for this claim is often not fully convincing (usually consisting only of the absence of visible "mold marks") or is not clearly stated.[13] Nevertheless, if further investigation confirms this assertion, it should not be taken for granted that the lost-wax process used in the making of these items was derived from the earlier lost-wax practice used in the Chinese Central Plains. From approximately 500 BCE onwards, profound artistic changes took place in Bronze Age China under the impact of the art of the Eurasian steppes and the Mediterranean world.[14] Around the same time, knowledge of new metallurgical techniques, such as cloisonné, strip-twisted wire, granulation, and mechanically linked loop-in-loop chains, was introduced into China through Inner Asia.[15] Was the lost-wax process also brought to China with these metalworking techniques as part of a kind of metalworkers' toolkit? Of course, Chinese workshops had already used lost-wax casting before this era, but that fact does not preclude the arrival of this skill in a different technological form. To verify such a hypothesis, the relevant items must be studied further, especially those with exotic motifs and those made for foreign consumption during the fifth to the third centuries BCE.

Even considering only the more closed or localized Chinese context, it is not completely certain whether all lost-wax castings have the three-dimensional interlaced structure discussed previously. For instance, Tan Derui has proposed that the six bronze musicians within the miniature house from Zhejiang Shaoxing Shizishan M306 (fig. 109), arguably dated to the fifth century BCE, were individually lost-wax cast.[16] This proposal, which is not completely convincing, is partially supported by the fact that no mold marks are visible from the front.[17] Lost-wax, though not proven here by the current evidence, still remains a possibility. The same uncertainty also applies to other artifacts, such as a *he* from the same tomb (fig. 110a) with individual figurines sitting atop its spout (fig. 110b) and the vessel lid (fig. 110c), none of which bear any visible mold marks on their surfaces.[18]

An Uncertain Correlation of Design and Technology

Another problem is that not all artifacts that have openwork interlace in high relief are lost-wax castings. As Helen Loveday commented,

> There has been a tendency, since the discovery of the bronzes from Xiasi and Leigudun, to assume that vessels with a complex structure involving openwork interlace, heavily-undercut relief elements, or parts rendered in the round must be further evidence for the use of an investment casting process.[19]

This statement is true; such alleged lost-wax examples range from the censer cup from Hubei Jiangling Yutaishan with comparatively simple openwork interlace (fig. 90)[20] to the more intricately structured lid crowns of the famous *pan long fang hu* from Henan Xinzheng Lijialou (fig. 87)[21] and the visually more convoluted vessel shell of the diatrete *ding* excavated at Shanxi Xinjiang Liuquan (fig. 88).[22] In fact, castings like these also include the lid appendages of the three large *ding* from Henan Xinzheng Lijialou,[23] a bronze *he** cup in the MEG (Musée d'ethnographie de Genève), Switzerland,[24] and a number of other censer cups and *fang hu* vessels.[25]

Among these vessels, the Liuquan *ding* (fig. 88) and the MEG *he**, both with an openwork outer case fixed over an undecorated inner shell, exhibit the most entangled interlace structures and so are the most likely to have been cast by the lost-wax process. However, by studying a group of Eastern Zhou diatrete vessels with similarly complex openwork interlace, Loveday convincingly reasoned that the Liuquan *ding* and the MEG *he**, unlike the other vessels she discussed (figs. 35, 36, 38, and 41), are likely to have been entirely cast by the section-mold process.[26] Indeed, the interlaces of the Liuquan *ding* (fig. 88b) and the MEG *he** convey a strong sense of two-dimensionality, which is distinct from the highly three-dimensional and interpenetrating interlaces of the lost-wax cast structures (figs. 35, 36, 38, and 41). Loveday's hypothesis is reinforced by the fact that these interlaces are flat-backed[27] and reminiscent of the square base of the Fengxiang burner (fig. 40) and many other section-mold cast structures.[28] As Loveday mentioned, given the "near-identical form of snake interlace" in the MEG *he** and the Liuquan *ding* (fig. 88b), as well as the adjacency of the latter's recovery site to the Houma foundry, the two vessels might have been products of Houma or related workshops.[29] That no hint of lost-wax production has ever been detected at Houma and that no lost-wax use has ever been confirmed on a casting of Houma style make it even less likely that the two vessels were cast by the lost-wax method. The conclusion that they were cast by the section-mold process is very plausible, especially considering the unparalleled transfer of decoration from model to mold at Houma, evidenced by a baked clay model there (fig. 13) and a corresponding bronze ornament not far distant (fig. 12).[30]

The absence of lost-wax evidence in the massive Houma debris is of exceptional interest. As discussed in chapter 2, it may have been the case that Houma founders were not interested in the lost-wax process, because they may have deemed it too inefficient. Whatever the reason, this is a useful clue to the casting method of many stylistically Houma or related designs with complicated openwork interlace,[31] such as a gold dagger hilt in the British Museum, said to be from Shanxi Hunyuan,[32] an iron dagger with a similar gold hilt from the tomb Yimencun M2 of Shaanxi Baoji (fig. 111),[33] comparable bronze daggers in the Guyuege collection and the Palace Museum, Taipei[34] and the formerly Frederick M. Mayer Collection, New York,[35] and a large number of luxurious garment hooks represented by several gold examples in the privately held Mengdiexuan collection, Hong Kong.[36] These finely manufactured artifacts, especially the gold hilt of the Yimencun iron dagger (fig. 111b), have in the past been widely considered as evidence of lost-wax use on the basis of their visually intricate shapes.[37] Close examination, however, reveals section-mold casting marks, usually too faint to easily detect, on many of these items.[38] In any case, the style of these objects is reminiscent of Houma, and Houma casters made equally complicated designs without using the lost-wax method (e.g., see figs. 12, 13), so there is no reason to suppose that they would have made these—and only these—objects by lost-wax.

As discussed, in addition to openwork interlace, other designs may have been cast by the lost-wax process. However, the argument that "it must be lost-wax because it is unimaginable that it might have been made in another way" is a dangerous one, because an expert caster might have been able to devise many intricate methods of fabrication that have simply eluded modern scholars. It is therefore worthwhile to do a critical review of lost-wax proposals made after the 1960s, when a consensus had been reached that most early Chinese bronzes were cast by the section-mold process. Because space prevents me from discussing all such proposals exhaustively, I will focus on two important groups of artifacts that have been dubiously labeled as lost-wax.

The first group comprises bronze vessels inlaid with precast copper, represented by the *lei* (fig. 112) from Henan Gushi Hougudui M1, a tomb dated no earlier than the end of the sixth century BCE.[39] To date, several dozen similar copper-inlaid bronzes have been discovered over a large area pertaining to Bronze Age China.[40] A few are early, produced during the late second millennium BCE,[41] but most are dated within a two-century span from the Late Spring and Autumn to the early Warring States periods.[42] Initially such copper inlays were believed to have been hammered into the prepared depressions of the finished bronzes.[43] Subsequent studies have indicated that the inlay materials were actually precast and later placed within the mold assembly of the bronze to be cast.[44] In the case of the Hougudui *lei* (fig. 112), there are more than 160 pieces of copper inlay.[45] How could the placement of the massed copper inlaid decorations within the mold assembly have been controlled during the casting process? How did casters prevent the shift of such a large number of painstakingly arrayed copper inlays through the powerful and rapid flow of the molten bronze? Li Jinghua has proposed a method involving an ingenious use of the lost-wax process to solve this problem. This method uses viscous wax to adhere copper inlays onto the inner core and fills all of the interspaces between these inlays with wax, until they are all "locked" in place and ready for the investment casting.[46] Although there is no evidence to support this method, Li considered it the only technically practical way to produce the cast copper-inlay vessels. Its application also reveals a new type of lost-wax casting in the Eastern Zhou period.[47]

As Li also argued, "when making the cast copper decorations, the lost-wax process was already developed at a high level."[48] In other words, Li saw his lost-wax proposal as fitting in terms of timing into the manufacture of the Eastern Zhou copper inlaid bronzes. Indeed, when Li finished his article in 2004, the earliest known cast-inlay process came from Henan Xichuan Xiasi M2,[49] a tomb dated to the second quarter of the sixth century BCE, where the lost-wax process was already verified to have been in use. Recent discoveries from Shandong Zaozhuang Xulou present an even earlier appearance of the cast-inlay process and confirm its use in Eastern Zhou no later than the first quarter of the sixth century BCE.[50] The current data therefore indicate roughly simultaneous rises of the cast-inlay process and the lost-wax process, posing the possibility that the two methods might be technologically related. Copper inlaid bronzes from Xulou, however, do not support this hypothesis.[51] For instance, the "王"-shaped copper ornamentations of the Xulou *gui* (M1: 6, fig. 113a) were each fixed in place by a metal spacer[52] at the back (fig. 113b).[53] Similar features can also be found in many other cases, such as the scar left by a spacer (fig. 114b), visible on another *gui* vessel from the same tomb (M1: 44, fig. 114a).[54]

Such metal spacers, much improved on the Hougudui *lei* (fig. 112b), were noted by Li Jinghua.[55] But Li did not even mention the function of these spacers in his proposed procedure, in which copper inlays could have been just as easily "locked" in place by wax. If so, why did the casters of the Hougudui *lei* bother to use so many spacers? Moreover, it is also questionable whether it was practical to use the lost-wax process to manufacture copper inlaid bronzes because the oxidization of copper is likely to occur during the removal of wax, when the precast copper inlays experience a high firing temperature. The oxidized copper would loosen the corresponding inlaid pieces and thereby cause their dislocation.[56] The lost-wax process does not allow founders to access the interior of the investment mold to fix these displaced inlays; for these reasons, it is not reasonable to assume that the Eastern Zhou copper inlaid bronzes were cast using the lost-wax method.[57]

The second group is a series of bronzes, usually inlaid with gold and/or silver, unearthed from the Hebei Pingshan Sanji Gongshe M1, a tomb of King Cuo of the state of Zhongshan, who died soon after 310 BCE.[58]

According to Tan Derui, the table base with four phoenix-like birds and four dragon-like beasts intertwined (fig. 14), the stand in the shape of a tiger devouring a deer, the pair of winged felines, and the lamp with fifteen oil-burning saucers were all cast using the lost-wax process.[59] Tan did not give specific reasons for his assertion, but with regard to the table base (fig. 14), he particularly mentioned that the twelve individual birds, beasts, and reclining deer below were precast separately by the lost-wax process and then joined to the assembly with rivets.[60] Jenny So has concurred with this opinion, commenting on the table base,

> This is the only extant bronze object of its kind from the Eastern Zhou period. The intricacy with which the creatures are interwoven suggests that the base might have been made by lost-wax casting. If so, the workmanship equals that of Zeng *zun-pan* of about 433 BC, but it also illustrates an exploration of the lost-wax casting technique drastically different from that intimately associated with Chu bronze masters in the south.[61]

Further investigation, however, indicates that the lost-wax process was probably not used in the manufacture of these Zhongshan bronzes, which introduced an extraordinary amount of separate casting and assembly.[62] For instance, the most intricately structured piece among all the Zhongshan bronzes, the aforementioned table base (fig. 14), is composed of 78 pieces cast from 188 mold pieces, which were joined together through 22 instances of interlock casting and 48 instances of soldering.[63] Contrary to Tan Derui's expectations, it is most likely that the individual animals of the table base were not cast by lost-wax but were formed by assembling a large number of section mold cast components.[64] It seems that lost-wax casting, already practiced in Chinese workshops centuries earlier, either was unknown at Zhongshan[65] or simply was not of interest to the Zhongshan metalworkers.

Here it is worthwhile to stress that separate casting and joining do not in theory rule out the lost-wax process. For instance, according to the archaeological report, the openwork drum stand from the tomb of

Marquis Yi of Zeng (fig. 115) is composed of sixteen serpents rendered as twenty-two physical components that were soldered together and adjoined to the base of the stand.[66] It is therefore believed that the lost-wax process was not used in the fabrication of this remarkable masterpiece.[67] One cannot, however, assume that the twenty-two different components were all cast in section molds. It would have been very difficult to remove mold sections from the spiraling components, especially the spiraling horns of the two dragons at the top next to the socket.[68] It is therefore possible that the stand was made in several stages. (1) It was designed by a craftsman adept at woodcarving; the craftsman made a wooden model in twenty-two separate pieces, so that the model could be taken apart. (2) When the craftsman was satisfied with the design, he took the wooden model apart. He then copied each of the wooden components in wax. To do so, he presumably had to copy freehand; that might explain why the final bronze components do not fit together perfectly. (3) The craftsman used each wax model to cast a bronze component. (4) He assembled the stand and soldered all its bronze components together. This is, of course, simply conjecture, in contrast to the unexamined assumption that the drum stand's maker only used section molds without any application of the lost-wax process. In a similar case, the long-curved necks of the aforementioned Qin waterfowl often comprise different metallurgically joined portions,[69] but this feature cannot be taken as proof against the use of lost-wax casting—for both section-mold and lost-wax processes, it would have been easier to cast such outstretched bending necks in a reasonably greater number of pieces (although the difficulty of joining them would have correspondingly increased).[70] In any case, regarding the lost-wax process as a master key that automatically and effortlessly solves every problem of shape constraint is unrealistic, especially considering the matter of "air pockets" that may occur during casting.[71]

Technological Variations of Lost-wax?

It is also necessary to address the presumed technological variations of the lost-wax process in Bronze Age China. The first variation is the so-called "lost-lead" approach, initially proposed by Li Zhiwei to explain the casting of the chime-bells from the tomb of Marquis Yi of Zeng.[72] In fact, lost-lead is very close to lost-wax, with the same "lost-model" principle, except that it uses lead or a related alloy instead of wax as the material for the working model. Later, Li announced unrealistically that early Chinese bronzes with an intricate shape, such as the four-ram *zun* from Hunan Ningxiang (fig. 116) and many bronze vessels from the Anyang tomb of Fuhao, had all been cast by the lost-lead process.[73] Li explained away the frequently seen section-mold cast marks with the unconvincing suggestion that an indirect lost-lead method, like the indirect lost-wax process, had been used in China since the Anyang period.[74] The main reason for Li's proposal is that lead or a related alloy that is relatively easy to carve, shape, or melt is discovered much more often (e.g., in the form of funerary *mingqi* vessels) than wax in Bronze Age China.[75] It is common knowledge, however, that wax, an organic material that decomposes easily, rarely survives for thousands of years. Even so, reliable evidence of wax in Bronze Age China has been discovered.[76]

Li's lost-lead theory is largely accepted by the bronze experts Noel Barnard[77] and Zhang Guangyuan.[78] For instance, Zhang Guangyuan has revised "lost-lead" into "lost-pewter" (pewter is the alloy of lead and tin) and applied this process to explain the casting of the crown of the famous *pan long fang hu* from Henan Xinzheng Lijialou (fig. 87b).[79] Zhang's main rationale is that a delicate model in wax would not have been tough enough to support the investment mold, especially in the summer when wax could easily melt in the prevailing ambient conditions.[80] As discussed, it is highly doubtful whether the strongly two-dimensional crown of the Xinzheng *pan long fang hu* was really made by the "lost-model" or investment casting process. Even if it were, there is still little reason to suspect that wax, the most commonly used material for investment casting worldwide, failed to work solely in China. In other words, Zhang's skepticism of wax as a casting material is answered by the presence of intricately shaped lost-wax castings throughout the world (e.g., the visually more fragile lost-wax cast votive rafts in the pre-Columbian New World[81]). It is widely known that wax, when mixed with other materials (e.g., tallow, resin) and processed in an appropriate manner, can be both flexible and strong at the same time.[82]

The second type of lost-wax variation, though still not completely verified, is the "lost-cord" or "burn-out" process variously named by Noel Barnard and Tan Derui.[83] As Lothar von Falkenhausen commented,

> In a [...] somewhat imaginative vein, Noel Barnard has speculated that the many portions of ancient Chinese bronzes that appear to imitate the shapes of rope or cord were cast from molds produced by coating actual pieces of cord with clay, removing the cord, and then firing the clay. [...] If there is any reality to Barnard's proposed "lost-cord" casting, it must have been limited to small portions of bronzes otherwise produced by the piece-mold technique.[84]

Because the "lost-cord" or "burn-out" proposal is relevant to the technological origin of early Chinese lost-wax casting, I have discussed it in detail in chapter 8. The foregoing discussion indicates that cord or rope, unlike wax, which could melt away, requires incineration into an ashen state and a subsequent clearing from the mold. In fact, the removal of cord material is not as easy as it sounds, because powdery ashes within the mold are especially difficult to clear out and would cause serious physical damage to the casting.

The third variation of the lost-wax process that deserves comment here is the "lost-wax and lost-textile casting" proposed by Joseph Ternbach and Emma Bunker to explain the coarse, loosely woven positive fabric pattern that occurs on the back of many late Eastern Zhou and Han belt plaques made of gold, silver, or gilded bronze, such as the plaque in the Metropolitan Museum of Art, New York (fig. 117).[85] As Bunker described it,

> This particular casting technique is a variation of the indirect lost-wax method. An open-section mold was formed by pressing a model of the plaque to be produced into a piece of clay. After the clay dried, wax was poured into the open-section mold and a piece of textile, probably woven hemp, was pressed onto the wax. After the textile-backed wax model hardened, it was removed. It was then packed in clay to make a mold, which was baked at a high temperature. Both the wax and the textile reinforcement were burned out of the mold and replaced by molten metal, which was poured into the mold during the casting process.[86]

Simply put, according to Bunker, the wax and the textile, one of which melted away and the other burned out, were both replaced by molten metal within the mold.[87] The reason for using fabric reinforcement was to allow for a thinner casting and thus to reduce the precious metal in use, because a fragile wax model of very little thickness could in this case be more safely peeled off from the open-section "mother" mold.[88] This approach, Tan Derui agreed, shares the rationale behind Tan's "burn-out" process: that a casting model incapable of being liquefied by heat must be incinerated and thoroughly cleared from the mold.[89] One advantage of the "lost-wax and lost-textile casting," if it did exist, would have been that models in wax could be shaped in sequence from the same permanent "mother" mold, facilitating mass production and replication. There is now a consensus that most of these luxury plaques with textile impressions were of Chinese manufacture, made to supply their nomadic neighbors.[90] According to Bunker, the proposed "lost-wax and lost-textile casting," unprecedented in history, was inspired by contemporaneous Chinese lacquer practices, in which objects "were lacquered over a simple wooden core, or reinforced with lacquer-impregnated fabric applied to the core."[91]

Though supported by some scholars,[92] the "lost-wax and lost-textile casting" theory has recently been questioned by Odagi Harutaro,[93] Luo Feng,[94] and especially Zhou Weirong and Huang Wei.[95] As Zhou and Huang argued, to reinforce the fragile wax model, the cloth should cover its front side, but the fabric texture, on the contrary, always occurs on the reverse.[96] This argument is ultimately inadequate, because the cloth, were it applied in this manner, would not only prevent the achievement of ornamental precision but also contaminate the original pattern with the impression of the fabric surface. In fact, a textile piece affixed to the back of a fragile wax design as a support could indeed facilitate the removal of the design from the "mother" mold, or at least provide a firm hold for the separation of the two.[97] Yet considering that almost all the plaques would not have been difficult to cast from a simple two-piece mold, this argument raises the question: why would Chinese craftsmen have adopted the seemingly more cumbersome "lost-wax and lost-textile casting" method?[98] Based on Odagi Harutaro' study of the Qin metalsmith's tomb at Xi'an dated to the late Warring States period,[99] Luo Feng proposed that many of the presumed "lost-wax and lost-textile" cast plaques could easily have been manufactured from two-piece molds; the corresponding mold pieces would have been taken from an original design with a sheet of cloth on its reverse.[100] Luo, however, failed to explain the function of the cloth attached to the original model. Inspired by similar textile impressions found on the undecorated surfaces of some Eastern Zhou gold coins cast by the section-mold process and slightly later cylindrical architectural tiles,[101] Zhou and Huang further pointed out that the addition of a textile piece made it easier to separate the mold pieces from the model (e.g., for metal casting) or the shaped product from the mold (e.g., for tile making).[102] These insightful ideas have greatly weakened the credibility of the "lost-wax and lost-textile" theory. Like the aforementioned "lost-lead/pewter" or "burn-out"/"lost-cord" process, "lost-wax and lost-textile" casting, though possible in theory, has not been historically verified in any convincing way.

FINAL WORDS

With this study of lost-wax casting in Bronze Age China, I have sought to raise a host of intriguing questions that are not confined solely to this narrowly defined topic but are related to broader issues of technology, art, and invention. For instance, what has been learned about technology, technological transfer, and technological innovations? How do they correlate with art, artistic diffusion, and artistic creation? How is it possible to better understand the intricate interplay between design and technology, and between tradition and invention? The difficulty and the possibility of answering these questions have been best articulated by Arthur Steinberg:

> Techniques, like iconographic or symbolic systems, do not generally develop in cultural isolation but are spread from place to place by itinerant craftsmen, traders, and travelers, and are then superimposed or added to indigenous motifs and techniques. It is often difficult to unscramble the new mix into its original constituent parts of local materials, ways of working them, themes, and motifs, and to analyze what has happened to them when the new stimuli have arrived. To a large extent this process of diffusion leads to homogenization. [...] Fortunately for those of us interested in studying some of these processes at earlier times and in different places this tendency towards homogenization occurred much more slowly and was less complete, so that we can still detect specific technological styles operating at certain times and watch them being replaced or altered by new ones.[103]

The use of lost-wax casting in China is a case that corresponds well to Steinberg's general remarks.[104] I believe that future discoveries will enrich the broad picture I have drawn here and provide more answers to the aforementioned questions in the context of Bronze Age China.

NOTES

1. See Shao, "Qinshihuang Lingyuan Chutu Caihui Qingtong Shuiqin Zhizuo Gongyi ji Xiangguan Wenti Yanjiu," 146–152.
2. Shao, "Qinshihuang Lingyuan Chutu Caihui Qingtong Shuiqin Zhizuo Gongyi ji Xiangguan Wenti Yanjiu," 147.
3. This feature refers to the tiny spaces within an entirely cast portion of the waterfowl. For instance, the multi-layers of feathers of the No. 21 crane were cast as a whole, with interspaces from layer to layer measuring only two to three millimeters (see fig. 105b), making it nearly impossible to withdraw mold sections from the model were the section-mold process in use. For details, see Shao, "Qinshihuang Lingyuan Chutu Caihui Qingtong Shuiqin Zhizuo Gongyi ji Xiangguan Wenti Yanjiu," 147–148.
4. Shao, "Qinshihuang Lingyuan Chutu Caihui Qingtong Shuiqin Zhizuo Gongyi ji Xiangguan Wenti Yanjiu," 65–66.
5. Ibid., 39, 152.
6. Ibid., 152.
7. For details, see Ibid., 147, 153–154.
8. For images, see Ibid., 37, figs. 3-14, 3-17. For discussion, see Ibid., 152–155.
9. It is still debated whether the long-curved necks were cast by the lost-wax or the section-mold process.
10. In a controversial article on the "sudden and short-lived surge of sculpture-making in third-century BC China," Lucas Nickel wrote, "The sculptures found in the tomb of the First Emperor of China stand isolated in the history of Chinese art. Both before and after the Qin Dynasty, as well as in contemporary tombs, there are no human or animal figures comparable to the terracotta warriors and horses, officials, dancers and acrobats. Although they are without parallel in Chinese art, the sculptures fit well in the wider Asian context of the third century BC, a time when Hellenistic art was at its height in Central and Western Asia. During the lifetime of the First Emperor, the Graeco-Bactrian kingdom of Diodotus encompassed all of Central Asia from the Jaxartes to the Indus, including the Ferghana Valley, and Hellenistic culture (and Hellenistic sculpture) flourished not only in distant Europe, but within reach of the growing Chinese empire. Knowledge of Hellenistic sculpture was accessible to

the Chinese at the time, and may very well have been the stimulus for the sudden and temporary rise of sculpture in China" ("The First Emperor and Sculpture in China," 442).

11. For the dramatic changes in art and technology of northwestern China after the Asian campaigns of Alexander the Great, see So and Bunker, *Traders and Raiders on China's Northern Frontier*, 56–62.
12. See White and Bunker, *Adornment for Eternity*, 41; Tan, *Canlan de Zhongguo Gudai Shila Zhuzao*, 64–65.
13. E.g., see White and Bunker, *Adornment for Eternity*, 84, 109; Tan, *Canlan de Zhongguo Gudai Shila Zhuzao*, 65.
14. For details, see Bagley, *Gombrich among the Egyptians*, 157–173.
15. So and Bunker, *Traders and Raiders on China's Northern Frontier*, 73–75.
16. Tan, *Canlan de Zhongguo Gudai Shila Zhuzao*, 43–47. For the burial context of this artifact, see Mou, "Shaoxing 306 hao Zhanguo mu Fajue Jianbao."
17. Detailed analysis of this object by myself and Su Rongyu, one of the most authoritative experts in the technical study of Chinese bronzes, has not detected any section-mold cast mark on these figurines in the direction of the opening (e.g., see fig. 109c), although it remains unclear whether relevant traces are present at the back of these figures. In any case, the absence of mold marks itself does not constitute sufficiently decisive evidence to confirm an object as having been cast by the lost-wax process. It is worth mentioning that clay, part of the core material, is exposed in the ear of one figurine of the miniature house (fig. 109d). But the use of cores is compatible with both the lost-wax and section-mold processes. In other words, the exposure of the clay core alone does not suggest either lost-wax or section-mold casting. The only thing I can say for certain is that the entire object was not cast in a single piece, because clear traces indicate that all the small figurines were precast and later joined to the house model (fig. 109b). My analysis of this house model was made possible by Li Yuxin, curator of the Zhejiang Provincial Museum, on November 9, 2015.
18. For detailed discussion, see Tan, *Canlan de Zhongguo Gudai Shila Zhuzao*, 44.
19. Loveday, "Diversity in Eastern Zhou Bronze Casting," 119.
20. See Tan, *Canlan de Zhongguo Gudai Shila Zhuzao*, 62–63.
21. Zhang proposed that the slightly three-dimensional interlace crown of this *fang hu* lid was cast by a variation of the lost-wax process ("Zhongguo Zuizao 'Shilafa,'" 64–69). This opinion has been partly acknowl-

edged by the National History Museum of History in Taiwan, where this artifact resides; the museum states on its current official website, "this vessel was cast by the lost wax casting process that was developed in the middle and late Spring and Autumn periods, making it, regardless of style, decoration and craftsmanship, of epoch-making significance in the history of bronze development" (translated by Notis and Wang, "Ancient Chinese Bronze Casting Methods," 1744.)

22. See Zhongguo, *Zhongguo Wenwu Jinghua*, 257.
23. E.g., see Peng, "The Lost-wax Casting in Bronze Age China," 306, fig. 7.22. For discussions, see Sun, *Xinzheng Yiqi*, 47–51; So, *Eastern Zhou Ritual Bronzes from the Arthur M. Sackler Collections*, 119–120; Henan and Taibei, *Xinzheng Zhenggong Damu Qingtongqi*, 79–86.
24. See Loveday, "Diversity in Eastern Zhou Bronze Casting," 105, fig. 3; 115, fig. 8.
25. E.g., see Peng, "The Lost-wax Casting in Bronze Age China," 332–333, figs. 9.14–9.16.
26. For details, see Loveday, "Diversity in Eastern Zhou Bronze Casting," 101, 119–126.
27. Ibid., 113, 125.
28. In addition, Loveday has noticed the presence of such filler metal between the open spaces and long and narrow raised marks visible in the interlace, which show a preference for the section-mold theory ("Diversity in Eastern Zhou Bronze Casting," 121–123).
29. Loveday, "Diversity in Eastern Zhou Bronze Casting," 101, 125, and 131.
30. For detailed discussion, see Bagley, "Anyang Mold-making and the Decorated Model," 50.
31. The items in the following list, with "dragons, granulated texturing, raised commas, and interlace [...] so closely resemble the same motifs on clay models of harness fittings from Houma" that they are reasonably expected to have been made in Houma or related foundries. For details, see So, *Eastern Zhou Ritual Bronzes from the Arthur M. Sackler Collections*, 43–44.
32. Artifact in the British Museum, London (acc. 177693). For images, see So, *Eastern Zhou Ritual Bronzes from the Arthur M. Sackler Collections*, 43, fig. 63; Lee, "Chengqi zhi Dao—Zhongguo Xianqin zhi Handai dui Huangjin de Renshi yu Gongyi Jishu Yanjiu," 44, fig. 8. For more information, see So, *Eastern Zhou Ritual Bronzes from the Arthur M. Sackler Collections*, 43–44.

33. See Tian, "Baojishi Yimencun Erhao Chunqiumu Fajue Jianbao," 4–5, and a series of articles compiled in Baoji, *Qinmu Yizhen.*
34. See Li, "Yimencun Jin, Yu Qi Wenshi Yanjiu," 16–17.
35. See So and Bunker, *Traders and Raiders on China's Northern Frontier,* 125–126.
36. For two gold garment hooks in the Mengdiexuan collection (acc. H061 and H043), see White and Bunker, *Adornment for Eternity,* 82, Plate 2.
37. E.g., see Zhang, "Qinqi Sanlun—Yimen Chunqiumu Jige Wenti Qiantan," 21; Huang and Li, "Jinyu Tongmeng," 49; Lee, "Chengqi zhi Dao," 15.
38. For instance, Su Rongyu has observed clear section-mold casting marks from the gold dagger hilt in the British Museum (personal communication, June 2015); W. T. Chase, likewise, "examined some gold-handled daggers [...] Mold marks had been finished off pretty well, but were still clearly discernible" (personal communication, August 2018); Emma Bunker also detected the same type of marks on at least three gold garment hooks from the Mengdiexuan collection, either by visual inspection or microscopic examination (see White and Bunker, *Adornment for Eternity,* 83).
39. See Henan, *Gushi Hougudui Yihao Mu,* 99–101.
40. Su and Wang, "Zaozhuang Xulou Chutu Zhuxiang Hongtong Qingtongqi Tanlun," 1.
41. Two early such samples, one dagger axe *ge* and one broad-axe *yue,* are in the Palace Museum of Beijing and the Asian Art Museum of San Francisco, respectively, and have been dated to the Anyang period (c. 1200–1050 BCE). For details, see Jia and Hu, "Dui Gudai Qingtongqi Hongtong Qianxiang de Yanjiu," 28.
42. Su and Wang, "Zaozhuang Xulou Chutu Zhuxiang Hongtong Qingtongqi Tanlun," 1.
43. See Li, "Gushi Hougudui Qingtong Zhuxiang Hongtong Huawen Gongyi Tantao," 134.
44. For details, see Jia and Hu, "Dui Gudai Qingtongqi Hongtong Qianxiang de Yanjiu," 32–34; Jia et al., "Zeng Hou Yi Qingtongqi Hongtong Huawen Zhuxiangfa de Yanjiu," 86.
45. Li, "Gushi Hougudui Qingtong Zhuxiang Hongtong Huawen Gongyi Tantao," 137.
46. For details, see Ibid., 140–142.
47. Ibid., 142.
48. Ibid., 140.
49. So, *Eastern Zhou Ritual Bronzes from the Arthur M. Sackler Collections,* 34.

50. Zaozhuang et al., "Shandong Zaozhuang Xulou Dongzhou Mu Fajue Jianbao," 26–27.
51. For detailed discussions, see Su and Wang, "Zaozhuang Xulou Chutu Zhuxiang Hongtong Qingtongqi Tanlun"; Hu et al., "Zaozhuang Shi Xulou Cun Hongtong Zhuxiang Qingtongqi Wenshi Xiangzhu Tezheng."
52. This device, called *tusun* ("protruding tenons") by Su and Wang when they noticed copper inlays from some Zaozhuang vessels, with a thickness similar to the wall of bronze, could be viewed as spacers themselves ("Zaozhuang Xulou Chutu Zhuxiang Hongtong Qingtongqi Tanlun," 7–9).
53. See Su and Wang, "Zaozhuang Xulou Chutu Zhuxiang Hongtong Qingtongqi Tanlun," 8.
54. Ibid., 8; Meyers, "Applications of X-ray radiography in the Study of Archaeological Objects," 91.
55. Li Jinghua called spacers with a square section *xindian* (literally, "core-cushion"), and those in a tapered shape *xincheng* (literally, "core-support"). See Li, "Gushi Hougudui Qingtong Zhuxiang Hongtong Huawen Gongyi Tantao," 136–139.
56. Tan Derui, oral communication, Shanghai, May 6, 2015.
57. In comparison, the procedure described by So on the production of a copper-inlaid *hu* makes more sense: "the motifs were first cast in copper, then inserted into the mold and held in place with spacers. [...]When molten bronze was poured into the space between the mold and core, it flowed around the spacers, thus locking, or 'casting,' the copper motifs into place" (*Eastern Zhou Ritual Bronzes from the Arthur M. Sackler Collections*, 257); see also Le Bas et. al., "Chinese Bronze Vessels with Copper Inlaid Decor and Pseudo-Copper Inlay of Ch'un-Ch'iu and Chan-Kuo Times - Part One," 159, and Barnard, "Chinese Bronze Vessels with Copper Inlaid Decor and Pseudo-Copper Inlay of Ch'un-Ch'iu and Chan-Kuo Times - Part Two," 189.
58. So, *Eastern Zhou Ritual Bronzes from the Arthur M. Sackler Collections*, 60–61.
59. For images, see Peng, "The Lost-wax Casting in Bronze Age China," 341–42, figs. 9.28–9.30. For discussion, see Tan, *Canlan de Zhongguo Gudai Shila Zhuzao*, 58–60.
60. See Tan, *Canlan de Zhongguo Gudai Shila Zhuzao*, 59.
61. So, *Eastern Zhou Ritual Bronzes from the Arthur M. Sackler Collections*, 64.

62. Su, "Zhanguo Zhongshan Wang Cuo Mu Qingtong Qiqun Zhuzao Gongyi Yanjiu," 264–266.
63. Hua, *Zhongguo Gudai Jinshu Jishu*, 200; Su, "Zhanguo Zhongshan Wang Cuo Mu Qingtong Qiqun Zhuzao Gongyi Yanjiu," 232–233, 262.
64. For details, see Hua, *Zhongguo Gudai Jinshu Jishu*, 200; Su, "Zhanguo Zhongshan Wang Cuo Mu Qingtong Qiqun Zhuzao Gongyi Yanjiu," 227–233.
65. Su, "Zhanguo Zhongshan Wang Cuo Mu Qingtong Qiqun Zhuzao Gongyi Yanjiu," 266.
66. Hubei, *Zeng hou Yi Mu*, 478.
67. See Hua, *Zhongguo Gudai Jinshu Jishu*, 200; Su, "Zhanguo Zhongshan Wang Cuo Mu Qingtong Qiqun Zhuzao Gongyi Yanjiu," 266.
68. The following section owes much to Robert Bagley's insightful comments (October 1, 2016).
69. See Shao, "Qinshihuang Lingyuan Chutu Caihui Qingtong Shuiqin Zhizuo Gongyi ji Xiangguan Wenti Yanjiu," 37, figs. 3-14, 3-17.
70. The incorporation of additional separately cast pieces would increase the difficulty in joining.
71. For details, see Bagley, *Shang Ritual Bronzes in the Arthur M. Sackler Collections*, 44–45; Loveday, "Diversity in Eastern Zhou Bronze Casting," 119.
72. For details, see Li, "Zeng Hou Yi Bianzhong ji Zun, Zunzuo Zhuzao Fangfa Xintan," 73–92.
73. Li, "Gudai Qingtong Zhuzao Fangfa de Zai Tantao," 95–96, 98–99.
74. Ibid., 97–98.
75. For details, see Li, "Zeng Hou Yi Bianzhong ji Zun, Zunzuo Zhuzao Fangfa Xintan"; Li, "Gudai Qingtong Zhuzao Fangfa de Zai Tantao"; Li, "Zeng Hou Yi Bianzhong Zhuzao Fangfa zhi Wojian"; Li, "Youguan Zeng Hou Yi Zun Pan Zhuzao Fangfa de Zhengming."
76. As Li also mentioned, a drop of wax has been detected with a musical instrument *sheng* from the tomb of Marquis Yi of Zeng ("Zeng Hou Yi Bianzhong Zhuzao Fangfa zhi Wojian"). For more recent discoveries of beeswax in Neolithic and Bronze Age China, see McGovern et al., "Fermented Beverages of Pre- and Proto-Historic China"; Luo et al., "Discovery of Beeswax as Binding Agent on a 6th-century BC Chinese Turquoise-inlaid Bronze Sword."
77. See Barnard, "The Entry of Cire-Perdue Investment Casting, and Certain Other Metallurgical Techniques (Mainly Metalworking) into South China and Their Progress Northwards," part 1.

78. See Zhang, "Zhongguo Zuizao 'Shilafa,'" 65–69.
79. Ibid., 65–68.
80. Ibid., 66–67.
81. For image of such a metal votive raft with mutiple figures (10×19.5×10.1 cm, Pasca, Cundinamarca, 600–1600 CE), see Económica, *The Art of Gold, the Legacy of Pre-Hispanic Colombia: Collection of the Gold Museum in Bogotá*, 212–213.
82. See Hua, "Zhongxifang Shilafa zhi Tongyi," 87–88.
83. For details, see Barnard, "The Entry of Cire-Perdue Investment Casting"; Barnard, "The Entry of Cire-Perdue Investment Casting, and Certain Other Metallurgical Techniques (Mainly Metalworking) into South China and Their Progress Northwards"; Tan, "Zhongguo Gudai Shila Zhuzao Qiyuan Wenti de Sikao"; Tan, "Zhongguo Zaoqi Shila Zhuzao Wenti de Guancha yu Sikao"; Hua, *Zhongguo Gudai Jinshu Jishu*, 179–180.
84. Falkenhausen, *Suspended Music*, 116.
85. For details, see Bunker and Ternbach, "A Variation of the 'Lost-wax' Process," 41–42; Bunker, "Lost Wax and Lost Textile," 222–224; White and Bunker, *Adornment for Eternity*, 41–42.
86. White and Bunker, *Adornment for Eternity*, 41. For detailed illustrations of this process, see Bunker and Ternbach, "A Variation of the 'Lost-wax' Process," 42.
87. White, "Early East Asian Metallurgy," 222.
88. Bunker and Ternbach, "A Variation of the 'Lost-wax' Process," 41; Bunker, "Lost Wax and Lost Textile," 222; Linduff, "Production of Signature Artifacts for the Nomad Market," 93–94.
89. Tan, "Zhongguo Gudai Shila Zhuzao Qiyuan Wenti de Sikao," 46.
90. For details, see Linduff, "Production of Signature Artifacts for the Nomad Market," 94; Luo, "Zhongyuan Zhizao," 60–62.
91. Bunker, "Lost Wax and Lost Textile," 223; White and Bunker, *Adornment for Eternity*, 41.
92. E.g., see Tan, "Zhongguo Gudai Shila Zhuzao Qiyuan Wenti de Sikao," 46; Hua, *Zhongguo Gudai Jinshu Jishu*, 180.
93. For details, see Odagi, "Hoppō-keichōhōkei obikazariban no tenkai," 79–94.
94. For details, see Luo, "Zhongyuan Zhizao," 58–60.
95. For details, see Zhou and Huang, "'Shila Shizhi Fa' Shangque," 59–62.
96. Zhou and Huang, "'Shila Shizhi Fa' Shangque," 60.

97. Joseph Ternbach gave a detailed explanation for this point. For details, see Bunker and Ternbach, "A Variation of the 'Lost-wax' Process," 41.
98. See Luo, "Zhongyuan Zhizao," 59.
99. Odagi, "Hoppō-keichōhōkei obikazariban no tenkai," 79–94.
100. Luo, "Zhongyuan Zhizao," 59–60.
101. For images, see Zhou and Huang, "'Shila Shizhi Fa' Shangque," 60–61, figs. 5.1, 5.2, and 6.1, 6.2.
102. According to Zhou and Huang, attractive force would hinder such a separation if the mold surface were too smooth ("'Shila Shizhi Fa' Shangque," 61).
103. Steinberg, "Technology and Culture," 55–56.
104. Ibid., 56.

Chronology

Table 1. Chronology of Bronze Age China (rough dates until 771 BCE).

Erlitou Culture	1900–1500 BCE ?
Erligang Culture	1500–1300 BCE ?
Erligang-Anyang "Transition Period"	13th century BCE
Anyang Period (of the Yinxu Culture)	1200–1050 BCE
Zhou Dynasty	1050–256 BCE
Western Zhou	1050–771 BCE
Eastern Zhou	770–256 BCE
Spring and Autumn Period	770–476 BCE
Warring States Period	475–221 BCE

Chronology of Bronze Age China

Scholars still have not reached a consensus about the dates of a "Bronze Age" in China. The trouble lies mainly in the definition of the term "Bronze Age," an epoch first proposed by Christian Jürgensen Thomsen (1788–1865) in his famous three-age system, in which "weapons and cutting implements were made of copper or bronze."[1] Later, the archaeological ages of stone, bronze, and iron were elaborated by Gordon Childe as three technological stages of production. According to Childe, copper or bronze, as the major material of production in the Bronze Age, is expected to take over "stone, bone, and wood," and later to be replaced by iron.[2] However, bronze and copper, usually too expensive to be used as tools in agricultural or craft production worldwide, were not a primary cause that could decisively affect social transformation.[3] Throughout

Bronze Age China, metal has rarely been discovered in use as productive tools.[4] In fact, as advances in knowledge or production technology, the materials of stone, bronze, and iron "showed no simple correlation with economic or social structures," and the urban revolution, or the rise of cities, civilizations, and stratified societies, was "more social than technological."[5] As Robert Bagley stated,

> when archaeologists today use the term "Bronze Age," in most parts of the world they use it only as a convenient label for some part of a local archaeological sequence, not as an acknowledgement of some special role played by metal. In China, however, the label is meaningful, for metal did play a special role, and the term "Bronze Age" can be defined in a way that firmly connects it with the rise of stratified societies.[6]

Though it is conventionally believed that the Chinese Bronze Age started around or before 2000 BCE,[7] and bronze was in use in China even earlier, the transition from primitive, small-scale metallurgy to large-scale metal industry in China seems to have only occurred around the sixteenth century BCE, when the term "Bronze Age" can most fairly be applied to a society we would like to call civilized.[8] Since iron-making technology had already been known in China since the early first millennium BCE, and within a few centuries the production of cast-iron implements had attained an industrial level,[9] K.C. Chang proposed that the Chinese Bronze Age was replaced by Iron Age around 500 BCE.[10] However, this replacement was a slow and prolonged course, as bronze remained the primary material for ceremonial utensils (mainly vessels and chime bells) and weaponry for an extensive period, and ritual and military affairs were of the greatest importance for the founding elites of the Chinese society. As Lothar von Falkenhausen mentioned,

> Even though, therefore, much of the period treated in the present book (*Chinese Society in the Age of Confucius [1000–250 BC]*) might well be said to constitute China's "iron Age," it is not, in fact, so designated by Chinese archaeologists. Referring to the entire Zhou period as part of the Bronze Age is justifiable because of

> the eminent cultural importance of bronze objects throughout this period.[11]

Following Falkenhausen's scheme, I regard the Chinese Bronze Age as terminated by the unification of China by the First Emperor of Qin, certainly a convenient endpoint. The Bronze Age in this book therefore refers to the period from around the sixteenth century to 221 BCE. The dates of the "Early Bronze Age" are usually given from the beginning of this era to the founding of Western Zhou dynasty around 1050 BCE, again a convenient endpoint;[12] the rest, from around 1050 to 221 BCE, can therefore be called the "Late Bronze Age" of China.

NOTES

1. Quoted in Daniel, *The Origins and Growth of Archaeology*, 94; Chang, "The Chinese Bronze Age," 35.
2. For details, see Childe, "Archaeological Ages as Technological Stages," 9–23; Chang, "The Chinese Bronze Age," 35–36.
3. Bagley, "Shang Archaeology," 136.
4. Chang, "The Chinese Bronze Age," 40.
5. Bagley, "Shang Archaeology," 136.
6. Ibid., 137.
7. E.g., Chang, "The Chinese Bronze Age," 40; Falkenhausen, *Chinese Society in the Age of Confucius*, 3, note 4.
8. Bagley, "Shang Archaeology," 137–138.
9. Falkenhausen, *Chinese Society in the Age of Confucius*, 3, note 4.
10. Chang, "The Chinese Bronze Age," 35, 40.
11. Falkenhausen, *Chinese Society in the Age of Confucius*, 3, note 4.
12. Bagley, "Shang Archaeology," 138.

Figures

Figure 1. Administrative Divisions of Contemporary China.

Source. © 2016 Maps of the World | Maps of all regions, countries and territories of the World, with the map title slightly modified (Accessed September 5, 2019): http://www.maps-of-the-world.net/maps/maps-of-asia/maps-of-china/large-detailed-administrative-divisions-map-of-china-2011.jpg.

Note. To correspond Chinese administrative divisions to the physiographic macroregions of China Proper, see "Cities and the Hierarchy of Local Systems" (p. 9) by William Skinner, in Arthur P. Wolf. ed. *Studies in Chinese Society*. Stanford: Stanford University Press, 1978.

Figure 2a. Bronze statuette of a horse cast with its base. H. 7.9 cm. Greek [Geometric], circa 750-700 BCE.

Source. Artifact in the J. Paul Getty Museum (acc. 85. AB.445); reproduced in *Enduring Bronze: Ancient Art, Modern Views* (p. 56, fig. 32) by Carol Mattusch, 2014, Los Angeles: J. Paul Getty Museum.

Figure 2b. The lost-wax process to cast the bronze statuette of Figure 2a.

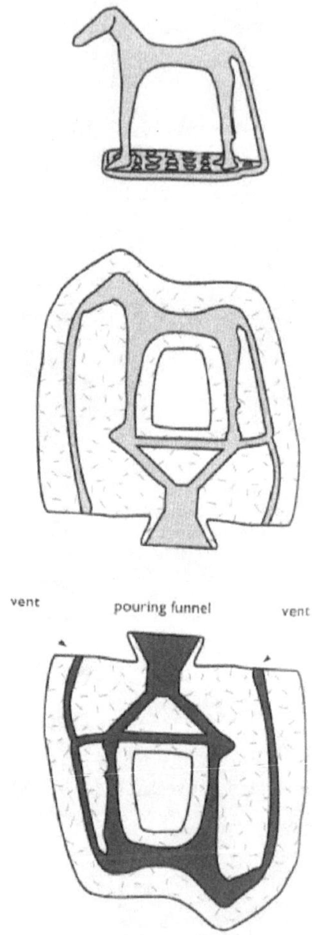

Source. Drawing by Sean Hemingway; reproduced in *Enduring Bronze: Ancient Art, Modern Views* (p. 56, figs. 33) by Carol Mattusch, 2014, Los Angeles: J. Paul Getty Museum.

Note. The lost-wax process illustrated by the diagram to cast the bronze horse (from top to bottom): (1) wax model for horse; (2) section through investment mold (clay surrounding wax model) with added wax funnel, gates (pouring channels), and air vents at each side; (3) mold with poured bronze replacing the wax model (script adapted from Carol Mattusch, *Enduring Bronze: Ancient Art, Modern Views,* p. 56).

Figure 3a. Riace Warrior A, one of a pair of full-size Greek bronze statues found in the sea at Riace Marina in 1972. H. 198 cm. Circa 5th century BCE.

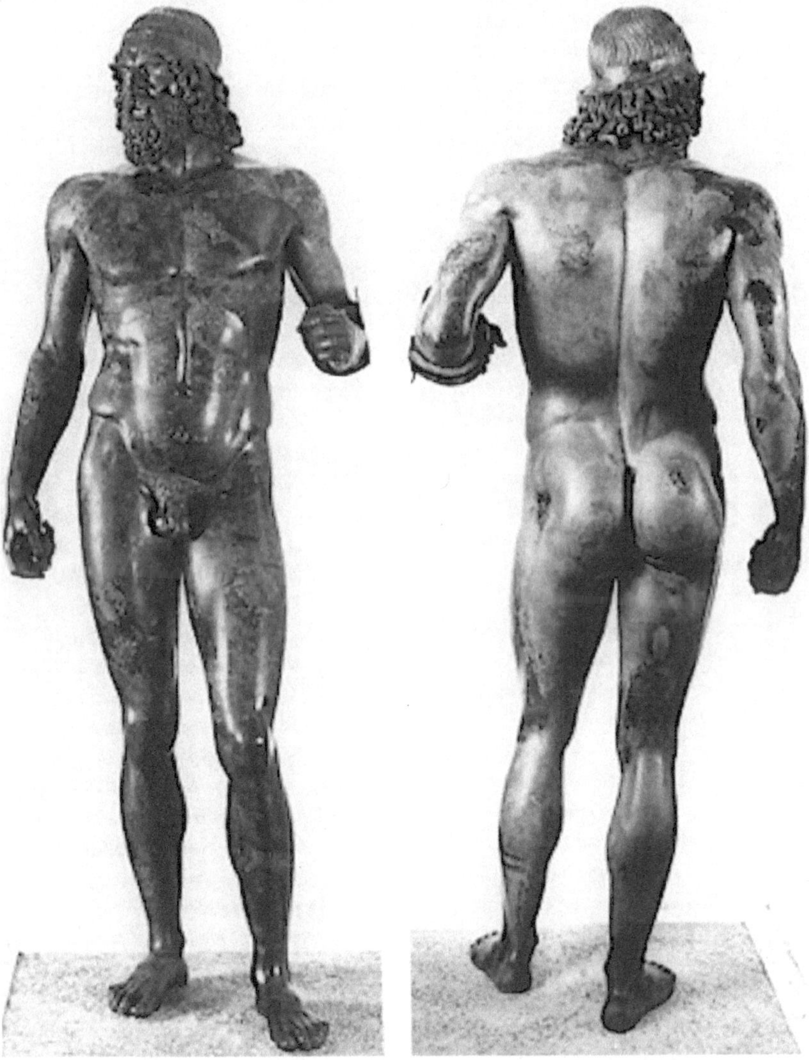

Source. Artifact in the in the Museo Nazionale della Magna Grecia, Reggio Calabria, Italy. After *Classical Bronzes: The Art and Craft of Greek and Roman Statuary* (p. 66, fig. 2.18: left) by Carol Mattusch, 1996, Ithaca, NY: Cornell University Press.

Figure 3b. Riace Warrior B, one of a pair of full-size Greek bronze statues found in the sea at Riace Marina in 1972. H. 197 cm. Circa 5th century BCE.

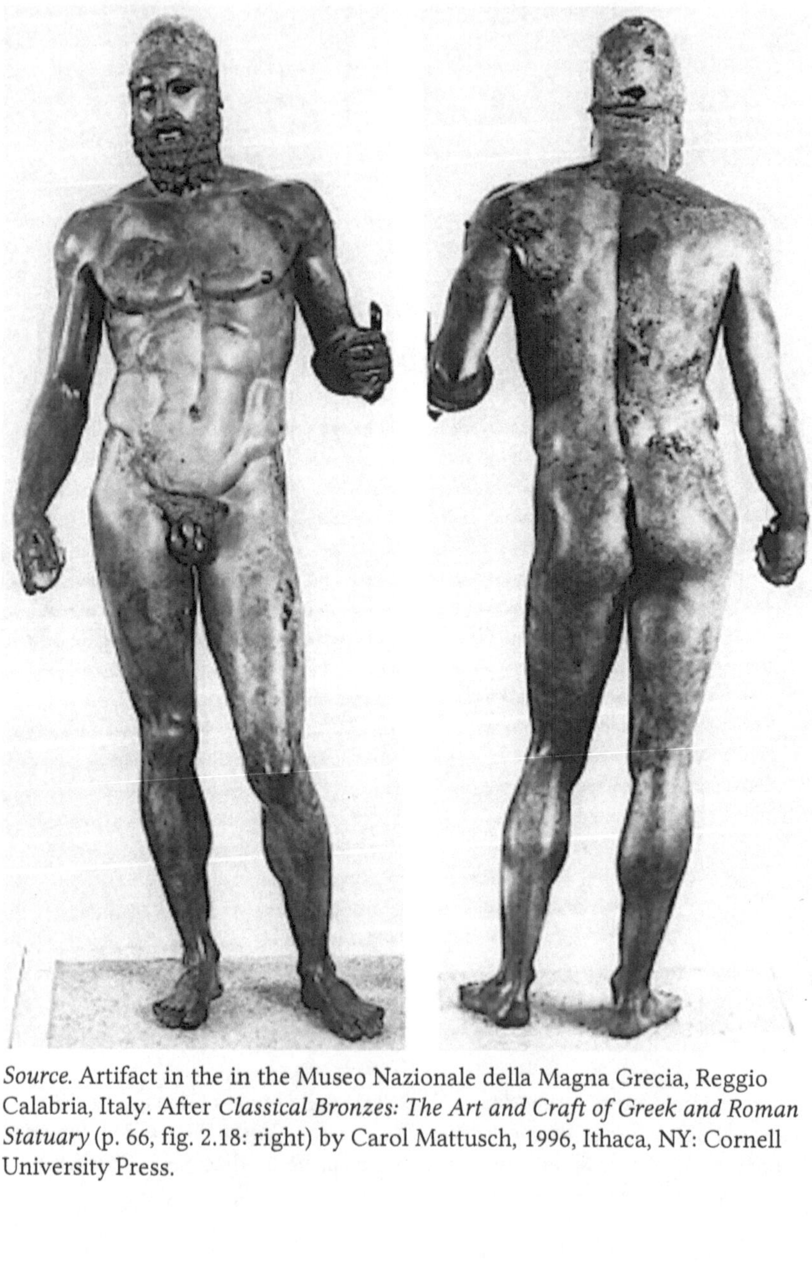

Source. Artifact in the in the Museo Nazionale della Magna Grecia, Reggio Calabria, Italy. After *Classical Bronzes: The Art and Craft of Greek and Roman Statuary* (p. 66, fig. 2.18: right) by Carol Mattusch, 1996, Ithaca, NY: Cornell University Press.

Figure 4. Seima-Turbino bronze dagger from the Rostovka burial near Omsk (in the Irtysh Basin) with figurines of a horse and a skier on the handle.

Source. After *Ancient Metallurgy in the USSR: The Early Metal Age* (p. 228, Plate 22) by EvgeniĬ Nikolaevich Chernykh, 1992, New York: Cambridge University Press.

Figure 5. The reconstruction by Maryon and Plenderleith for the casting of a bronze *zun* vessel: (A) General view of the interior of mold. (B) Wax model with walls for first part of piece-mold in position. (C) First part of piece-mold completed with mortises cut; boundaries of second part in position.

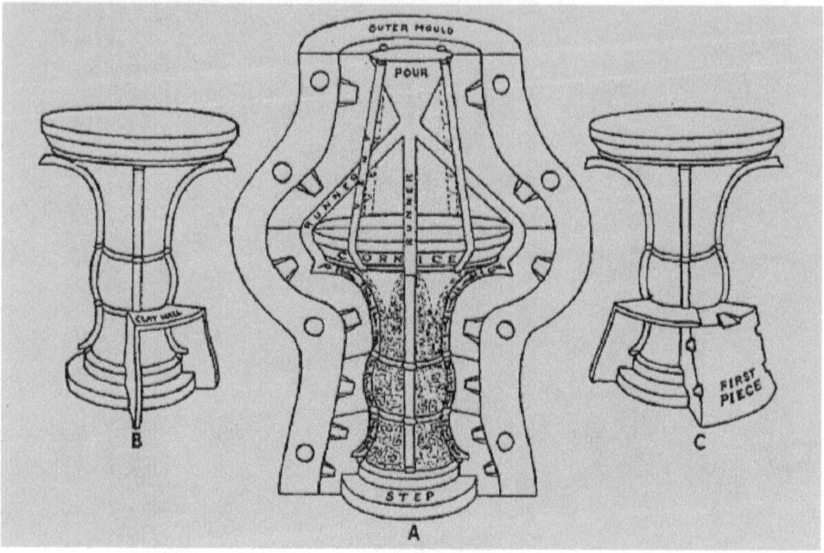

Source. Image and script after "Fine Metal-work" (p. 628, fig. 425) by Herbert Maryon and Harold Plenderleith, 1954. In *A History of Technology*, edited by Charles Singer et al., 623-662. Oxford: Oxford University Press.

Note. The modelled ornament has been omitted in (B) and (C). Vessel hollow-based up to level of first piece of mold in (C) (script adapted from Maryon and Plenderleith, "Fine Metal-work," p. 628).

Figure 6. The cross section of a leg from a broken bronze *ding* in the Freer Gallery of Art, Washington, D.C.

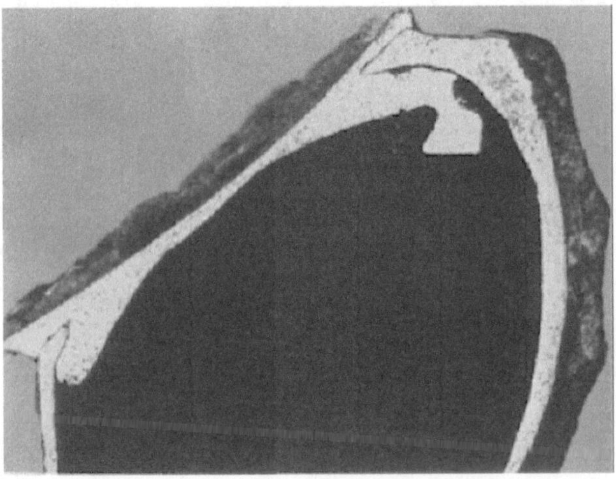

Source. After *The Freer Chinese Bronzes, vol II: Technical Studies* (p. 79, fig. 64) by Rutherford J. Gettens, 1969. Washington: Smithsonian Institution.
Note. The cross section of the *ding* leg shows how parts of a bronze are made to interlock through the casting process. Light and dark areas indicate metal pieces and clay core, respectively (script adapted from Rutherford J. Gettens, *The Freer Chinese Bronzes, vol II: Technical Studies*, p. 79).

Figure 7. Diagram showing the relationship between a *ding* (Erligang style) and the mold used to cast it.

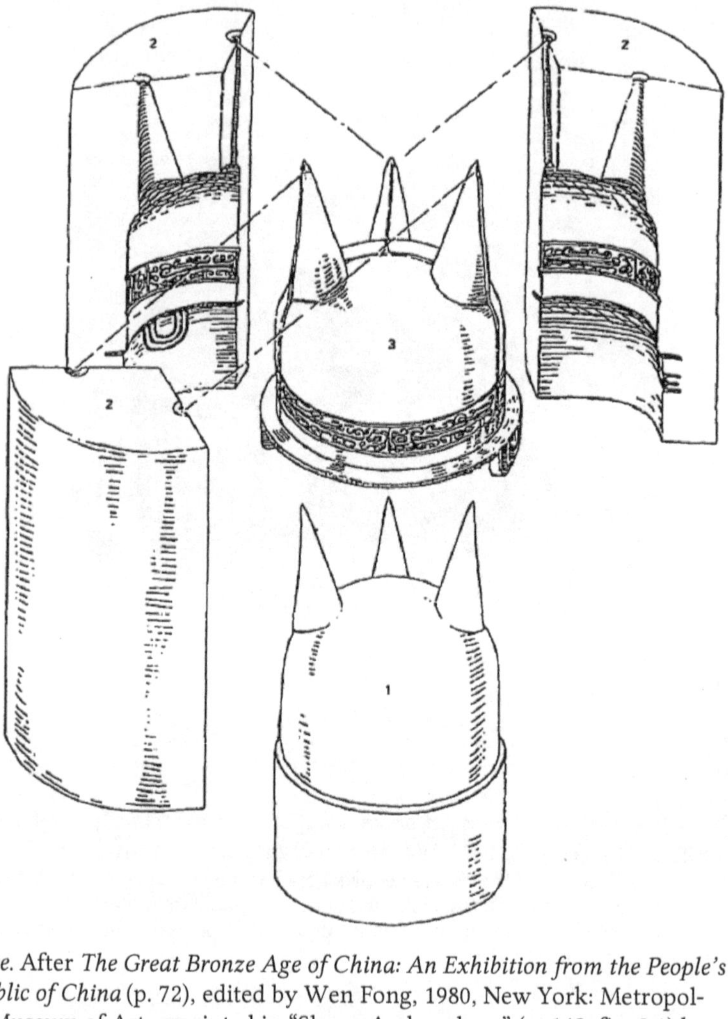

Source. After *The Great Bronze Age of China: An Exhibition from the People's Republic of China* (p. 72), edited by Wen Fong, 1980, New York: Metropolitan Museum of Art; reprinted in "Shang Archaeology" (p. 143, fig. 3.1) by Robert Bagley, 1999, in *The Cambridge History of Ancient China*, edited by Michael Loewe and Edward L. Shaughnessy, 124–231. Cambridge and New York: Cambridge University Press.

Note. 1) clay core; 2) clay mold sections (these were formed on an undecorated model, and the decoration was then carved into them); 3) finished bronze, showing mold marks along the lines where the mold sections met (script adapted from Robert Bagley, "Shang Archaeology," p. 143).

Figure 8a. Bronze *jue* from Erlitou (3rd stratum). H. 22.5 cm. Circa 16th century BCE.

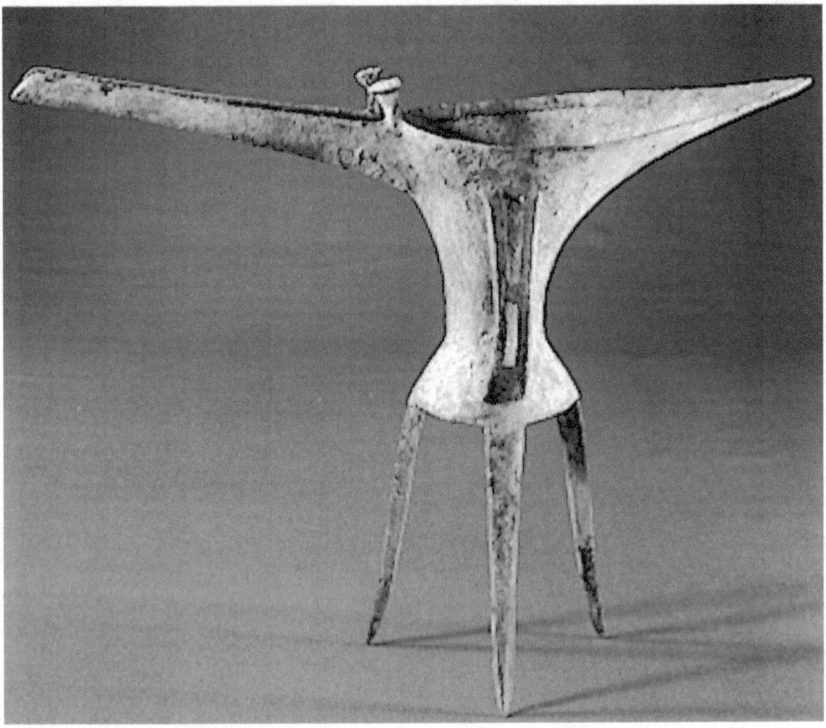

Source. After *Zhongguo qingtongqi quanji*, vol. 1 (Plate 7), edited by Zhongguo qingtongqi quanji bianji weiyuanhui, 1993, Beijing: Wenwu Chubanshe.

Figure 8b. Rubbing of the surface decoration of the bronze *jue* shown in Figure 8a.

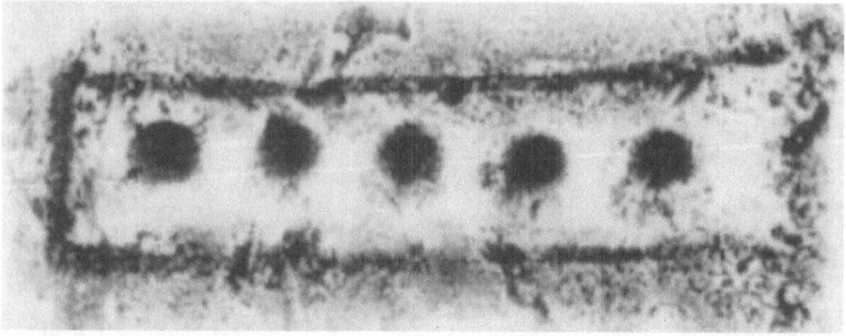

Source. After *The Great Bronze Age of China: An Exhibition from the People's Republic of China* (p. 17), edited by Wen Fong, 1980, New York: Metropolitan Museum of Art; reprinted in "Shang Ritual Bronzes: Casting Technique and Vessel Design" (p.11, fig. 9b) by Robert Bagley, *Archives of Asian Art* 43 (1990): 7–20.

Figure 9. Bronze *ding* from Huangpi Panlongcheng, Hubei. H. 54 cm. Circa 15th century BCE.

Source. After *Zhongguo gu qingtongqi xuan*, no. 2., 1976, Beijing: Wenwu Chubanshe; reprinted in "Shang Ritual Bronzes: Casting Technique and Vessel Design" (p.14, fig. 15) by Robert Bagley, *Archives of Asian Art* 43 (1990): 7–20.

Figure 10. Clay pattern block from the Houma foundry site, Shanxi. Width 32.8 cm. Late 6th or early 5th century BCE.

Source. Artifact in the Houma Jinguo Gudu Bowuguan ("Houma Museum of Ancient Jin Capital"), Shanxi; reprinted in *The Chinese Exhibition: A Pictorial Record of the Exhibition of Archaeological Finds of the People's Republic of China* (catalogue no. 132), 1975, Kansas City, Mo.: The Nelson Gallery Atkins Museum.

Figure 11a. Bronze *hu* with dragon interlace. H. 44.8 cm, greatest circumference 83.5 cm. First half of 5th century BCE.

Source. Freer Gallery of Art, Smithsonian Institution, Washington, D.C.: Purchase — Charles Lang Freer Endowment, F1957.22.

Figure 11b. Ornament in the main register of the bronze *hu* shown in Figure 11a .

Source. Freer Gallery of Art, Smithsonian Institution, Washington, D.C.: Purchase — Charles Lang Freer Endowment, F1957.22.

Figure 12. Bronze openwork ornament from Henan Sanmenxia. Diameter 7.5 cm. Circa 5th century BCE.

Source. After *Kaogu jinghua* (p. 219), Beijing: Kexue Chubanshe, 1993; reprinted in "Anyang Mold-making and the Decorated Model" (p. 61, fig. 17) by Robert Bagley, *Artibus Asiae* 69 (2009): 39–90.

Figure 13a. Baked clay model for an openwork ornament, unearthed at the Houma foundry site, Shanxi. Diameter of decorated area 7 cm. Circa 5th century BCE.

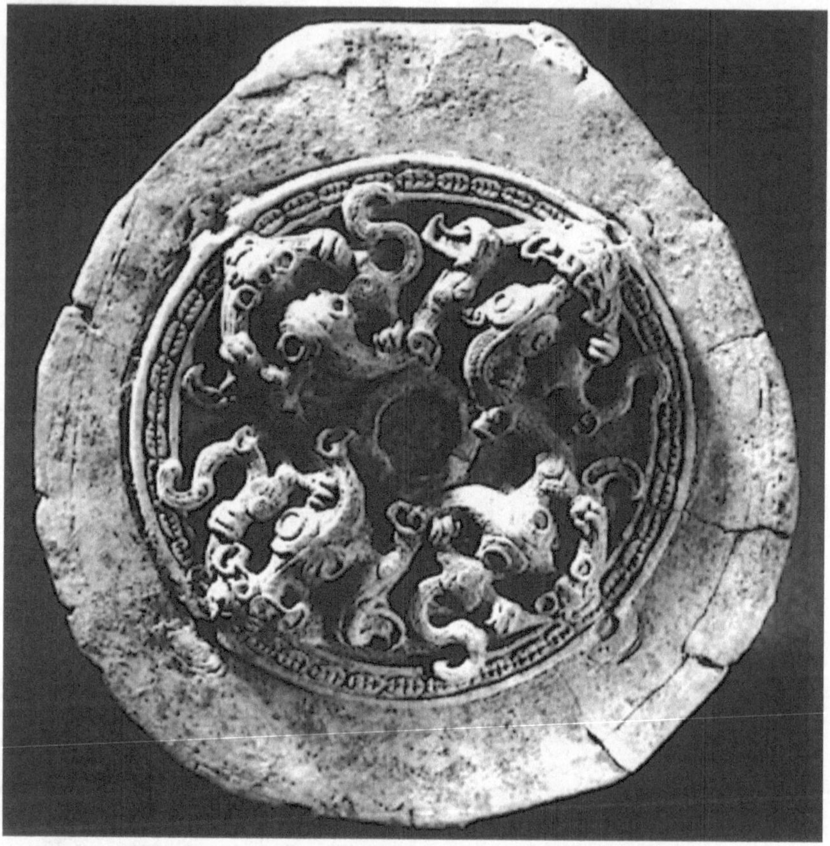

Source. Photograph courtesy of the Institute of Archaeology of Shanxi Province, reprinted in "Anyang Mold-making and the Decorated Model" (p.61, fig. 18a) by Robert Bagley, *Artibus Asiae* 69 (2009): 39–90.

Figure 13b. The decorated part of the model shown in Figure 13a.

Source. Drawing by Li Xiating; reprinted in "Anyang Mold-making and the Decorated Model" (p.61, fig. 18b) by Robert Bagley, *Artibus Asiae* 69 (2009): 39–90.

Figure 14a. Table base from Pingshan Sanji Gongshe M1, Hebei. Bronze inlaid with gold and silver. Late 4th century BCE. H.36.2cm.

Source. After *Zhongguo qingtongqi quanji*, vol. 9 (Plate 165), edited by Zhongguo qingtongqi quanji bianji weiyuanhui, 1997, Beijing: Wenwu Chubanshe.

Figure 14b. Detail of the table base of Figure 14a.

Source. After *Zhongguo qingtongqi quanji*, vol. 9 (Plate 167), edited by Zhongguo qingtongqi quanji bianji weiyuanhui, 1997, Beijing: Wenwu Chubanshe.

Figure 15a. Bronze *zun* of the *zun-pan* set from the tomb of Marquis Yi of Zeng (d. 433 BCE), Suizhou, Hubei province. H. 30.1 cm.

Source. After *Zhongguo qingtongqi quanji*, vol. 10 (Plate 140), edited by Zhongguo qingtongqi quanji bianji weiyuanhui, 1996, Beijing: Wenwu Chubanshe.

Figure 15b. Bronze *pan* of the *zun-pan* set from the tomb of Marquis Yi of Zeng (d. 433 BCE), Suizhou, Hubei province. H. 23.5 cm.

Source. After *Zhongguo qingtongqi quanji*, vol. 10 (Plate 141), edited by Zhongguo qingtongqi quanji bianji weiyuanhui, 1996, Beijing: Wenwu Chubanshe.

Figure 16. Rim appendage of the *zun* (top view) shown in Figure 15. Diameter 25 cm.

Source. After *Zhongguo qingtongqi quanji*, vol. 10 (Plate 138), edited by Zhongguo qingtongqi quanji bianji weiyuanhui, 1996, Beijing: Wenwu Chubanshe.

Figure 17. Detail of structure on the rim appendage of the *zun* shown in Figure 15.

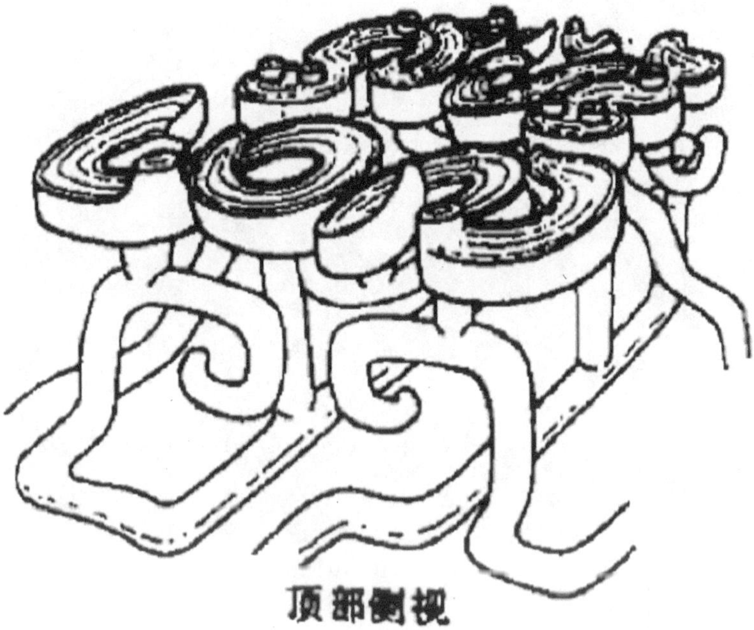

Source. After "Zenghou Yu zun-pan he shilafa de qiyuan" (p. 356, fig. 14) by Hua Jueming and Jia Yunfu, *Ziran Kexueshi Yanjiu* 4 (1983): 352–359.

Figure 18. Detail of the rim appendage of the *zun* in Figure 15, showing traces of a so-called "runner" thought to have been removed from individual pattern elements (Perspective 1).

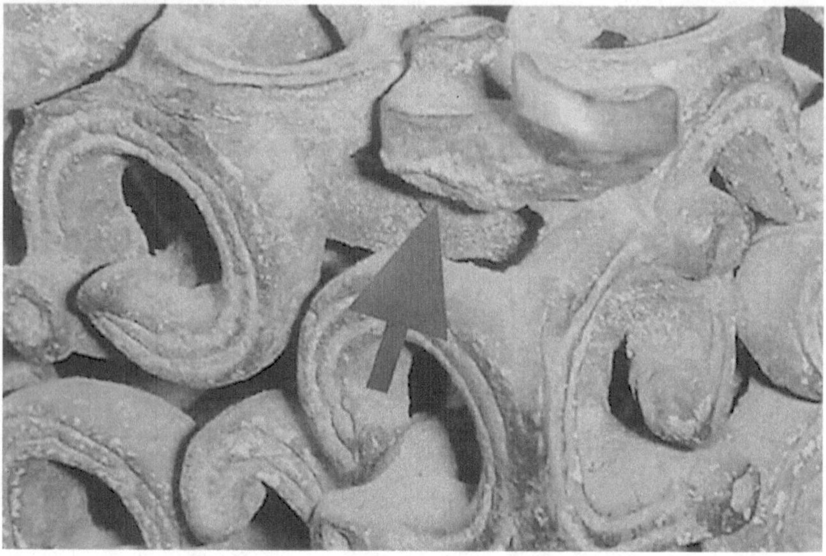

Source. After "Zai Lun 'Shila Gongyi Bushi Zhongguo Qingtong Shidai de Xuanze'" (p. 43, fig. 2) by Zhou Weirong et al., *Nanfang Wenwu* 2 (2007): 41–48,132.

Figure 19. Casting procedure proposed by Perspective 2: use of the model (center) to produce five mold sections.

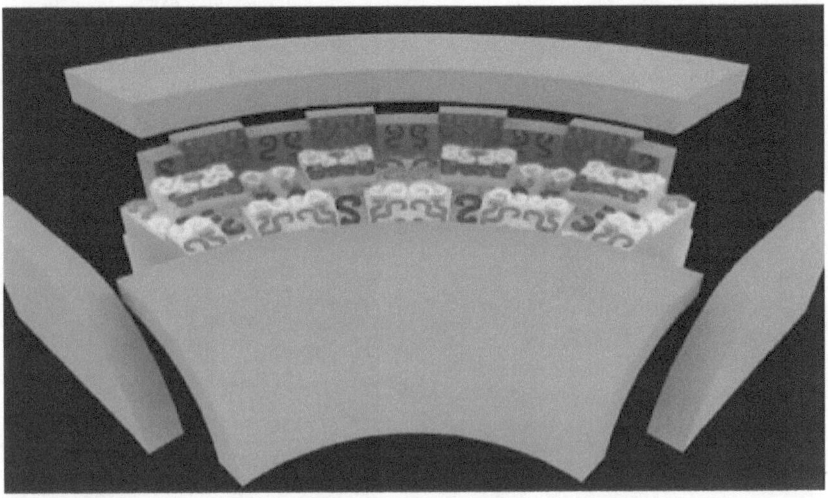

Source. After "Sanwei Donghua li de Gudai Chuantong Gongyi" (p. 116, fig.5) by Wang Wei et al., *Jianghan Kaogu* 3 (2010): 113–122.

Figure 20a. Pattern elements before being decorated (Perspective 2).

Source. After "Zhongguo Qingtong Shidai Loukong Qingtongqi Nifan Zhuzuo Gongyi Yanjiu" (p. 282, fig. 3.1) by Wang Jinchao, 2009, in *Qingtong Wenhua Yanjiu.* Vol. 6, 278–287, 190. Tongling: Huangshan Shushe.

Figure 20b. Pattern elements after being decorated (Perspective 2).

Source. After "Zhongguo Qingtong Shidai Loukong Qingtongqi Nifan Zhuzuo Gongyi Yanjiu" (p. 282, fig. 3.2) by Wang Jinchao, 2009, in *Qingtong Wenhua Yanjiu*. Vol. 6, 278–287, 190. Tongling: Huangshan Shushe.

Figure 21. Illustration of the double-layered inner mold assembly (Perspective 2).

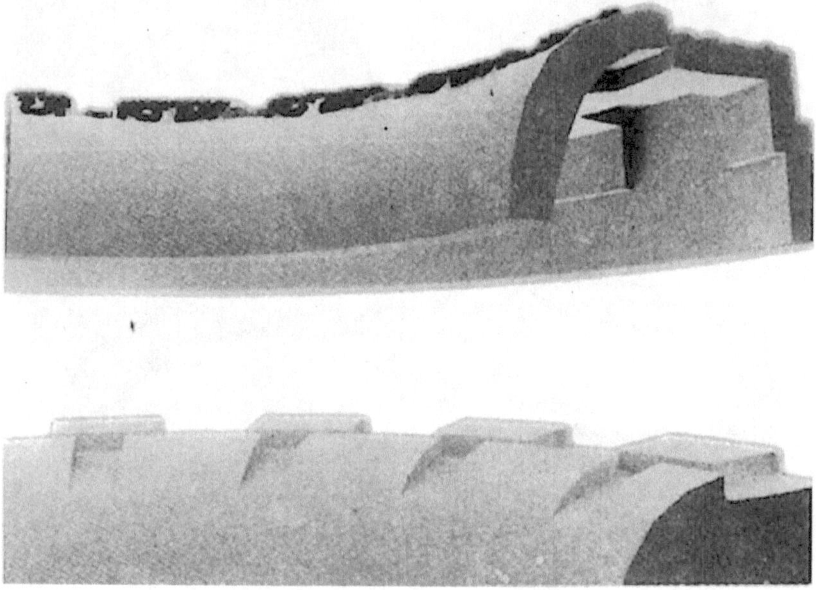

Source. After "Zhongguo Qingtong Shidai Loukong Qingtongqi Nifan Zhuzuo Gongyi Yanjiu" (p. 283, fig. 4.2) by Wang Jinchao, 2009, in *Qingtong Wenhua Yanjiu.* Vol. 6, 278–287, 190. Tongling: Huangshan Shushe.

Figure 22. Use of a cylindrical tool to penetrate the outer layer of the inner mold and pierce the inner layer (Perspective 2).

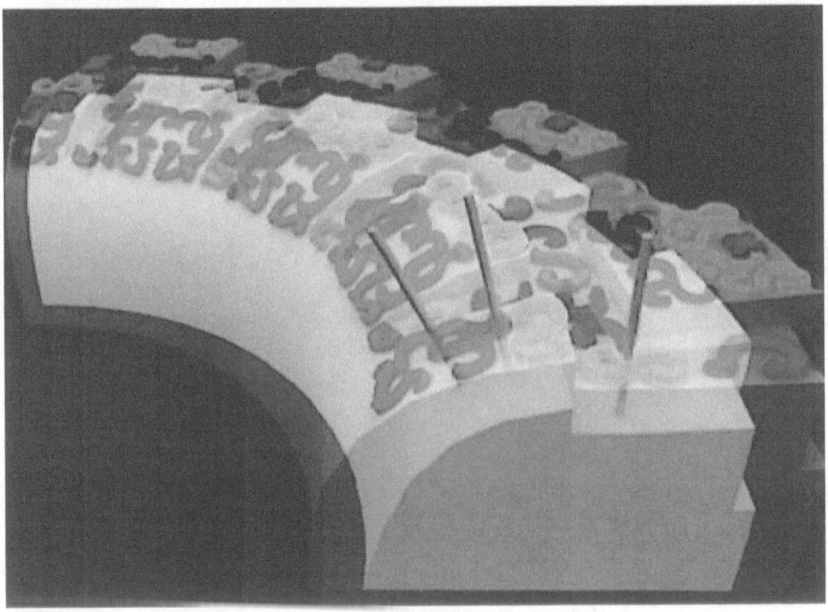

Source. After "Sanwei Donghua li de Gudai Chuantong Gongyi" (p. 118, fig. 10) by Wang Wei et al., *Jianghan Kaogu* 3 (2010): 113–122.

Figure 23. Connecting all the "pits" to create "channels" on the surface of the inner layer of the inner mold (Perspective 2).

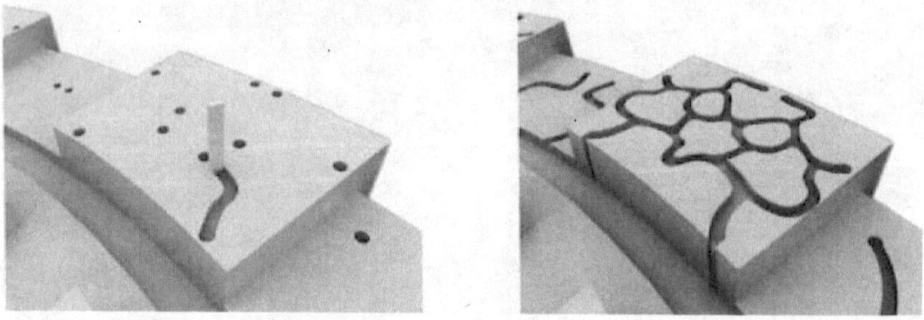

Source. After "Zhongguo Qingtong Shidai Loukong Qingtongqi Nifan Zhuzuo Gongyi Yanjiu" (p. 283, fig. 5.3) by Wang Jinchao, 2009, in *Qingtong Wenhua Yanjiu.* Vol. 6, 278–287, 190. Tongling: Huangshan Shushe.

Figure 24. Deformed pattern elements in the rim appendage of the *zun* shown in Figure 15.

Source. After “Zai Lun ‘Shila Gongyi Bushi Zhongguo Qingtong Shidai de Xuanze’” (p. 44, fig. 5) by Zhou Weirong et al., *Nanfang Wenwu* 2 (2007): 41–48,132.

Figure 25. The venting system attached to the wax model of a simple vessel in a modern bronze foundry in the City of New York, U.S.

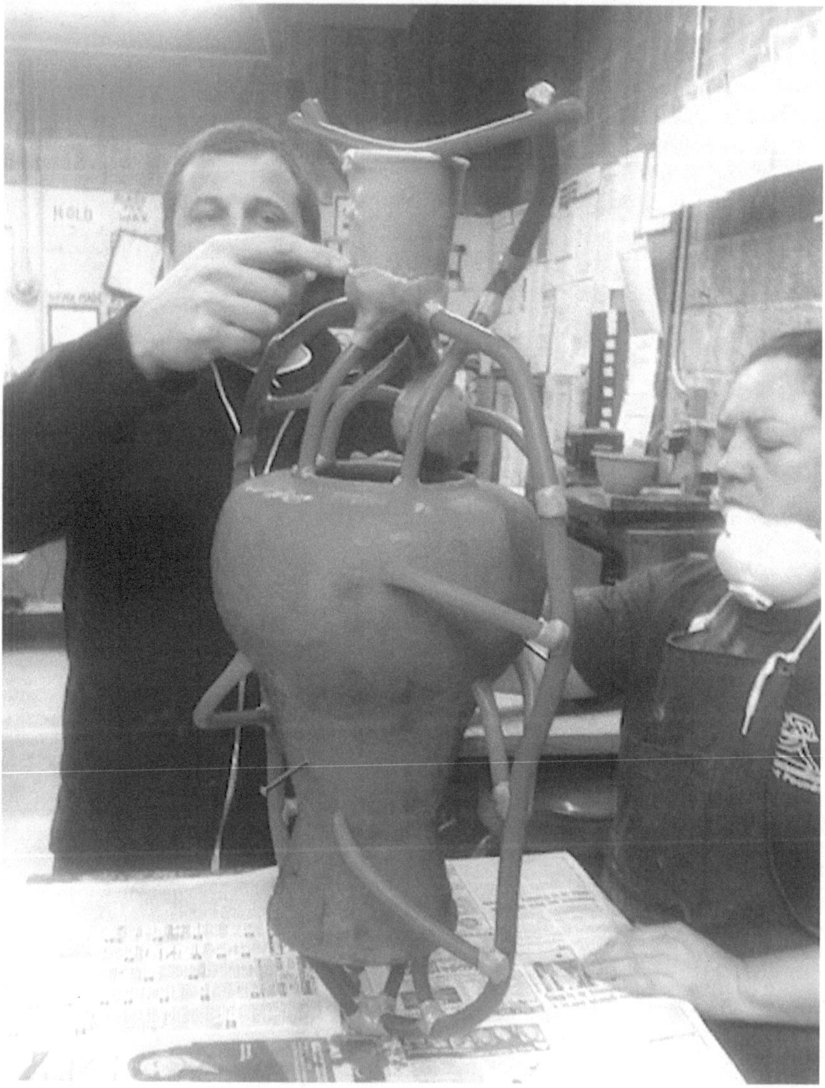

Source. Author's photograph.

Figure 26a. Bronze *jin* (altar table) from Henan Xichuan Xiasi M2. 2nd quarter of 6th century BCE. L.124 cm.

Source. After *Zhongguo qingtongqi quanji*, vol. 10 (Plate 76), edited by Zhongguo qingtongqi quanji bianji weiyuanhui, 1996, Beijing: Wenwu Chubanshe.

Figure 26b. Decoration of bronze *jin* (altar table) from Henan Xichuan Xiasi M2.

Source. After *Zhongguo qingtongqi quanji*, vol. 10 (Plate 77), edited by Zhongguo qingtongqi quanji bianji weiyuanhui, 1996, Beijing: Wenwu Chubanshe.

Figure 26c. The illustrated structure of bronze *jin* (altar table) from Henan Xichuan Xiasi M2.

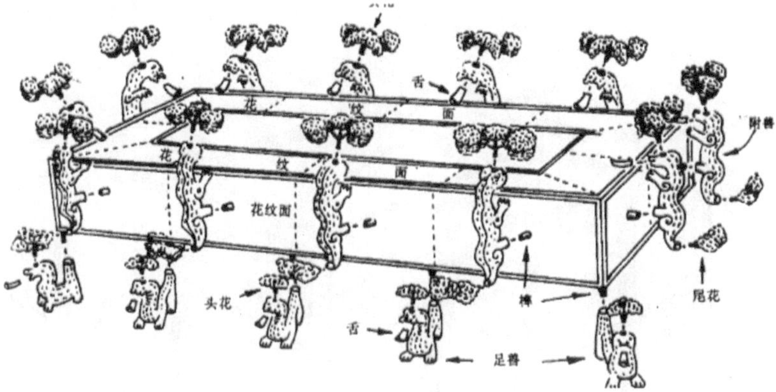

Source. After "Xichuan Chunqiu Chumu Tongjin Shila Zhuzaofa de Gongyi Tantao" (p. 40, fig. 2) by Li Jinghua, *Wenwu Baohu yu Kaogu Kexue* 6, no. 1 (1994): 39–45.

Figure 27a. The exterior of the Xiasi *jin* body in sections.

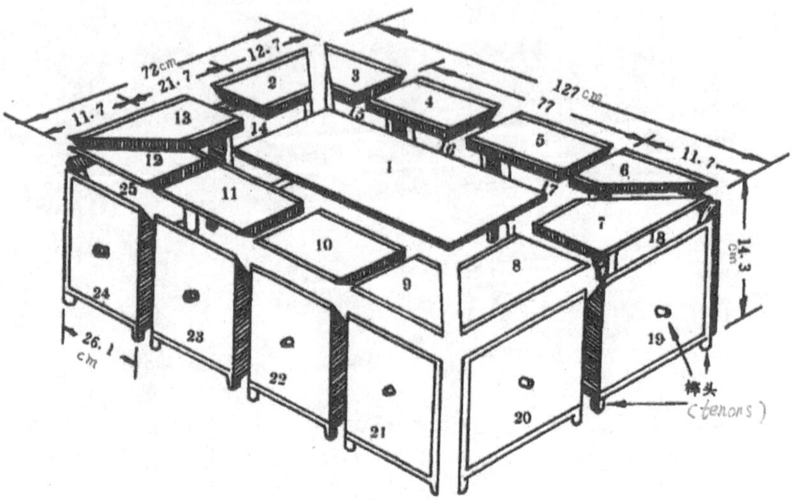

Source. After "Xichuan Chunqiu Chumu Tongjin Shila Zhuzaofa de Gongyi Tantao" (p. 40, fig. 3) by Li Jinghua, *Wenwu Baohu yu Kaogu Kexue* 6, no. 1 (1994): 39–45.

Figure 27b. The interior of the Xiasi *jin* body in sections.

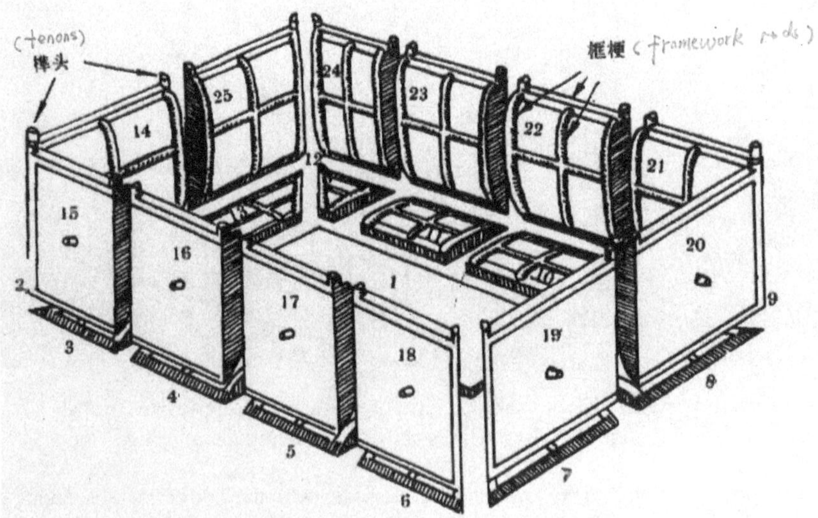

Source. After "Xichuan Chunqiu Chumu Tongjin Shila Zhuzaofa de Gongyi Tantao" (p. 40, fig. 4) by Li Jinghua, *Wenwu Baohu yu Kaogu Kexue* 6, no. 1 (1994): 39–45.

Figure 28. Rods and ornamental elements in the openwork structure of the Xiasi *jin* body.

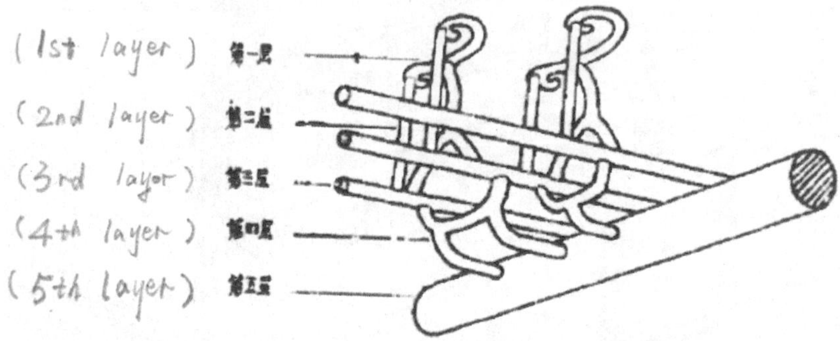

Source. After "Henan Xichuan Xiasi Chunqiu Yunwen Tong Jin de Zhuzao yu Xiufu" (p. 475, fig. 2) by Ren Changzhong and Wang Changqing, *Kaogu* 5 (1987): 474–478.

Note. The *jin*, with structure shown here right-side up (with the side of the table facing up), was actually cast upside down. Readers should mentally rotate the diagram to visualize the flow of bronze.

Figure 29. *Kuang geng* ("rods of the framework") and *gongxing geng* ("arched rods"), viewed from interior of the Xiasi *jin* body.

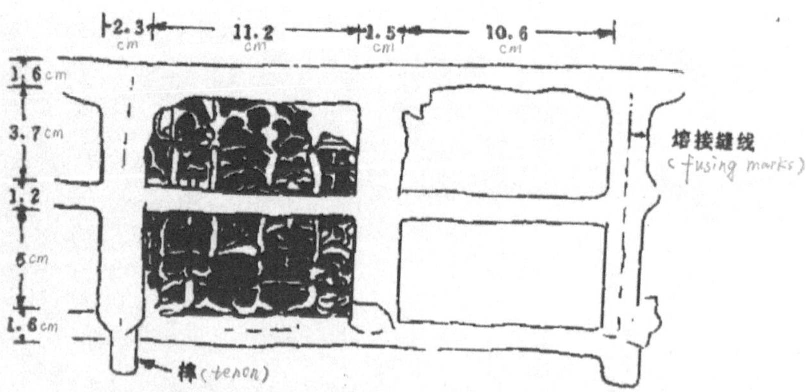

Source. After "Xichuan Chunqiu Chumu Tongjin Shila Zhuzaofa de Gongyi Tantao" (p. 50, fig.23.3) by Li Jinghua, 1994. In Li Jinghua, *Zhongyuan Gudai Yejin Jishu Yanjiu.* Zhengzhou: Zhongzhou Guji Chubanshe, 48–52.

Figure 30. The construction of *gongxing geng* ("arched rods") and *zhi geng* ("straight rods") within a single unit gridded by *kuang geng* ("rods of the framework"), viewed from the interior of the Xiasi *jin* body.

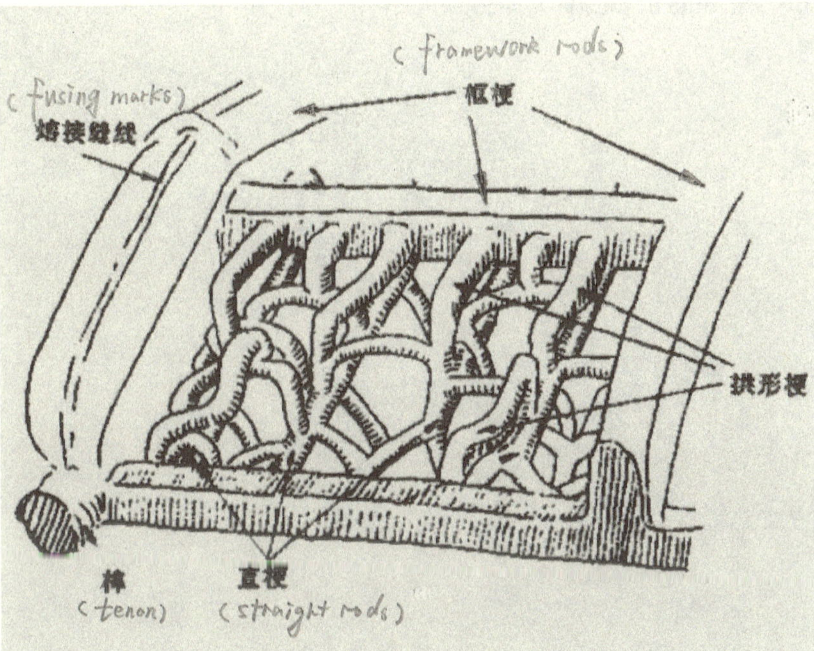

Source. After "Xichuan Chunqiu Chumu Tongjin Shila Zhuzaofa de Gongyi Tantao" (p. 50, fig. 23.2) by Li Jinghua, 1994. In Li Jinghua, *Zhongyuan Gudai Yejin Jishu Yanjiu.* Zhengzhou: Zhongzhou Guji Chubanshe, 48–52.

Figure 31. The construction of *gongxing* geng ("arched rods"), *zhi geng* ("straight rods"), *cheng geng* ("supporting rods"), and *huawen geng* ("the ornamental elements") from a single *kuang geng* ("rods of the framework"), openwork structure of the Xiasi *jin* body.

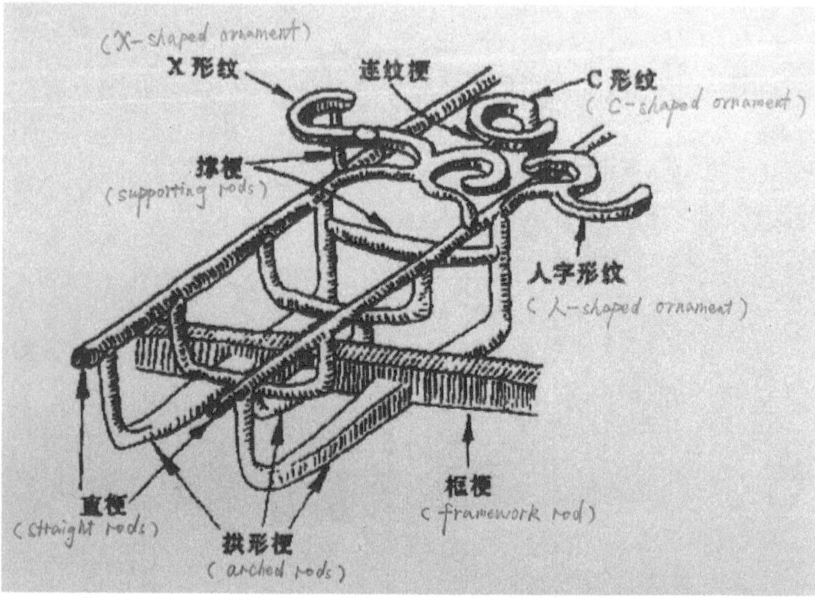

Source. After "Xichuan Chunqiu Chumu Tongjin Shila Zhuzaofa de Gongyi Tantao" (p. 50, fig.23.1) by Li Jinghua, 1994. In Li Jinghua, *Zhongyuan Gudai Yejin Jishu Yanjiu.* Zhengzhou: Zhongzhou Guji Chubanshe, 48–52.
Note. The *jin*, with structure shown here right-side up (with the side of the table facing up), was actually cast upside down. Readers should mentally rotate the diagram to visualize the flow of bronze.

Figure 32. The construction of *huawen geng* ("the ornamental elements") and *cheng geng* ("supporting rods") upon *zhi geng* ("straight rods"), viewed from the exterior of the Xiasi *jin* body.

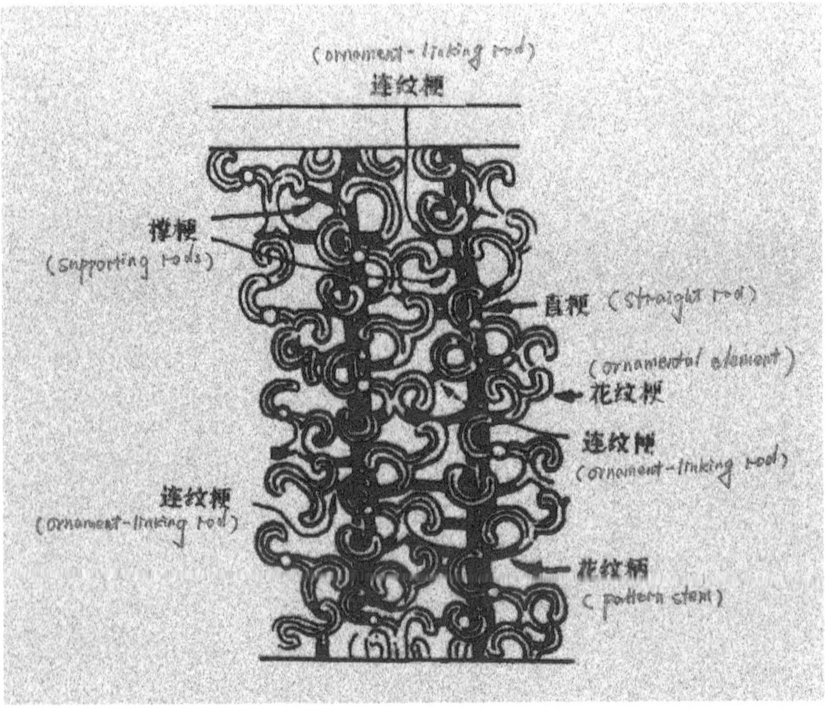

Source. After "Xichuan Chunqiu Chumu Tongjin Shila Zhuzaofa de Gongyi Tantao" (p. 42, fig. 11) by Li Jinghua, *Wenwu Baohu yu Kaogu Kexue* 6, no. 1 (1994): 39–45.

Figure 33. The interrelationship of bronze rods in profile, from the *huawen geng* ("the ornamental elements") to *kuang geng* ("rods of the framework").

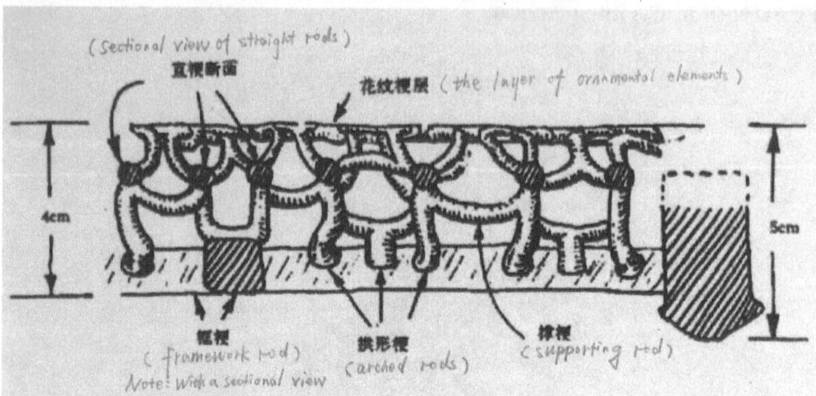

Source. After "Xichuan Chunqiu Chumu Tongjin Shila Zhuzaofa de Gongyi Tantao" (p. 41, fig. 9) by Li Jinghua, *Wenwu Baohu yu Kaogu Kexue* 6, no. 1 (1994): 39–45.

Figure 34a. The interior of the Xiasi *jin* body: area at one corner.

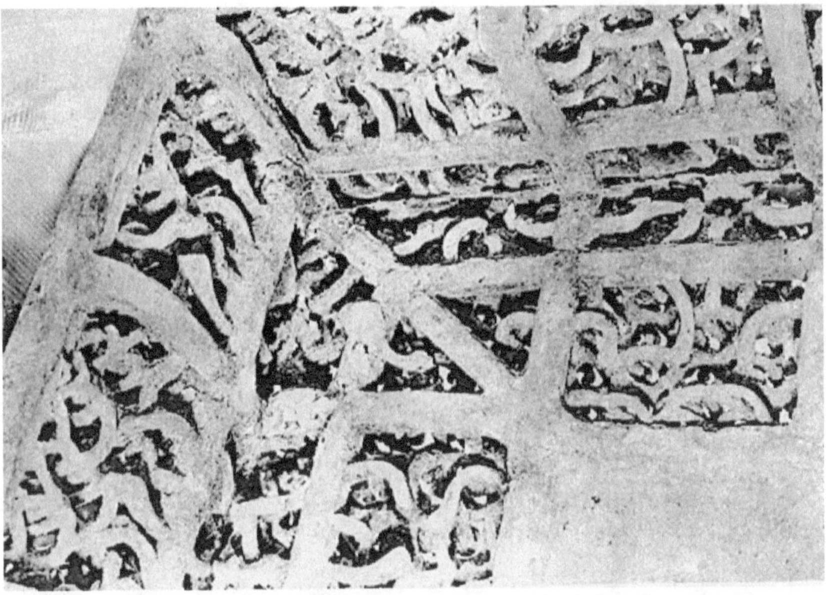

Source. After *Xichuan Xiasi Chunqiu Chumu* (Plate 114.3) by Henan Sheng Wenwu Yanjiusuo, Henan Sheng Danjiang Kuqu Kaogu Fajuedui, and Xichuan Xian Bowuguan, 1991, Beijing: Wenwu Chubanshe.

Figure 34b. The interior of the Xiasi *jin* body: the revealed inner structure before being restored.

Source. After "Henan Xichuan Xiasi Chunqiu Yunwen Tong Jin de Zhuzao yu Xiufu" (Plate 8.2) by Ren Changzhong and Wang Changqing, *Kaogu* 5 (1987): 474–478.

Figure 35a. Inscribed "Chen Zhang" *hu* from Jiangsu Xuyi. Mid-late 4th century BCE. H.24cm.

Source. After *Zhongguo qingtongqi quanji*, vol. 9 (Plate 122), edited by Zhongguo qingtongqi Quanji bianji weiyuanhui, 1997, Beijing: Wenwu Chubanshe.

Figure 35b. Openwork ornamentation of the "Chen Zhang" *hu* from Jiangsu Xuyi.

Source. After *Zhongguo qingtongqi quanji*, vol. 9 (Plate 123), edited by Zhongguo qingtongqi Quanji bianji weiyuanhui, 1997, Beijing: Wenwu Chubanshe.

Figure 35c. One of the mask motifs of the "Chen Zhang" *hu* from Jiangsu Xuyi.

Source. After "Nanjing Bowuyuan cang Zhanguo Toukong Panlongwen Qingtongqi" (p. 73, fig. 4.2) by Wan Li, *Dongnan Wenhua* 6 (2013): 72–76.

Figure 36a. The globular lidded tripod in The Metropolitan Museum of Art, New York. H. 21.9 cm. Warring States period.

Source. The Metropolitan Museum of Art, Rogers Fund, 1947 (acc. 47.27ab), reprinted in *Ancient Chinese Bronze Art: Casting the Precious Sacral Vessel* (p. 73, no. 35) by W. Thomas Chase, 1991. New York City: China House Gallery: China Institute in America.

Figure 36b. Openwork ornamentation of the tripod shown in Figure 36a, exhibiting "abstract undercuts"

Source. Author's photograph, courtesy of The Metropolitan Museum of Art, New York.

Figure 36c. The disputable "mold marks" (see arrow) of the tripod shown in Figure 36a.

Source. Author's photograph, courtesy of The Metropolitan Museum of Art, New York.

Figure 37a. Overhead view of the interior, the lower openwork ornamentation of the tripod shown in Figure 36a.

Source. Author's photograph, courtesy of The Metropolitan Museum of Art, New York.

Figure 37b. Details of the interior, the lower openwork ornamentation of the tripod shown in Figure 36a.

Source. Author's photograph, courtesy of The Metropolitan Museum of Art, New York.

Figure 38a. Incense burner found at Fengxiang, Shaanxi, H. 35.3 cm. Warring States period.

Source. After "Fengxiang Faxian Zhanguo Fengniao Xianhuan Tong Xunlu" (p. 57) by Jing Hongwei and Wang Zhouying, *Wenbo* 1 (1996): 57; reprinted in "Diversity in Eastern Zhou Bronze Casting: A Look at a Group of Openwork Vessels" (p. 112, fig. 6) by Helen Loveday, *Journal of East Asian Archaeology* 4, no. 1–4 (2002): 101–142.

Figure 38b. Sphere of the incense burner of Figure 38a.

Source. Photograph courtesy of Su Rongyu.

Figure 39a. Side view of the exterior, the square base of Fengxiang burner.

Source. Photograph courtesy of Su Rongyu.

Figure 39b. Top view of the exterior, the square base of Fengxiang burner. .

Source. Photograph courtesy of Su Rongyu.

Figure 40a. Bottom view of the interior, the square base of Fengxiang burner.

Source. Photograph courtesy of Su Rongyu.

Figure 40b. Side view of the interior, the square base of Fengxiang burner.

Source. Photograph courtesy of Su Rongyu.

Figure 41a. Bronze jar inlaid with turquoise. Winthrop Collection, Sackler Museum, Harvard Art Museums. H. 9.7 cm. Warring States period.

Source. Photograph courtesy of the President and Fellows of Harvard College. Harvard Art Museums/Arthur M. Sackler Museum, Bequest of Grenville L. Winthrop (acc.1943.52.70).

Figure 41b. Sphere of the bronze jar of Figure 41a shown in detail.

Source. Photograph courtesy of the President and Fellows of Harvard College. Harvard Art Museums/Arthur M. Sackler Museum, Bequest of Grenville L. Winthrop (acc.1943.52.70).

Figure 42a. Bronze *lei* excavated from Jingmen Zilinggang (M64.2), Hubei Province. H. 13.2 cm.

Source. After *Jingmen Zilinggang* (Color Plate 2.2) by Jingmen shi Bowuguan, 2008, Beijing: Wenwu Chubanshe.

Figure 42b. Structure of the bronze *lei* shown in Figure 42a.

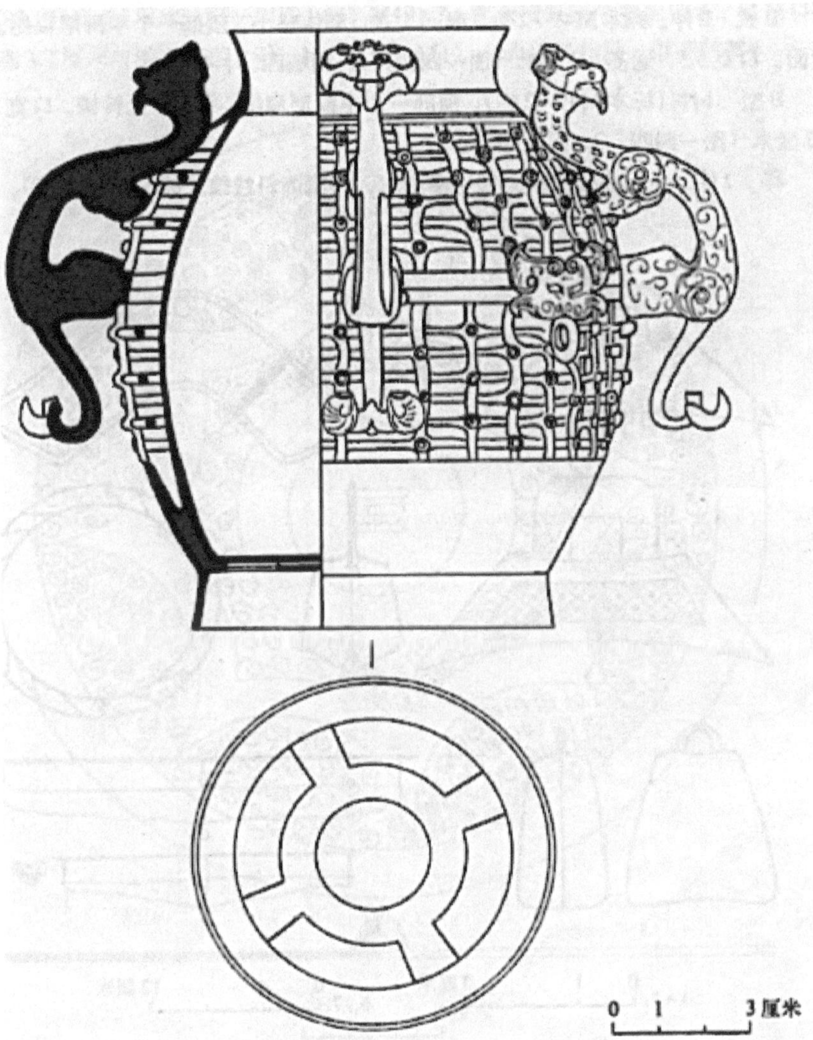

Source. After *Jingmen Zilinggang* (p. 155, fig. 143) by Jingmen shi Bowuguan, 2008, Beijing: Wenwu Chubanshe.

Figure 42c. Detail of openwork ornamentation, the bronze *lei* shown in Figure 42a.

Source. Taken from *Jingmen Zilinggang* (Color Plate 2.2) by Jingmen shi Bowuguan, 2008, Beijing: Wenwu Chubanshe.

Figure 43a. Bronze *bu* from Jiangsu Pixian Liulin. H. 7.1 cm. Warring States period.

Source. After "Nanjing Bowuyuan cang Zhanguo Toukong Panlongwen Qingtongqi" (p. 73, fig. 3) by Wan Li, *Dongnan Wenhua* 6 (2013): 72–76.

Figure 43b. Detail of openwork ornamentation, the bronze *bu* shown in Figure 43a.

Source. Photograph courtesy of Su Rongyu.

Figure 44a. Bronze *fang hu* cover from Jiangsu Pixian Liulin. Warring States period.

Source. After "Nanjing Bowuyuan cang Zhanguo Toukong Panlongwen Qingtongqi" (p. 72, fig. 1) by Wan Li, *Dongnan Wenhua* 6 (2013): 72–76.

Figure 44b. Interlace for one vertical wall of the bronze cover shown in Figure 44a.

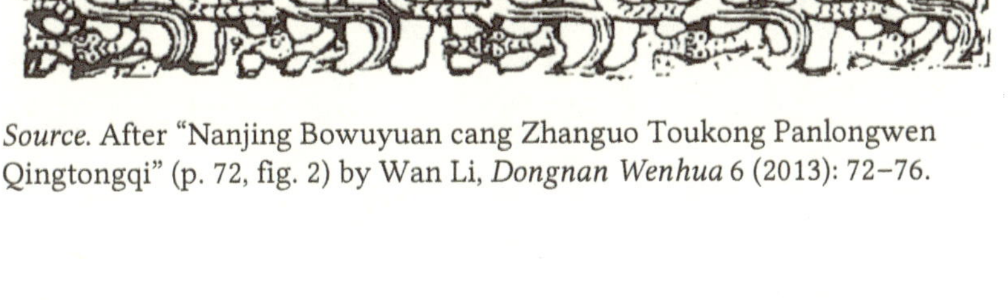

Source. After "Nanjing Bowuyuan cang Zhanguo Toukong Panlongwen Qingtongqi" (p. 72, fig. 2) by Wan Li, *Dongnan Wenhua* 6 (2013): 72–76.

Figure 44c. Traces of the interior of the bronze cover shown in Figure 44a.

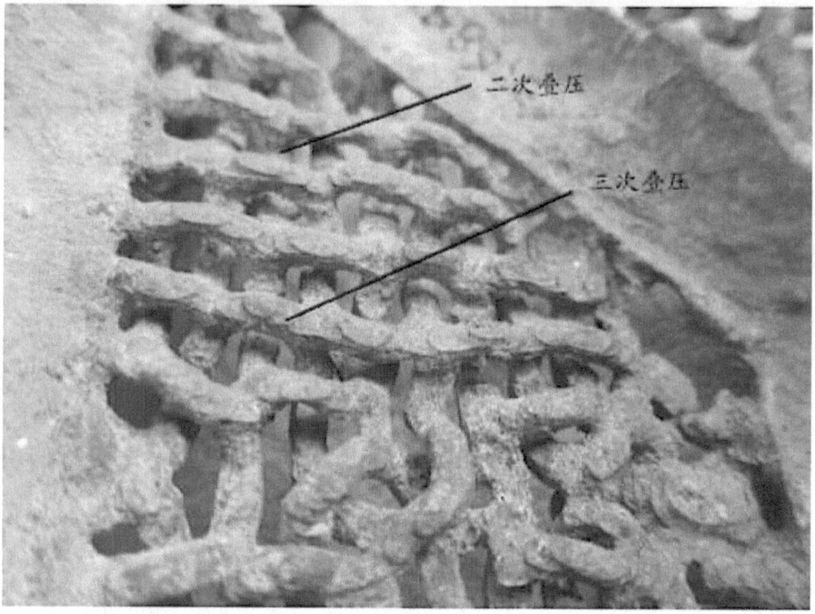

Source. After "Nanjing Bowuyuan cang Zhanguo Toukong Panlongwen Qingtongqi" (p. 75, fig. 6) by Wan Li, *Dongnan Wenhua* 6 (2013): 72–76.

Figure 45a. The exterior of trapezoidal slopes, the bronze *fang hu* cover of Figure 44a.

Source. Photograph courtesy of Su Rongyu.

Figure 45b. The Interior of trapezoidal slopes, the bronze *fang hu* cover of Figure 44a.

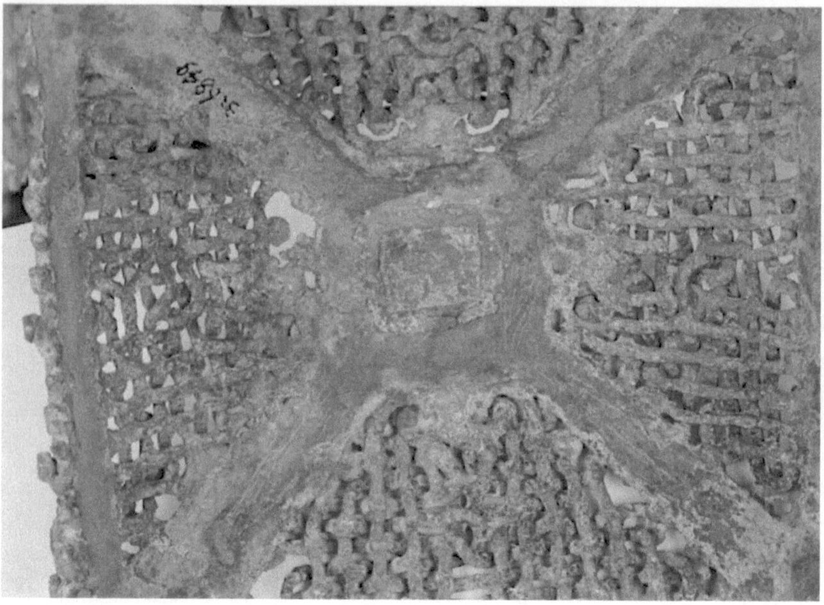

Source. Photograph courtesy of Su Rongyu.

Figure 46a. The double-tier bronze mirror with three-dimensional interlace in the Shanghai Museum, China. Warring States period. Dia. 10.2 cm. The exterior of back-tier of the mirror.

Source. After *Lianxing Shenye, Yingzhi Lianggong: Shanghai Bowuguan cang Tongjing Jingpin* (p. 95) by Shanghai Bowuguan, 2005, Shanghai: Shanghai Shuhua Chubanshe.

Figure 46b. The interior of back-tier of the mirror shown in Figure 46a.

Source. Author's photograph, courtesy of Shanghai Museum (acc. 75949).

Figure 47a. The so-called "mold marks" found on the exterior of back tier of the mirror shown in Figure 46a.

Source. After "Shanghai Bowuguan Wenwu Baohu yu Kaogu Kexue Shiyanshi Guance yu Jiance Baogao" (p. 50, fig. 1), 2005, appendix one to Li Chaoyuan. "Xinhuo Zhanguo Toukong Fuhejing Yanjiu." In *Lianxing Shenye, Yingzhi Lianggong: Shanghai Bowuguan cang Tongjing Jingpin*, edited by Shanghai Bowuguan, 50–52. Shanghai: Shanghai Shuhua Chubanshe.

Figure 47b. The so-called "mold marks" found on the interior of back tier of the mirror shown in Figure 46a.

Source. After "Shanghai Bowuguan Wenwu Baohu yu Kaogu Kexue Shiyanshi Guance yu Jiance Baogao" (p. 50, fig. 2), 2005, appendix one to Li Chaoyuan. "Xinhuo Zhanguo Toukong Fuhejing Yanjiu." In *Lianxing Shenye, Yingzhi Lianggong: Shanghai Bowuguan cang Tongjing Jingpin*, edited by Shanghai Bowuguan, 50–52. Shanghai: Shanghai Shuhua Chubanshe.

Figure 48. The small straight segments observed on the interior of back tier of the mirror shown in Figure 46a, respectively marked as A-A, B-B, and so on.

Source. After "Shanghai Bowuguan Wenwu Baohu yu Kaogu Kexue Shiyanshi Guance yu Jiance Baogao" (p. 50, fig. 3), 2005, appendix one to Li Chaoyuan. "Xinhuo Zhanguo Toukong Fuhejing Yanjiu." In *Lianxing Shenye, Yingzhi Lianggong: Shanghai Bowuguan cang Tongjing Jingpin*, edited by Shanghai Bowuguan, 50–52. Shanghai: Shanghai Shuhua Chubanshe.

Figure 49. The double-tier mirror with *loukong* openwork design in the Shanghai Museum. Dia. 7.1 cm. Warring States period.

Source. Author's photograph, courtesy of Shanghai Museum.

Figure 50a. The double-tier bronze mirror in the Japanese catalogue *Chugoku ōchō no sui*.

Source. After "Xinhuo Zhanguo Toukong Fuhejing Yanjiu" (p. 43, fig. 18) by Li Chaoyuan, 2005, in *Lianxing Shenye, Yingzhi Lianggong: Shanghai Bowuguan cang Tongjing Jingpin*, edited by Shanghai Bowuguan, 50–52. Shanghai: Shanghai Shuhua Chubanshe.

Figure 50b. The comparison of the *Chugoku* mirror, shown here, and the Shanghai Museum mirror (Figure 50c). Comparative analysis conducted by Hua Lin, who believes the two mirrors are identical in form and dimension.

Source. After "'Toukong Jiaolongwen Jing' yanjiu"(Plate 27) by Hua Lin, online research series 2013-2015 (Accessed March 6, 2018): https://tongjing.jimdo.com/%E5%9B%9B%E5%8F%B6%E7%BA%B9%E9%95%9C/.
Note. The auxiliary lines Hua Lin draws for the convenience of comparison, with corresponding areas marked by the same number.

Figure 50c. The comparison of the *Chugoku* mirror (Figure 50b) and the Shanghai Museum mirror, shown here. Comparative analysis conducted by Hua Lin, who believes the two mirrors are identical in form and dimension.

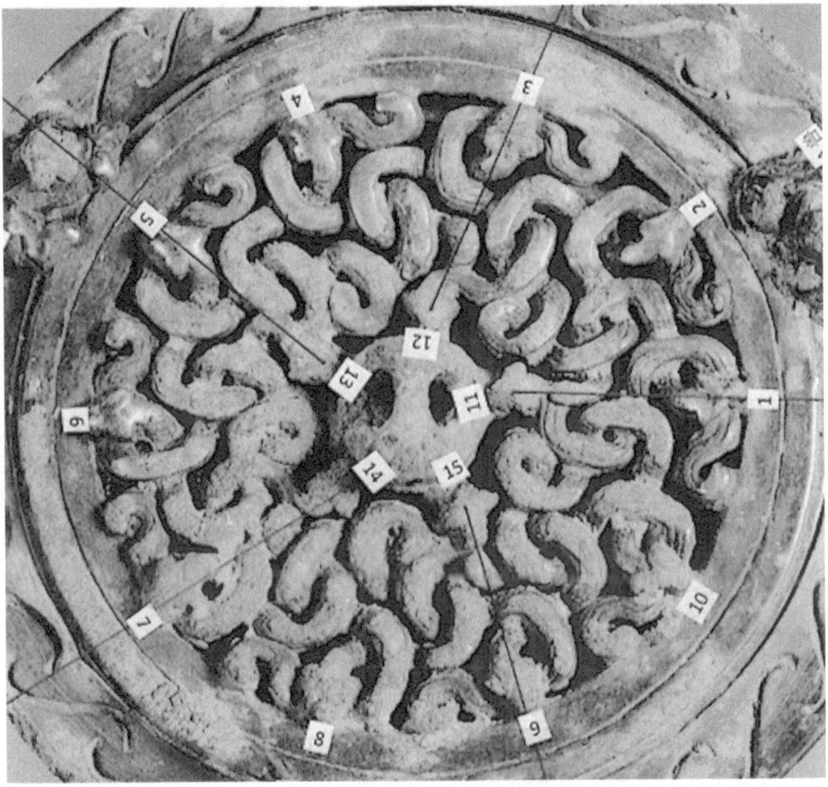

Source. After "'Toukong Jiaolongwen Jing' yanjiu"(Plate 27) by Hua Lin, online research series 2013-2015 (Accessed March 6, 2018): https://tongjing.jimdo.com/%E5%9B%9B%E5%8F%B6%E7%BA%B9%E9%95%9C/.
Note. The auxiliary lines Hua Lin draws for the convenience of comparison, with corresponding areas marked by the same number.

Figure 51a. The double-tier bronze mirror with three-dimensional interlace in the HUAM (Harvard University Art Museums). Dia. 13.9 cm. Warring States period. The exterior of back tier of the mirror.

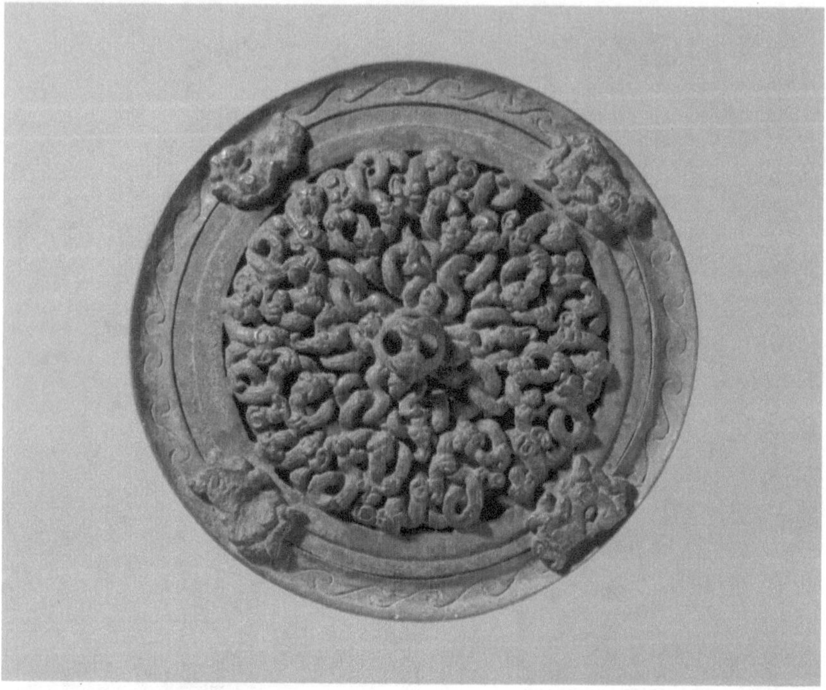

Source. Photograph courtesy of the President and Fellows of Harvard College. Harvard Art Museums/Arthur M. Sackler Museum, Bequest of Grenville L. Winthrop (acc. 1943.52.150).

Figure 51b. Three-dimensional interlace from the exterior of back tier of the mirror shown in Figure 51a.

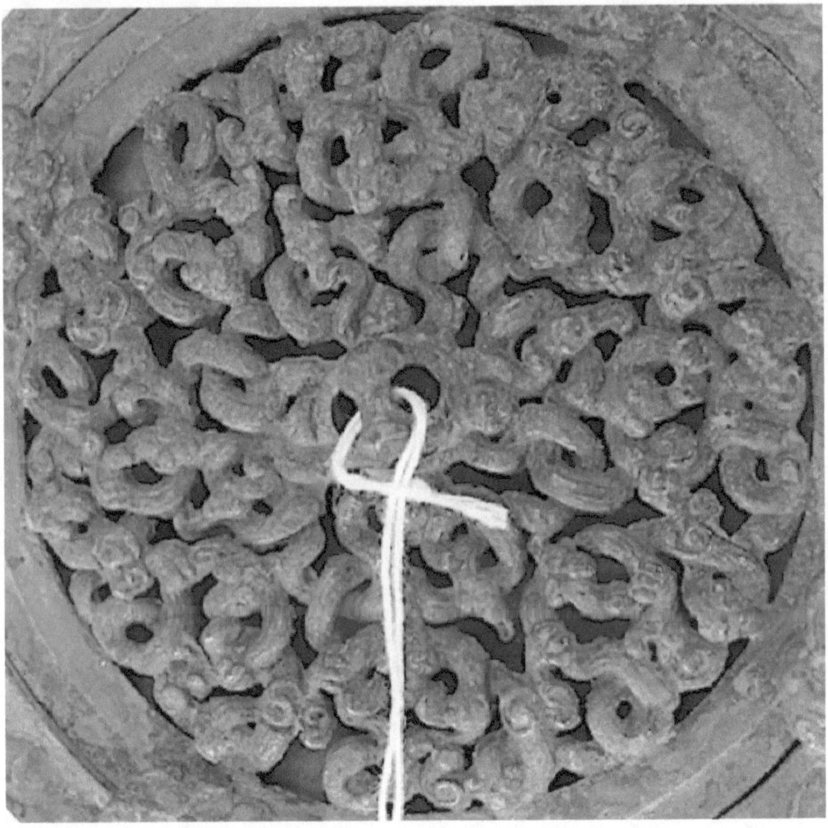

Source. Author's photograph, courtesy of Harvard Art Museums.

Figure 52a. The double-tier bronze mirror with three-dimensional interlace in the Shosoin collection, Nara, Japan. Warring States period.

Source. After "Zhongqiu Hua Yue Ye, Jiajing Gong Shang Shi" (p. 27, fig. 2) by Wang Wei, *Wenwu Jianding yu Jianshang* 9 (2011): 26–29.

Figure 52b. The double-tier bronze mirror with three-dimensional interlace in a private collection, China. Warring States period.

Source. After http://www.wenbo.cc/html/wb_jcdt/1196172353F72EB.asp (Accessed March 6, 2018).

Figure 53a. The unprovenanced "Chu Wang Yin Shen" *zhan* (officially labelled as "Grain Vessel [*dui*]") in The Metropolitan Museum of Art, New York, early 6th century BCE.

Source. Image courtesy of The Metropolitan Museum of Art, New York (acc. 1992.165.24a, b). Charlotte C. and John C. Weber Collection, Gift of Charlotte C. and John C. Weber, 1992.

Figure 53b. The disk-shaped crown on top of the vessel lid of the "Chu Wang Yin Shen" *zhan*.

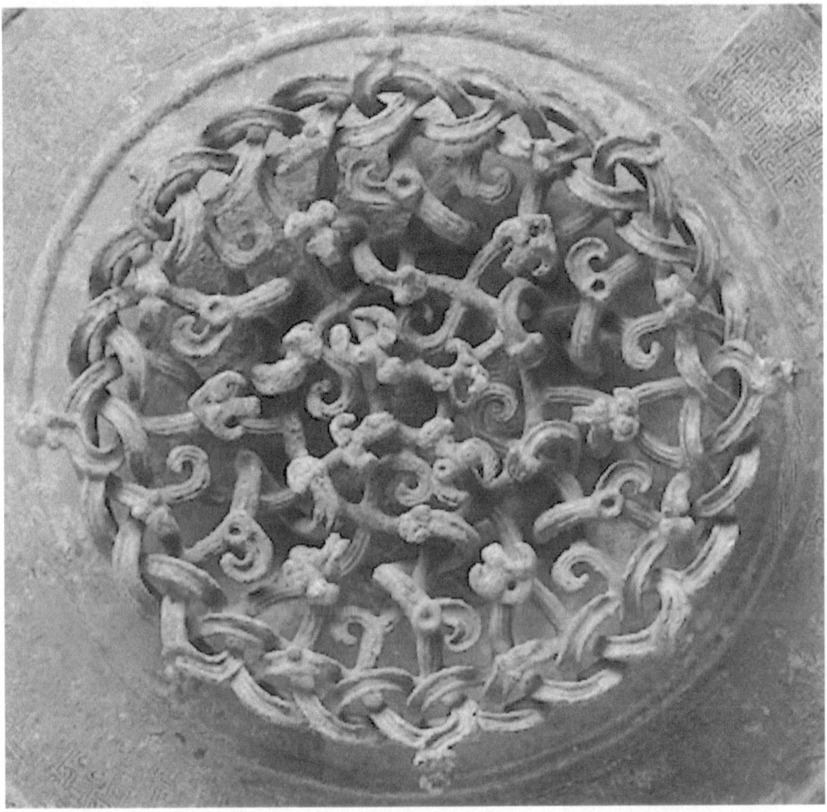

Source. Image courtesy of The Metropolitan Museum of Art, New York (acc. 1992.165.24a, b). Charlotte C. and John C. Weber Collection, Gift of Charlotte C. and John C. Weber, 1992.

Figure 53c. Zoomorphic heads and tails within the crown of the vessel lid of the "Chu Wang Yin Shen" *zhan*.

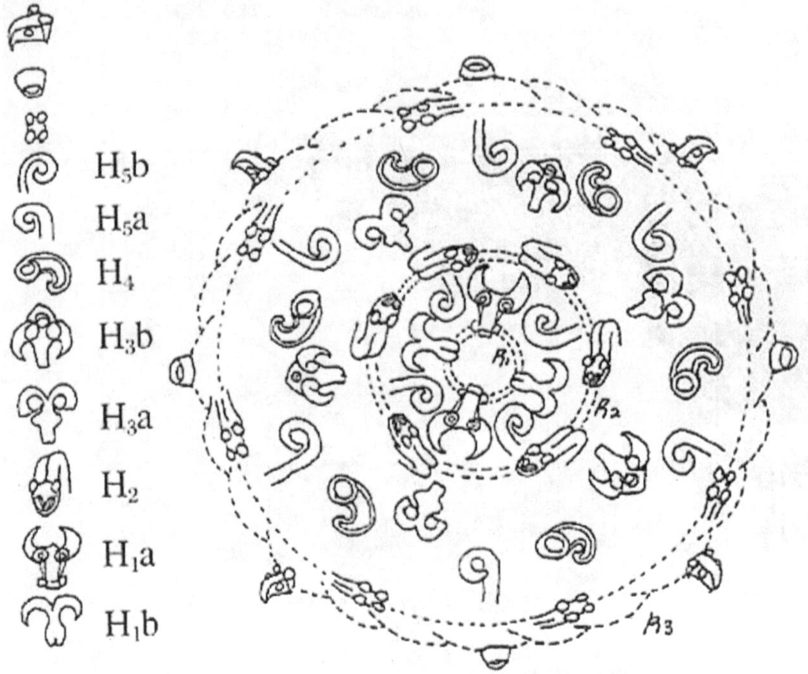

Source. After "Chu Wang Yu: Yizhi Zhongguo Zuizao de Shila Zhujian, Jianlun Shimofa Jiqi Gongyi Yanbian" (p. 83, fig. 5) by Hua Jueming and Xiao Huifang, 2005, in *Qingtong Wenhua Yanjiu.* Vol. 4. Tongling: Huangshan Shushe, 80–89.

Figure 54a. Structure of interlaced robs within the crown of the "Chu Wang Yin Shen" *zhan* lid, with corresponding photograph in Figure 54b for comparison.

Source. After "Chu Wang Yu: Yizhi Zhongguo Zuizao de Shila Zhujian, Jianlun Shimofa Jiqi Gongyi Yanbian" (p. 83, fig. 8) by Hua Jueming and Xiao Huifang, in *Qingtong Wenhua Yanjiu.* Vol. 4, Tongling: Huangshan Shushe, 2005, 80–89.

Figure 54b. Structure of interlaced robs within the crown of the "Chu Wang Yin Shen" *zhan* lid, with corresponding diagram in Figure 54a for comparison.

Source. Image courtesy of The Metropolitan Museum of Art, New York (acc. 1992.165.24a, b). Charlotte C. and John C. Weber Collection, Gift of Charlotte C. and John C. Weber, 1992.

Figure 55a. The innermost two rings, R1 and R2, from the crown of the "Chu Wang Yin Shen" *zhan* lid. Diagram with corresponding photograph of Figure 55c for comparison.

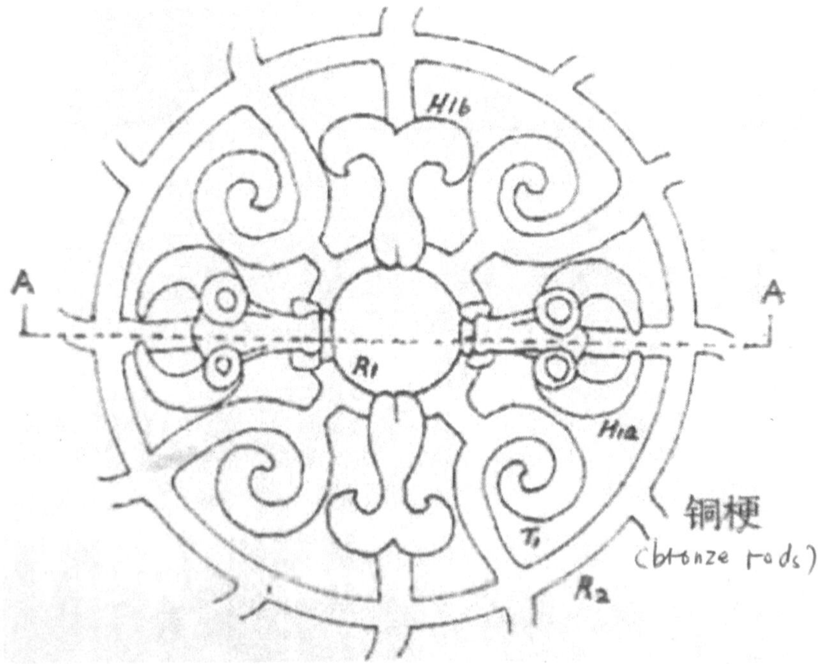

Source. After "Chu Wang Yu: Yizhi Zhongguo Zuizao de Shila Zhujian, Jianlun Shimofa Jiqi Gongyi Yanbian" (p. 83, fig. 4) by Hua Jueming and Xiao Huifang, in *Qingtong Wenhua Yanjiu*. Vol. 4, Tongling: Huangshan Shushe, 2005, 80–89.

Figure 55b. Detail of the innermost two rings shown in Figure 55a.

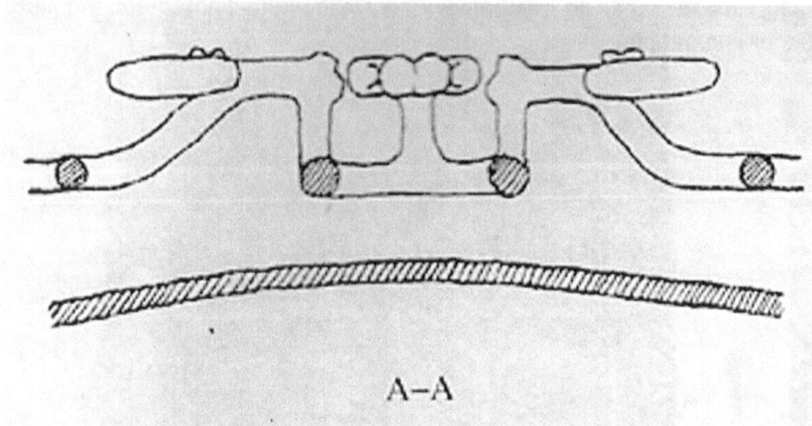

Source. After "Chu Wang Yu: Yizhi Zhongguo Zuizao de Shila Zhujian, Jianlun Shimofa Jiqi Gongyi Yanbian" (p. 83, fig. 4) by Hua Jueming and Xiao Huifang, in *Qingtong Wenhua Yanjiu.* Vol. 4, Tongling: Huangshan Shushe, 2005, 80–89.

Figure 55c. The innermost two rings marked as R1 and R2 in Figure 55a.

Source. Image courtesy of The Metropolitan Museum of Art, New York (acc. 1992.165.24a, b). Charlotte C. and John C. Weber Collection, Gift of Charlotte C. and John C. Weber, 1992.

Figure 56a. Side view of the crown of the "Chu Wang Yin Shen" *zhan* lid, with corresponding diagram of Figure 56b for comparison.

Source. Image courtesy of The Metropolitan Museum of Art, New York (acc. 1992.165.24a, b). Charlotte C. and John C. Weber Collection, Gift of Charlotte C. and John C. Weber, 1992.

Figure 56b. Side view of the crown of the "Chu Wang Yin Shen" *zhan* lid, with corresponding photograph of Figure 56a for comparison.

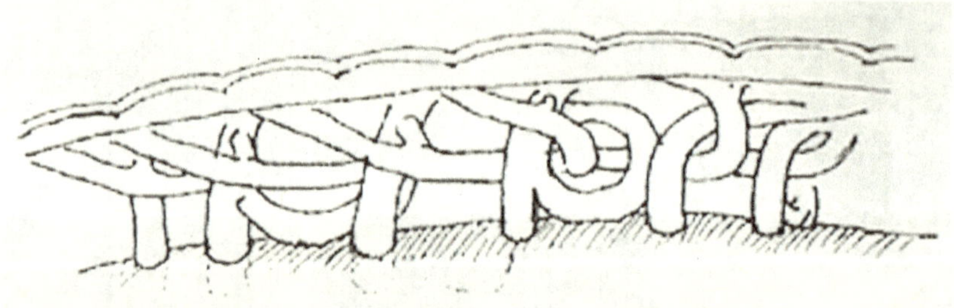

Source. After "Chu Wang Yu: Yizhi Zhongguo Zuizao de Shila Zhujian, Jianlun Shimofa Jiqi Gongyi Yanbian" (p. 83, fig. 6) by Hua Jueming and Xiao Huifang, in *Qingtong Wenhua Yanjiu.* Vol. 4, Tongling: Huangshan Shushe, 2005, 80–89.

Figure 57a. One of the interlaced appendages of the "Chu Wang Yin Shen" *zhan*.

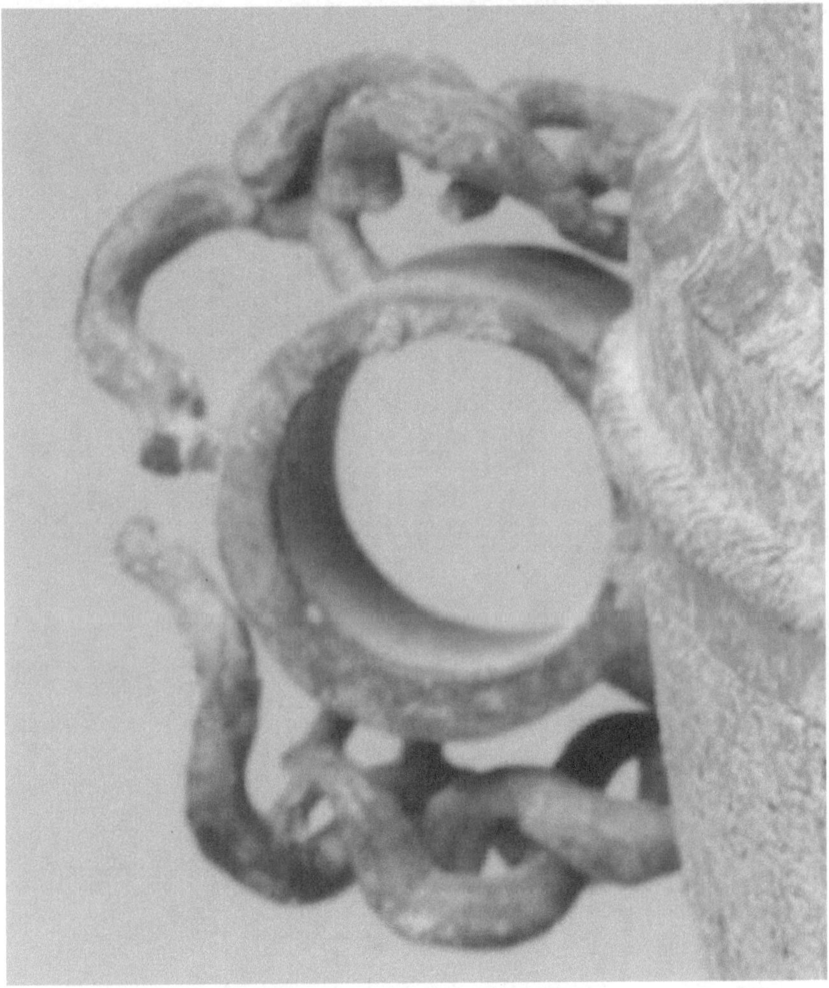

Source. Image courtesy of The Metropolitan Museum of Art, New York (acc. 1992.165.24a, b). Charlotte C. and John C. Weber Collection, Gift of Charlotte C. and John C. Weber, 1992.

Figure 57b. The cabriole leg of the "Chu Wang Yin Shen" *zhan*.

Source. Image courtesy of The Metropolitan Museum of Art, New York (acc. 1992.165.24a, b). Charlotte C. and John C. Weber Collection, Gift of Charlotte C. and John C. Weber, 1992.

Figure 58a. Bronze *zhan* from Henan Xichuan Xiasi M1. Second quarter of 6th century BCE.

Source. After *Eastern Zhou Ritual Bronzes from the Arthur M. Sackler Collections* (p. 121, fig. 10.4) by Jenny So, 1995, New York: Published by the Arthur M. Sackler Foundation in association with the Arthur M. Sackler Gallery, Smithsonian Institution: Distributed by Abrams.

Figure 58b. Illustration of the bronze *zhan* shown in Figure 58a.

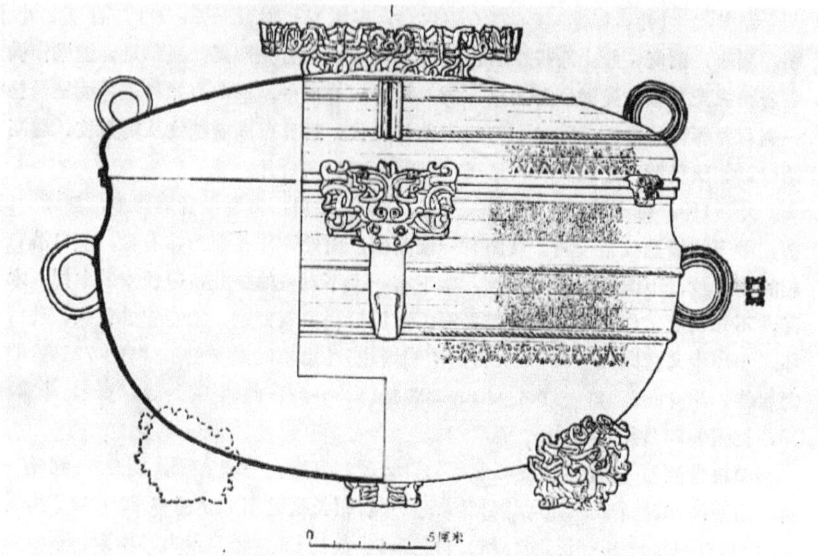

Source. After *Xichuan Xiasi Chunqiu Chumu* (p. 76, fig. 65), edited by Henan Sheng Wenwu Yanjiusuo et al., 1991, Beijing: Wenwu Chubanshe.

Figure 59a. The lid ornament of the bronze *zhan* shown in Figure 58a.

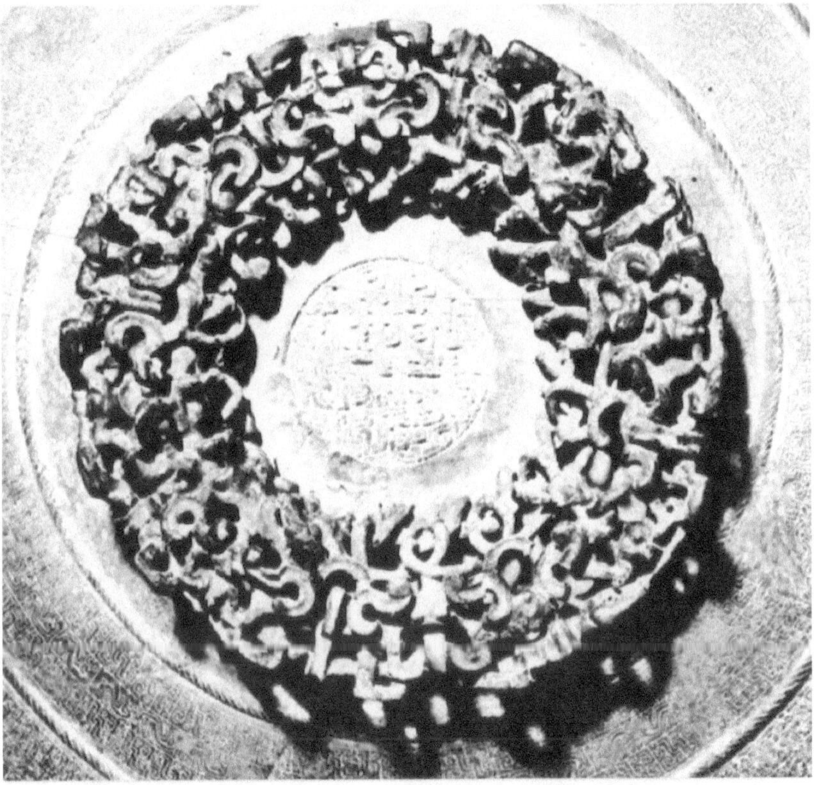

Source. After *Xichuan Xiasi Chunqiu Chumu* (Plate 32.2), edited by Henan Sheng Wenwu Yanjiusuo et al., 1991, Beijing: Wenwu Chubanshe.

Figure 59b. Illustration of the lid ornament shown in Figure 59a.

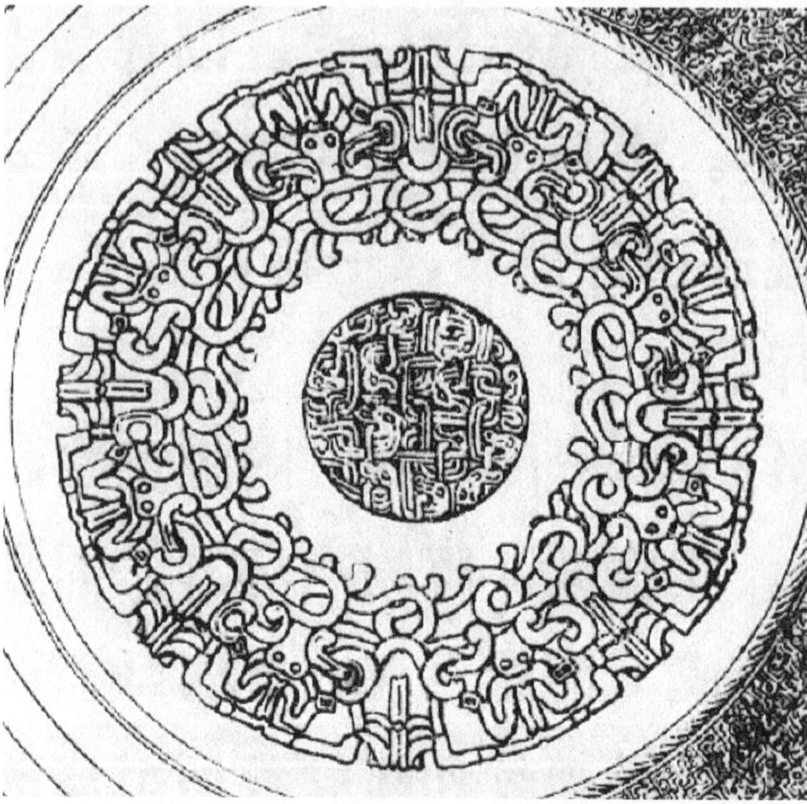

Source. After *Xichuan Xiasi Chunqiu Chumu* (p. 76, fig. 65), edited by Henan Sheng Wenwu Yanjiusuo et al., 1991, Beijing: Wenwu Chubanshe.

Figure 60a. Bronze *zhan* appendage (left side) from Henan Xichuan Xiasi M2. 2nd quarter of 6th century BCE.

Source. After "Chunqiu Shiqi Shilafa Zhuzao Gongyi Wenti Tantao" (p. 87, fig. 4), *Zhongyuan Wenwu* 6 (2006): 84–91.

Figure 60b. Bronze *zhan* appendage (right side) from Henan Xichuan Xiasi M2. 2nd quarter of 6th century BCE.

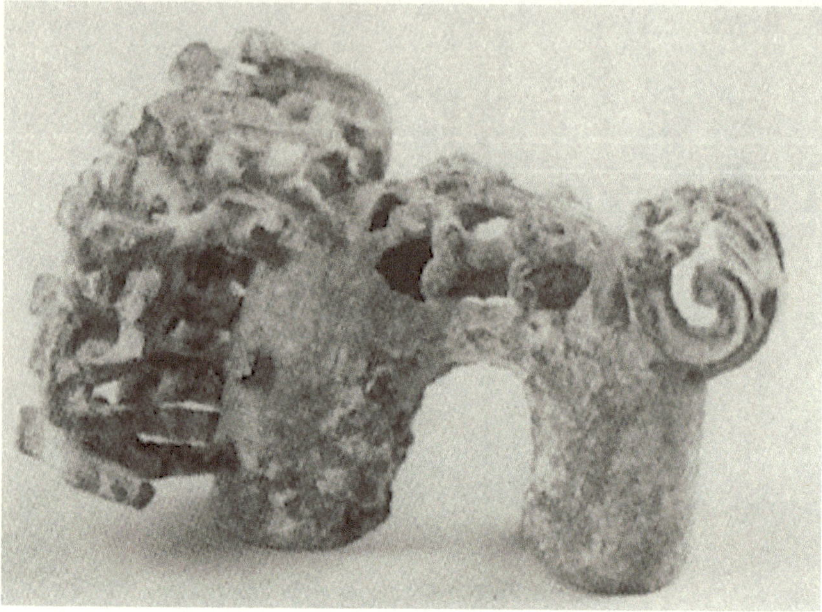

Source. After "Chunqiu Shiqi Shilafa Zhuzao Gongyi Wenti Tantao" (p. 87, fig. 5), *Zhongyuan Wenwu* 6 (2006): 84–91.

Figure 60c. Bronze *zhan* appendage (pattern surface) from Henan Xichuan Xiasi M2. 2nd quarter of 6th century BCE.

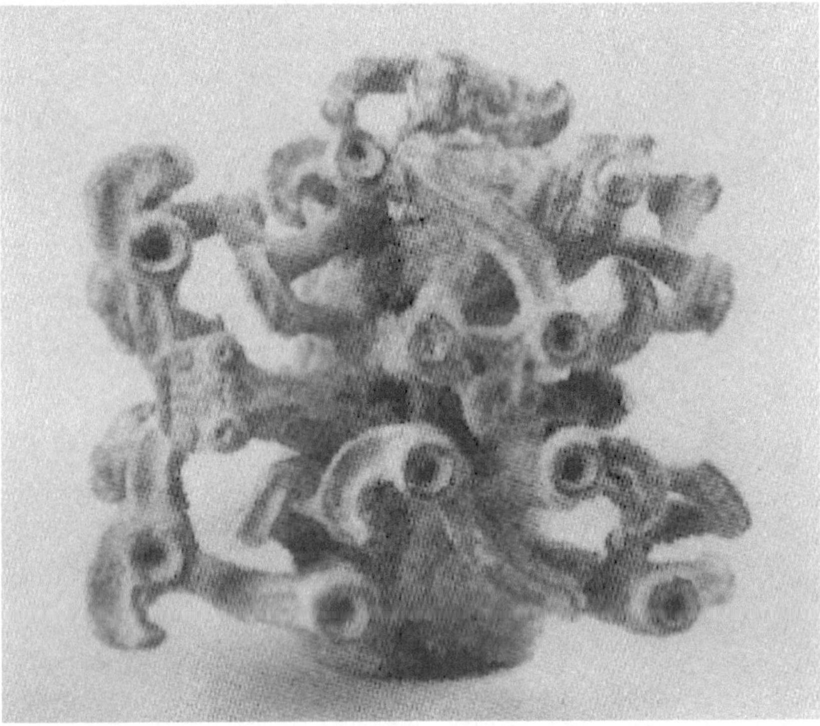

Source. After "Chunqiu Shiqi Shilafa Zhuzao Gongyi Wenti Tantao" (p. 87, fig. 6), *Zhongyuan Wenwu* 6 (2006): 84–91.

Figure 60d. Bronze *zhan* appendage (top) from Henan Xichuan Xiasi M2. 2nd quarter of 6th century BCE.

Source. After "Chunqiu Shiqi Shilafa Zhuzao Gongyi Wenti Tantao" (p. 87, fig. 7), *Zhongyuan Wenwu* 6 (2006): 84–91.

Figure 60e. Bronze *zhan* appendage (bottom) from Henan Xichuan Xiasi M2. 2nd quarter of 6th century BCE.

Source. After "Chunqiu Shiqi Shilafa Zhuzao Gongyi Wenti Tantao" (p. 87, fig. 8), *Zhongyuan Wenwu* 6 (2006): 84–91.

Figure 61a. Bronze *zhan* appendage from Henan Yexian, the tomb of Duke Ning of state Xu. Tomb dated to c. 547 BCE.

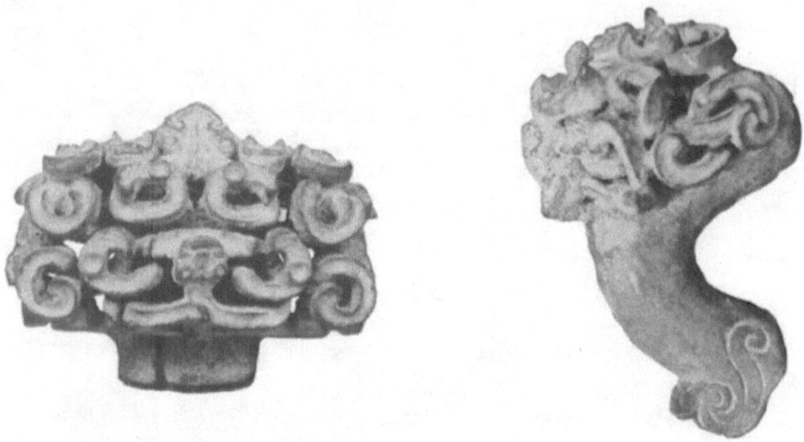

Source. After "Zhongguo Zaoqi Shila Zhujian—Xu Gong Ning Panhui Wen Toukong Shijian de Fuzhi Yanjiu" (p. 192, fig. 3) by Li Yuanzhi et al., *Tezhong Zhuzao ji Youse Hejin* 2 (2010): 192–194.

Figure 61b. Inner structure of the bronze *zhan* appendage shown in Figure 61a.

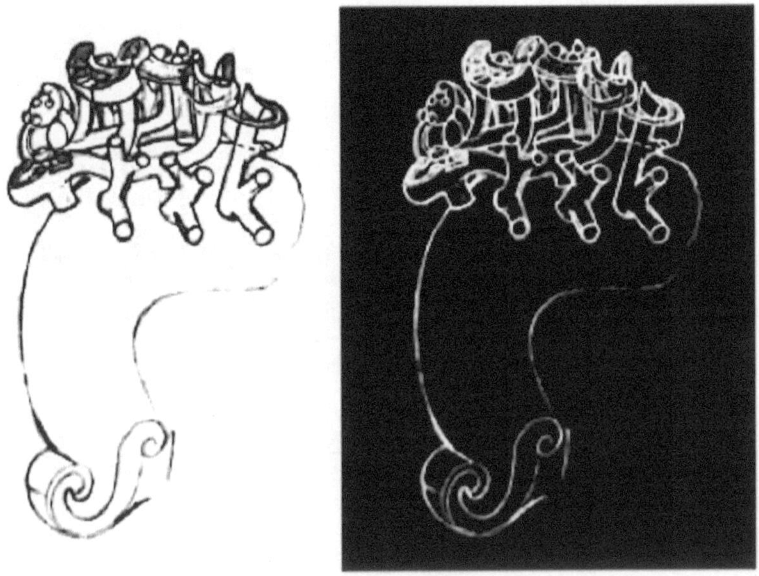

Source. After "Xu Gong Ning Panhuiwen Toukong Shijian Fuzhi Yanjiu" (p. 47, fig. 2) by Chen Hongliang et al., *Nanfang Wenwu* 4 (2009): 46–50.

Figure 62a. Ornamental elements of the bronze *zhan* appendage shown in Figure 61a.

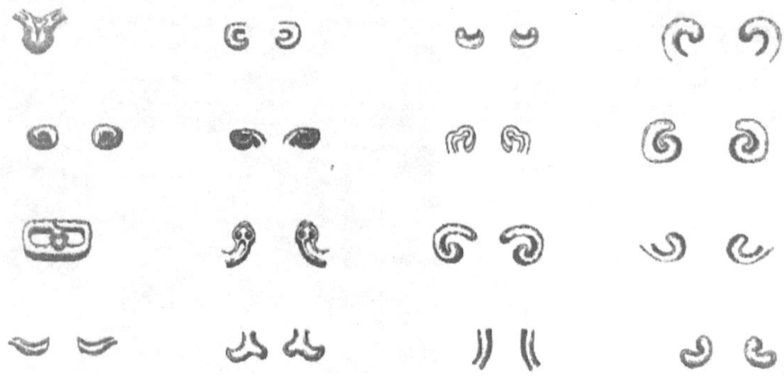

Source. After "Zhongguo Zaoqi Shila Zhujian–Xu Gong Ning Panhui Wen Toukong Shijian de Fuzhi Yanjiu" (p. 193, fig. 4) by Li Yuanzhi et al., *Tezhong Zhuzao ji Youse Hejin* 2 (2010): 192–194.

Figure 62b. The wooden template used by Chen Hongliang et al. for the reproduction of ornamental elements in wax.

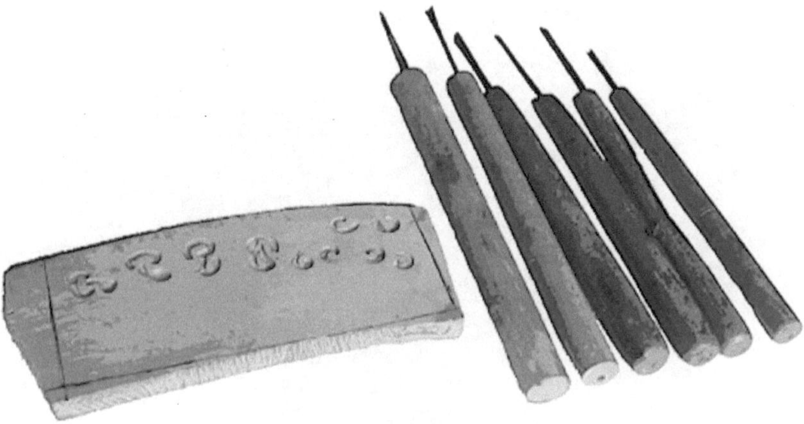

Source. After "Xu Gong Ning Panhuiwen Toukong Shijian Fuzhi Yanjiu" (p. 49, fig. 11) by Chen Hongliang et al., *Nanfang Wenwu* 4 (2009): 46–50.

Figure 63a. Reproduction of the bronze *zhan* appendage from Henan Yexian by the lost-wax process.

Source. After "Xu Gong Ning Panhuiwen Toukong Shijian Fuzhi Yanjiu" (p. 50, fig. 19) by Chen Hongliang et al., *Nanfang Wenwu* 4 (2009): 46–50.

Figure 63b. Reproduction of the bronze *zhan* appendage from Henan Yexian by Wang Jinchao's section-mold process.

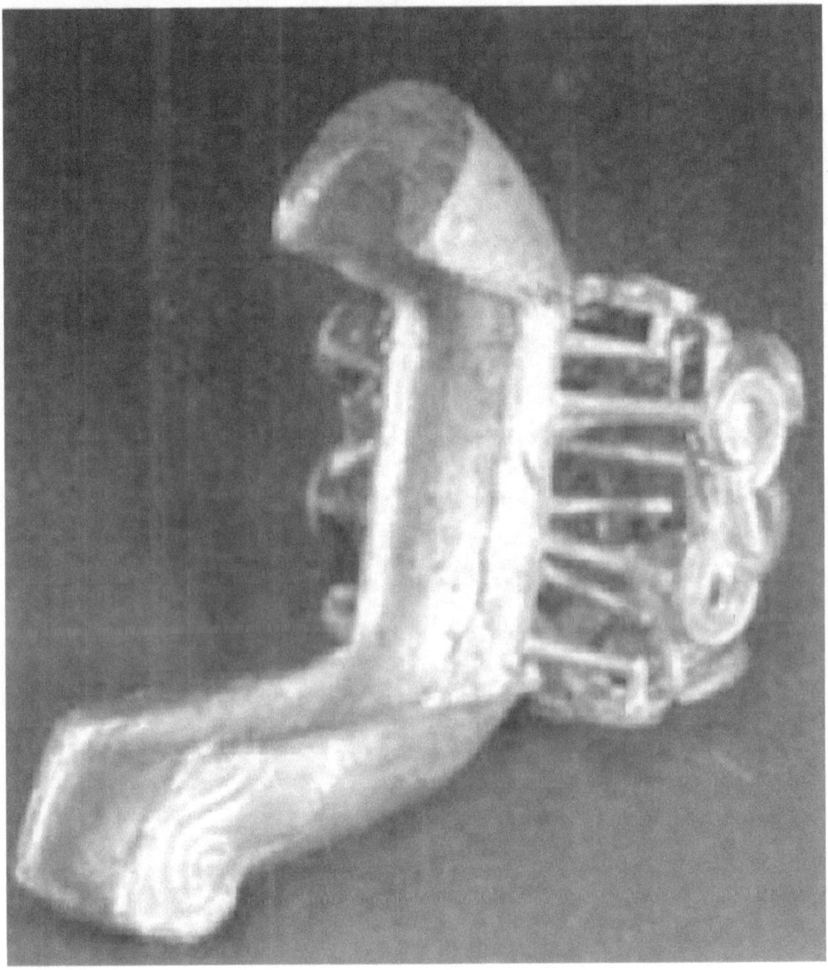

Source. After "Nifanfa Fuyuan Xugong Ning Loukong Panhuiwen Qingtong Shijian Yanjiu" (p. 36, fig. 4) by Wang Jinchao, *Nanfang Wenwu* 4 (2008): 34–36.

Figure 64. The inscribed "Yun Er" *zhan* from Hunan Yueyang Fengxingzuishan. Circa 6th century BCE. H. 17.4 cm.

Source. Photograph courtesy of Yueyang Municipal Museum.

Figure 65a. Crown of the vessel lid of the "Yun Er" *zhan*.

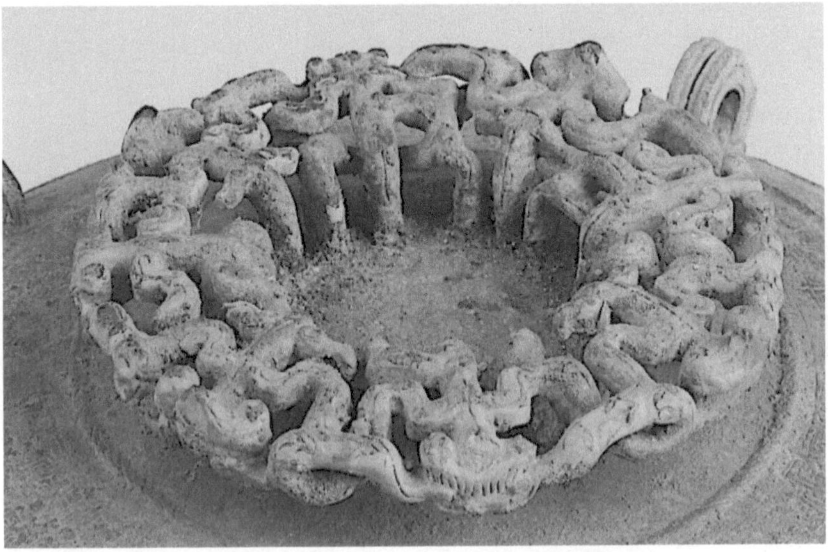

Source. Photograph courtesy of Yueyang Municipal Museum.

Figure 65b. Crown of the vessel lid of the "Yun Er" *zhan*: illustration.

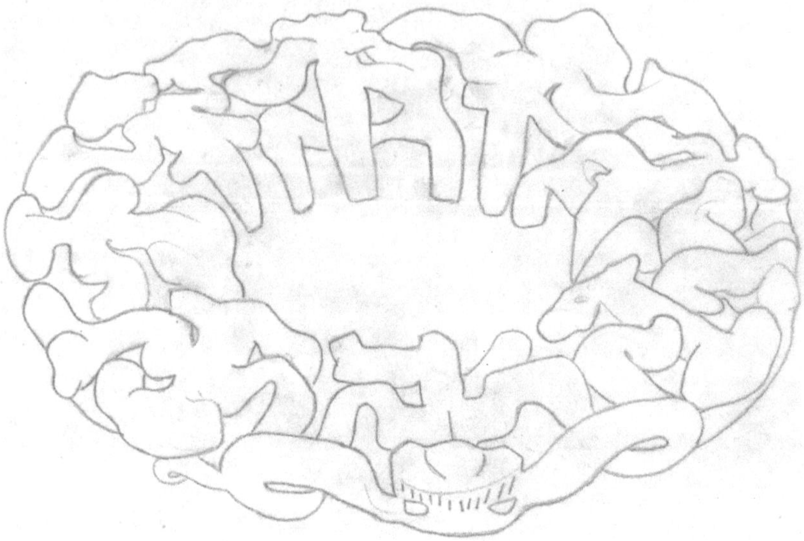

Source. Illustration by the author.

Figure 66a. Bronze *zhan* with openwork appendages from M38 of the Mingmenhuafu construction area, Nanyang, Henan Province.

Source. After "Henan Nanyang Chunqiu Chu Pengshe Mu Fajue Jianbao" (p. 21, fig. 51.1) by Nanyang shi Wenwu Kaogu Yanjiusuo, *Wenwu* 3 (2011): 4–31.

Figure 66b. Crown of the vessel lid of the bronze *zhan* shown in Figure 66a.

Source. Photograph courtesy of Ding Yao.

Figure 67a. The "Huang Zi" *zhan* in a private overseas collection.

Source. Photograph courtesy of a private overseas collection (owner and copyright holder of the artifact).

Figure 67b. The "Huang Zi" *zhan* in a private overseas collection: the openwork crown of the vessel lid.

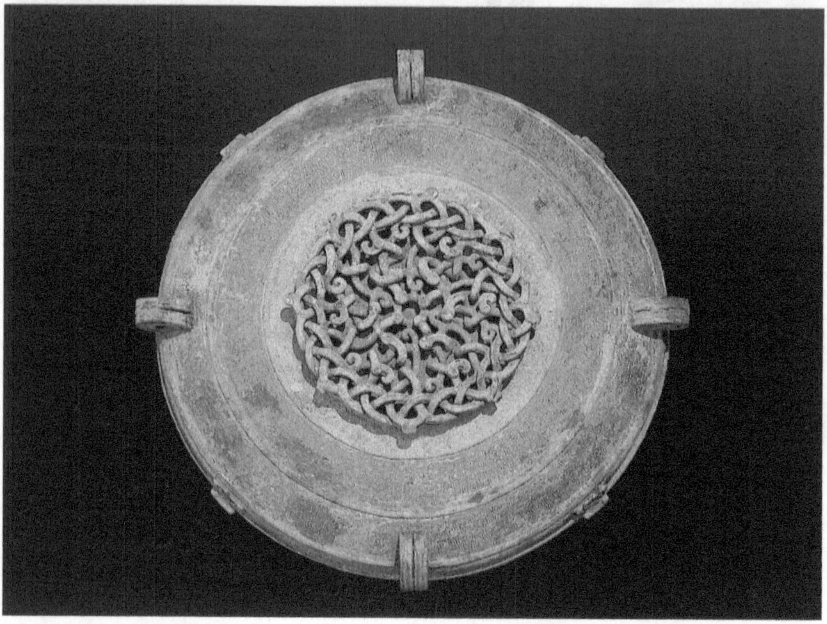

Source. Photograph courtesy of a private overseas collection (owner and copyright holder of the artifact).

Figure 68a. The "Huang Zi Lou" *zhan* in a private overseas collection.

Source. Photograph courtesy of a private overseas collection (owner and copyright holder of the artifact).

Figure 68b. The "Huang Zi Lou" *zhan* in a private overseas collection: the openwork crown of the vessel lid.

Source. Photograph courtesy of a private overseas collection (owner and copyright holder of the artifact).

Figure 69a. Bronze *pan* from Hubei Suizhou Wenfengta M33. Middle to Late Spring and Autumn period.

Source. Photograph courtesy of Huang Fengchun.

Figure 69b. One of the openwork appendages of the bronze *pan* shown in Figure 69a.

Source. Photograph courtesy of Huang Fengchun.

Figure 70a. Openwork appendages of the *pan* shown in Figure 69a: wrinkle marks on the surface of bronze rods.

Source. Photograph courtesy of Huang Fengchun.

Figure 70b. Openwork appendages of the *pan* shown in Figure 69a: small round, oval, or irregularly shaped slices laminated together on the surface of bronze rods.

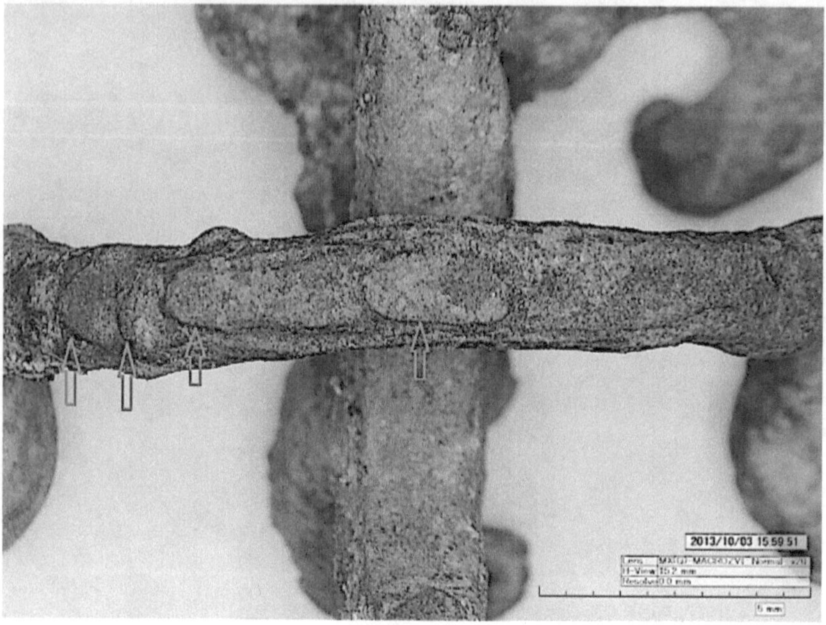

Source. Photograph courtesy of Huang Fengchun.

Figure 71a. The ripple-like marks on the bronze rods (within broken pieces) from a handle-shaped appendage of the *pan* of the Zeng *zun-pan* set.

Source. Photograph by Yung-ti Li and the author.

Figure 71b. Detail of the ripple-like marks shown in Figure 71a.

Source. Photograph by Yung-ti Li and the author.

Figure 72a. Tenuous raised mark along the middle of rods (within broken pieces) from a handle-shaped appendage of the *pan* of the Zeng *zun-pan* set.

Source. Photograph by Yung-ti Li and the author.

Figure 72b. Detail of the tenuous raised mark shown in Figure 72a.

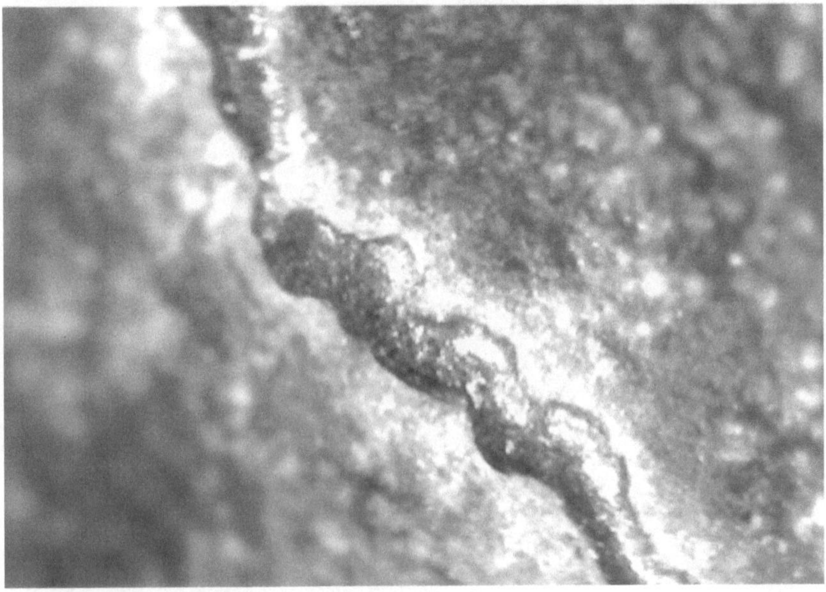

Source. Photograph by Yung-ti Li and the author.

Figure 72c. The "folded" bronze rod (within broken pieces) from a handle-shaped appendage of the *pan* of the Zeng *zun-pan* set.

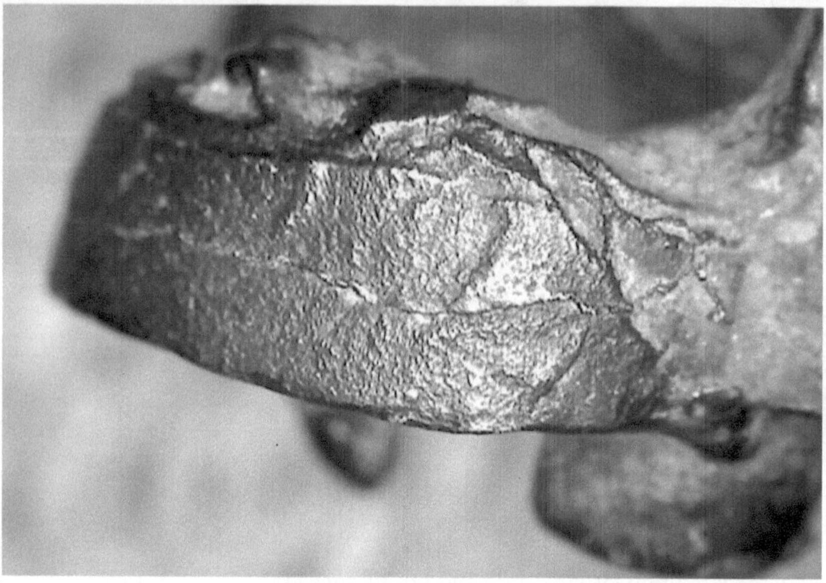

Source. Photograph by Yung-ti Li and the author.

Figure 72d. Fingerprints left on the bronze rod (within broken pieces) from a handle-shaped appendage of the *pan* of the Zeng *zun-pan* set.

Source. Photograph by Yung-ti Li and the author.

Figure 73a. Crown appendage to the lid of a bronze *hu* excavated from Henan Xichuan Xiasi M2: fragmentary piece from the bottom rim. Second quarter of 6th century BCE.

Source. After "Chunqiu Shiqi Shilafa Zhuzao Gongyi Wenti Tantao" (p. 88, fig. 11) by Zhao Shigang, *Zhongyuan Wenwu* 6 (2006): 84–91.

Figure 73b. Crown appendage to the lid of a bronze *hu* excavated from Henan Xichuan Xiasi M2: fragmentary piece from the bottom rim. Second quarter of 6th century BCE.

Source. After "Chunqiu Shiqi Shilafa Zhuzao Gongyi Wenti Tantao" (p. 88, fig. 12) by Zhao Shigang, *Zhongyuan Wenwu* 6 (2006): 84–91.

Figure 73c. Crown appendage to the lid of a bronze *hu* excavated from Henan Xichuan Xiasi M2: fragmentary piece from the wall. Second quarter of 6th century BCE.

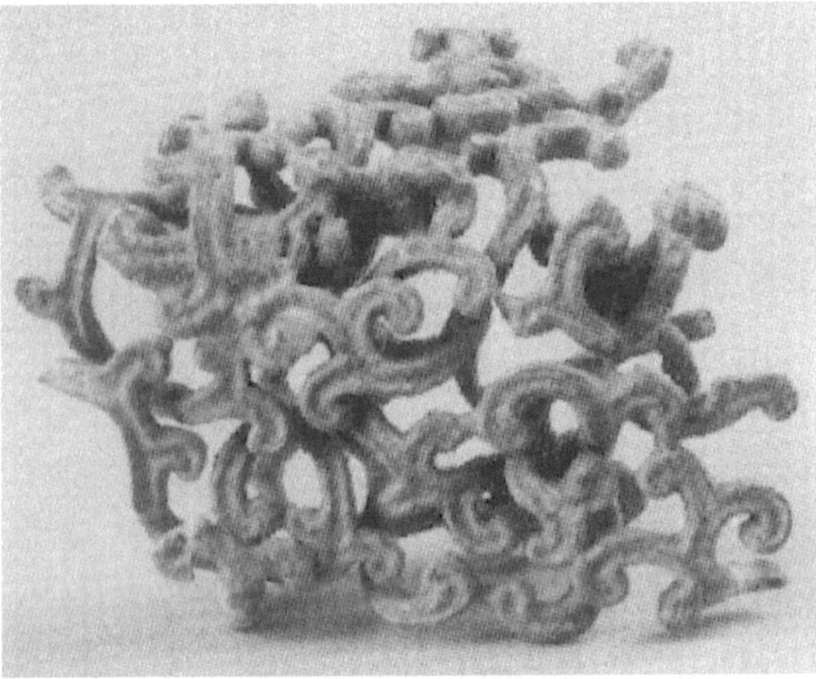

Source. After "Chunqiu Shiqi Shilafa Zhuzao Gongyi Wenti Tantao" (p. 88, fig. 13) by Zhao Shigang, *Zhongyuan Wenwu* 6 (2006): 84–91.

Figure 73d. Crown appendage to the lid of a bronze *hu* excavated from Henan Xichuan Xiasi M2: fragmentary piece from the wall. Second quarter of 6th century BCE.

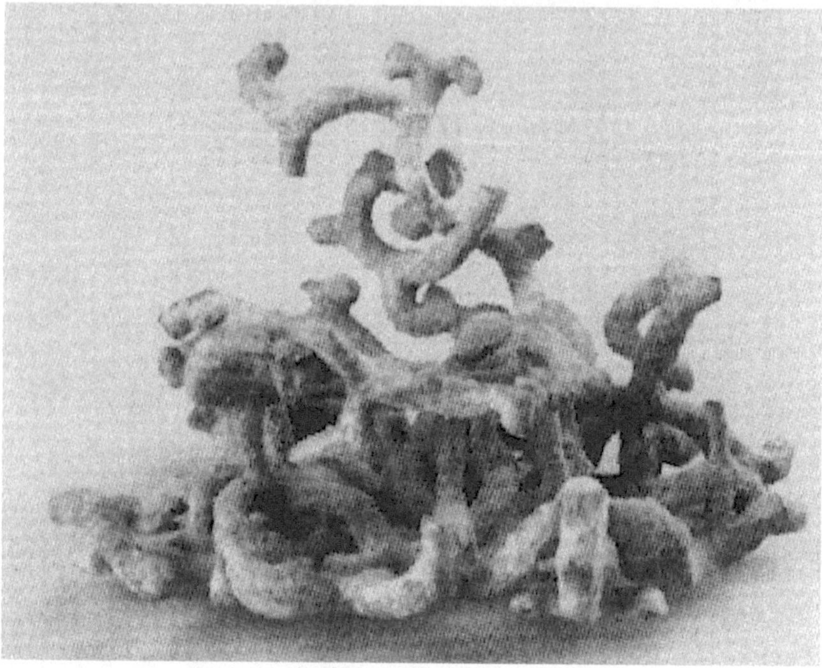

Source. After "Chunqiu Shiqi Shilafa Zhuzao Gongyi Wenti Tantao" (p. 89, fig. 14) by Zhao Shigang, *Zhongyuan Wenwu* 6 (2006): 84–91.

Figure 73e. Crown appendage to the lid of a bronze *hu* excavated from Henan Xichuan Xiasi M2: fragmentary piece from the wall. Second quarter of 6th century BCE.

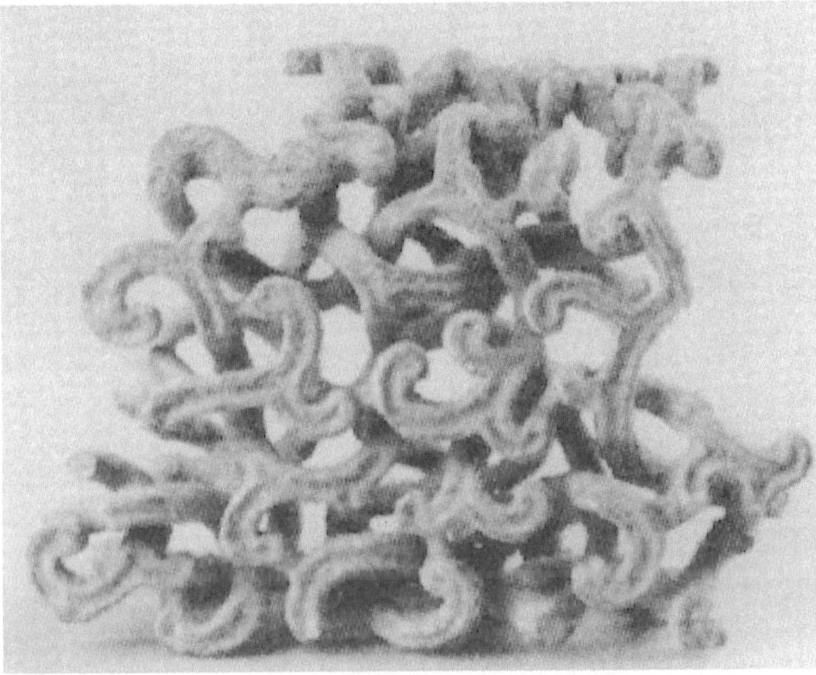

Source. After "Chunqiu Shiqi Shilafa Zhuzao Gongyi Wenti Tantao" (p. 89, fig.16) by Zhao Shigang, *Zhongyuan Wenwu* 6 (2006): 84–91.

Figure 73f. Crown appendage to the lid of a bronze *hu* excavated from Henan Xichuan Xiasi M2: fragmentary piece from the wall. Second quarter of 6th century BCE.

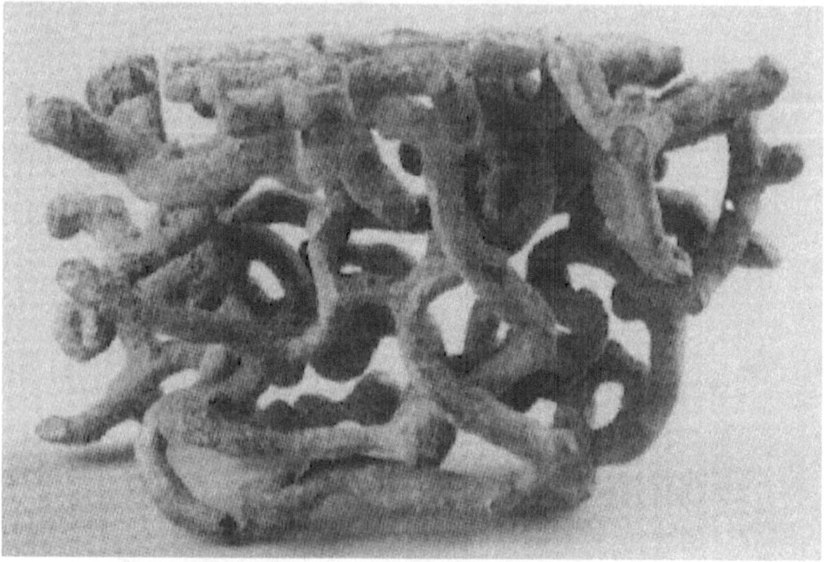

Source. After "Chunqiu Shiqi Shilafa Zhuzao Gongyi Wenti Tantao" (p. 89, fig.19) by Zhao Shigang, *Zhongyuan Wenwu* 6 (2006): 84–91.

Figure 74a. The *sheng ding* M2:38 excavated from Henan Xichuan Xiasi M2.

Source. After *Xichuan Xiasi Chunqiu Chumu* (Color Plate 6), edited by Henan Sheng Wenwu Yanjiusuo et al., 1991, Beijing: Wenwu Chubanshe.

Figure 74b. One of the beast-shaped figurines, the *sheng ding* M2:38 excavated from Henan Xichuan Xiasi M2.

Source. Taken from *Xichuan Xiasi Chunqiu Chumu* (Color Plate 6), edited by Henan Sheng Wenwu Yanjiusuo et al., 1991, Beijing: Wenwu Chubanshe.

Figure 75. The "Huang Zi Lou" *he* in a private overseas collection, associated with the Viscount Lou of Huang.

Source. Photograph courtesy of a private overseas collection (owner and copyright holder of the artifact).

Figure 76. Chariot fitting in the Nanjing Museum.

Source. After "Zhongqiu Hua Yue Ye, Jiajing Gong Shang Shi" (p.28, fig. 6) by Wang Wei, *Wenwu Jianding yu Jianshang* 9 (2011): 26–29.

Figure 77a. A pair of bronze *hu* connected to the same table. From the Suizhou Leigudun M1, tomb attributed to the Marquis Yi of Zeng and dated to circa 433 BCE.

Source. After *Zhongguo qingtongqi quanji*, vol. 10 (Plate 126), edited by Zhongguo qingtongqi quanji bianji weiyuanhui, 1996, Beijing: Wenwu Chubanshe.

Figure 77b. Detail of the beast-shaped handle of the paired *hu* shown in Figure 77a.

Source. After *Zhongguo qingtongqi quanji*, vol. 10 (Plate 126), edited by Zhongguo qingtongqi quanji bianji weiyuanhui, 1996, Beijing: Wenwu Chubanshe.

Figure 77c. Detail of the beast-shaped handle of the paired *hu* shown in Figure 77a.

Source. After *Zhongguo qingtongqi quanji*, vol. 10 (Plate 126), edited by Zhongguo qingtongqi quanji bianji weiyuanhui, 1996, Beijing: Wenwu Chubanshe.

Figure 78. Bronze harness ring. Iran, probably from Luristan. Early first millennium BCE. Dimensions: 9.4 x 8.51 cm.

Source. Image courtesy of The Metropolitan Museum of Art, New York (acc. 57.51.44). Cora Timken Burnett Collection of Persian Miniatures and Other Persian Art Objects, Bequest of Cora Timken Burnett, 1956.

Figure 79. Bronze chariot fitting from Shaanxi Chang'an Zhangjiapo M2. H.9cm.

Source. After *Zhangjiapo Xizhou Mudi* (p.212, fig.160: 6) by Zhongguo Shehui Kexueyuan Kaogu Yanjiusuo, 1999, Beijing: Zhongguo Dabaike Quanshu Chubanshe.

Figure 80. Bronze *hu* from a cache of Shaanxi Meixian Yangjiacun. Late Western Zhou. H.59.6cm.

Source. After "Shaanxi Meixian Yangjiacun Xi Zhou Qingtongqi Jiaocang Fajue Jianbao" (fig. 28) by Shaanxi Sheng Kaogu Yanjiusuo et al., *Wenwu* 6(2003): 4–42.

Figure 81. Bronze knife with openwork handle representing intertwined rope from Ningcheng Nanshangen, Inner Mongolia. L. 23.8 cm.

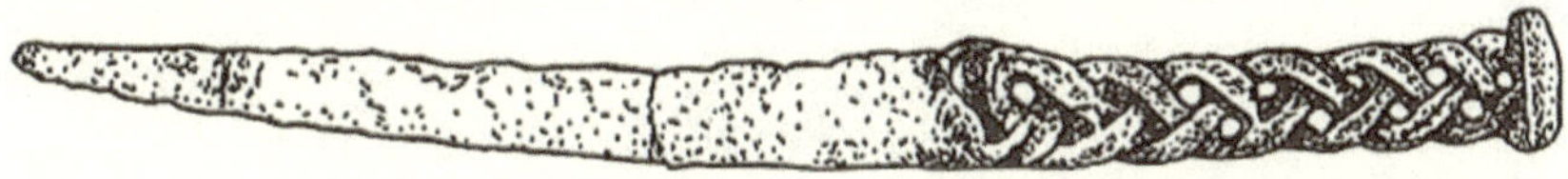

Source. After *Beifang Caoyuan Kaoguxue Wenhua Yanjiu: Qingtong Shidai zhi Zaoqi Tieqi Shidai* (p.189, fig.87: 1) by Wu'en Yuesitu, 2007, Beijing: Kexue Chubanshe.

Figure 82a. Bronze scabbard fitting decorated with snake interlaces and kneeling human figures from Beijing Fangshan Liulihe. Circa 10th century BCE.

Source. After *Beijing Wenwu Jingcui Daxi* (*Qingtongqi juan*, p.156, no. 139) edited by Mei Ninghua and Tao Xincheng, 2006, Beijing: Beijing Chubanshe.

Figure 82b. Bronze scabbard fitting decorated with snake interlaces and kneeling human figures in the Freer Gallery of Art, Washington, D.C. Circa 10th century BCE.

Source. Image courtesy of the Freer Gallery of Art, Smithsonian Institution, Washington, D.C.: Gift of Therese and Erwin Harris in honor of the 75th Anniversary of the Freer Gallery of Art, F1998.6.

Figure 83. Ke *bo* bronze bell from Shaanxi Fufeng Renjiacun. Late Western Zhou, 9th–8th century BCE. H. 62.5 cm.

Source. After *Zhongguo qingtongqi quanji*, vol. 5 (Plate 189), edited by Zhongguo qingtongqi quanji bianji weiyuanhui, 1997, Beijing: Wenwu Chubanshe.

Figure 84a. Bronze *you* with decoration of plume-interlaced birds from Anhui Tunxi. Middle Western Zhou period,10th century BCE. H.23.5 cm.

Source. After *Zhongguo qingtongqi quanji*, vol. 6 (Plate 119), edited by Zhongguo qingtongqi quanji bianji weiyuanhui, 1996, Beijing: Wenwu Chubanshe.

Figure 84b. Decoration of plume-interlaced birds of the bronze *you* shown in Figure 84a.

Source. After *Zhongguo qingtongqi quanji*, vol. 6 (Plate 120), edited by Zhongguo qingtongqi quanji bianji weiyuanhui, 1996, Beijing: Wenwu Chubanshe.

Figure 85. Bronze *gui*, likely from southeast China. Circa 8th–6th century BCE. H. 19.9 cm.

Source. Photograph courtesy of the President and Fellows of Harvard College. Harvard Art Museums/Arthur M. Sackler Museum, Anonymous gift (acc. 1944.57.5).

Figure 86. Bronze *hu* from Shanxi Houma Shangmacun. Early Eastern Zhou, 6th century BCE. H. 84.5 cm.

Source. After *Zhongguo meishu quanji: Gongyi meishu bian* 4: *Qingtongqi 2* (no. 11), edited by Li Xueqin, 1986, Beijing: Wenwu Chubanshe.

Figure 87a. Bronze *Pan long fang hu* from Henan Xinzheng Lijialou. One of a pair, currently in the National Museum of History (Taipei, China). H.90 cm. Early 6th century BCE.

Source. After *Zhongguo qingtongqi quanji*, vol. 7 (Plate 23), edited by Zhongguo qingtongqi quanji bianji weiyuanhui, 1998, Beijing: Wenwu Chubanshe.

Figure 87b. Detail of the ornamental crown, *Pan long fang hu* from Henan Xinzheng Lijialou.

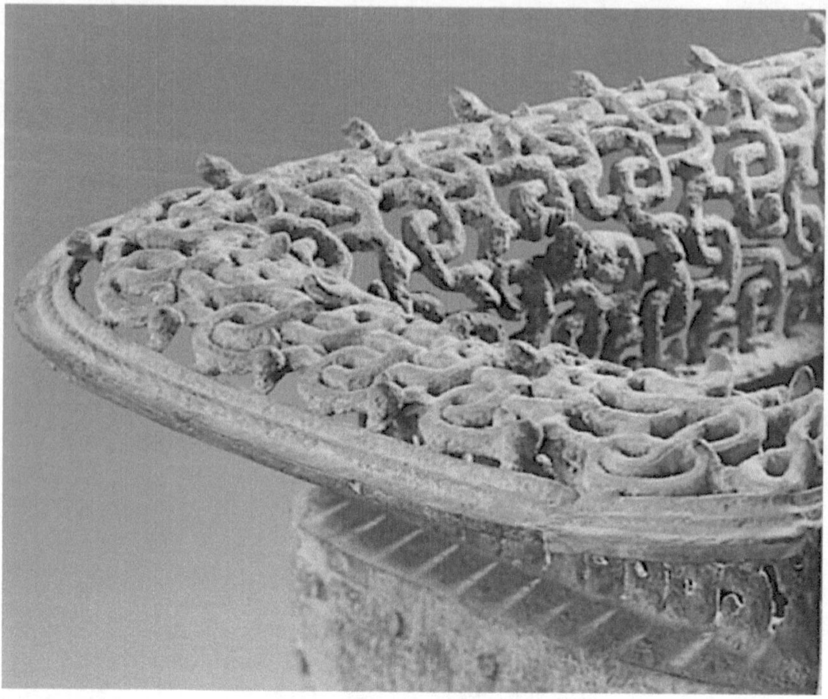

Source. After "Ancient Chinese Bronze Casting Methods: The Dilemma of Choice" (p. 1744, fig. 2) by Michael Notis and Dongning Wang, *MRS (Materials Research Society) Advances* 2, no. 33–34 (2017): 1743–1768.

Figure 88a. Bronze *ding* from Shanxi Xinjiang Liuquan. Late 6th or early 5th century BCE. H.24 cm.

Source. After *Zhongguo qingtongqi quanji*, vol. 8 (Plate 6), edited by Zhongguo qingtongqi quanji bianji weiyuanhui, 1995, Beijing: Wenwu Chubanshe.

Figure 88b. Interlace of the bronze *ding* shown in Figure 88a.

Source. After *Zhongguo qingtongqi quanji*, vol. 8 (Plate 7), edited by Zhongguo qingtongqi quanji bianji weiyuanhui, 1995, Beijing: Wenwu Chubanshe.

Figure 89. Carved wooden box from Hubei Jiangling Yutaishan M471. H.20.9 cm. Warring States period. H. 20.9 cm.

Source. After *Jiangling Yutaishan Chu mu* (Plate LIX) by Hubei Jingzhou Diqu Bowuguan, 1984, Beijing: Wenwu Chubanshe.

Figure 90a. Bronze cup from Hubei Jiangling Yutaishan. Middle Warring States period. H. 12.7cm.

Source. After *Zhongguo qingtongqi quanji*, vol. 10 (Plate 80), edited by Zhongguo Qingtongqi quanji bianji weiyuanhui, 1996, Beijing: Wenwu Chubanshe.

Figure 90b. Illustration of the bronze cup of Figure 90a.

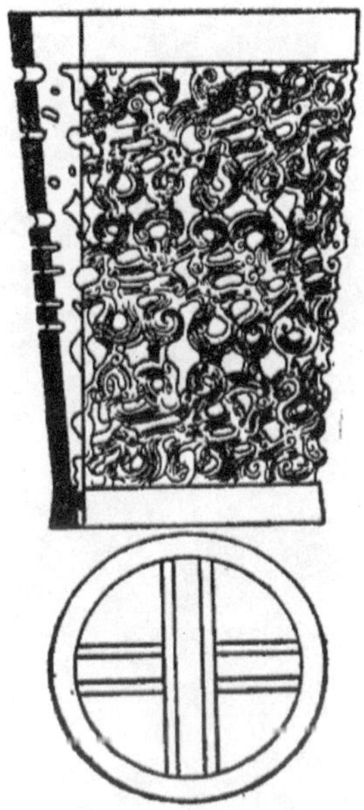

Source. After *Canlan de Zhongguo Gudai Shila Zhuzao* (p. 64) by Tan Derui, 1989, Shanghai: Shanghai Kexue Jishu Wenxian Chubanshe.

Figure 91a. Unprovenanced bronze *zhan* formerly in the Sackler Gallery of Art, Washington, D.C. Height to rim: 11.0 cm; height to top of lid: 16.3cm. Late 7th or early 6th century BCE. Side view.

Source. Artifact used to be in the Sackler Gallery of Art, Smithsonian Institution, Washington, D.C. (acc. V-292; perhaps sold at Christie's in 2009, information courtesy of Jennifer E. Berry). After *Eastern Zhou Ritual Bronzes from the Arthur M. Sackler Collections* (p. 118), 1995, New York: Published by the Arthur M. Sackler Foundation in association with the Arthur M. Sackler Gallery, Smithsonian Institution: Distributed by Abrams.

Figure 91b. Top view of the bronze *zhan* shown in Figure 91a .

Source. Artifact used to be in the Sackler Gallery of Art, Smithsonian Institution, Washington, D.C. (acc. V-292; perhaps sold at Christie's in 2009, information courtesy of Jennifer E. Berry). After *Eastern Zhou Ritual Bronzes from the Arthur M. Sackler Collections* (p. 119), 1995, New York: Published by the Arthur M. Sackler Foundation in association with the Arthur M. Sackler Gallery, Smithsonian Institution: Distributed by Abrams.

Figure 92a. Bronze *ding* with openwork lid crown from the cemetery of Dangyang Zhaojiahu, Hubei Province. H. 35.7 centimeters. Middle Spring and Autumn period (circa 670–570 BCE).

Source. After the official website of Yichang Municipal Museum, Hubei (Accessed March 8, 2018): http://www.ycbwg.com/view.asp?id=788.

Figure 92b. Openwork lid crown of the bronze *ding* shown in Figure 92a.

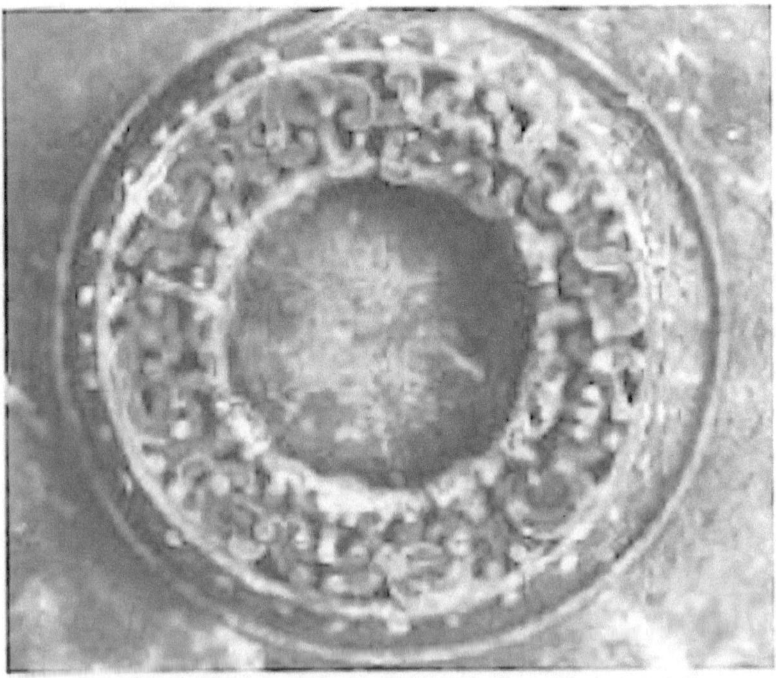

Source. After the official website of Yichang Municipal Museum, Hubei (Accessed March 8, 2018): http://www.ycbwg.com/view.asp?id=788.

Figure 93a. The double-owl bronze *you* in the Shanghai Museum. Anyang period. H.20.5 cm.

Source. After *Xia Shang Zhou Qingtongqi Yanjiu: Shanghai Bowuguan Cangpin* (p. 312, Plate 153) by Chen Peifen, 2004, Shanghai: Shanghai Guji Chubanshe.

Figure 93b. Swing handle of the double-owl bronze *you* in the Shanghai Museum.

Source. After *Xia Shang Zhou Qingtongqi Yanjiu: Shanghai Bowuguan Cangpin* (p. 312, Plate 153) by Chen Peifen, 2004, Shanghai: Shanghai Guji Chubanshe.

Figure 93c. Diagram of swing handle, the double-owl bronze *you* in the Shanghai Museum.

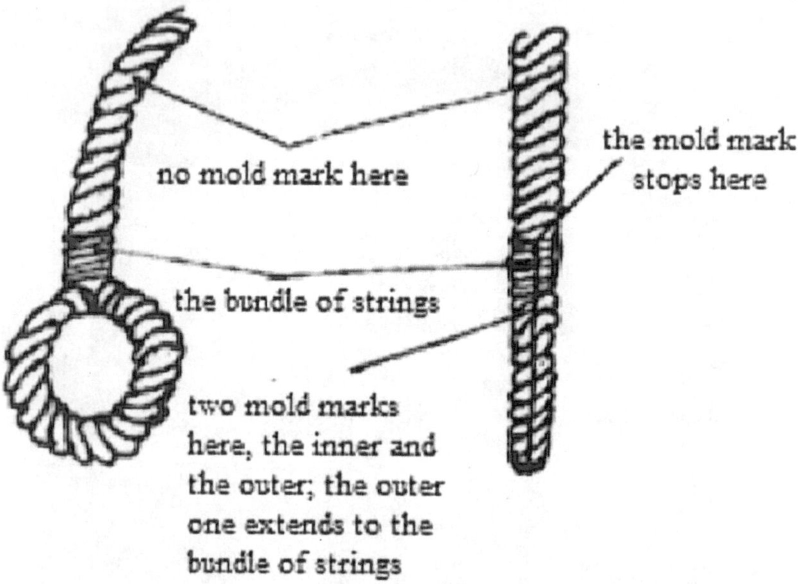

Source. After "Zhongguo Zaoqi Shila Zhuzao Wenti de Guancha yu Sikao" (fig. 5) by Tan Derui, *Nanfang* Wenwu 2 (2007): 36–40.

Figure 94a. One of the three bronze *yan* from Anhui Tongling in the Spring and Autumn Period. H. 58.2 cm.

Source. After "Bianxing Kuiwen Fenti Yan—Cunzai Fenshifa de you yi Shengdong Lizheng" (p. 293, Color Plate) by Tan Derui, *Tezhong Zhuzao ji Youse Hejin* 3 (2010): 293–294.

Figure 94b. Swing handle of the bronze *yan* shown in Figure 94a.

Source. After "Bianxing Kuiwen Fenti Yan—Cunzai Fenshifa de you yi Shengdong Lizheng" (p. 294, fig. 2) by Tan Derui, *Tezhong Zhuzao ji Youse Hejin* 3 (2010): 293–294.

Figure 94c. Swing handle of the bronze *yan* shown in Figure 94a.

Source. After "Zai Lun 'Shila Gongyi Bushi Zhongguo Qingtong Shidai de Xuanze'" (fig. 12.1) by Zhou Weirong et al., *Nanfang Wenwu* 2 (2007): 41–48, 132.

Figure 94d. Swing handle of the bronze *yan* shown in Figure 94a.

Source. After "Zai Lun 'Shila Gongyi Bushi Zhongguo Qingtong Shidai de Xuanze'" (fig. 12.2) by Zhou Weirong et al., *Nanfang Wenwu* 2 (2007): 41–48, 132.

Figure 95a. Bronze tiger from Jiangxi Xin'gan Dayangzhou. Circa 1300 BCE.

Source. After *Xin'gan Shangdai Da Mu* (Plate 49) by Jiangxisheng Wenwu Kaogu Yanjiusuo et al., 1997, Beijing: Wenwu Chubanshe.

Figure 95b. Decoration of a bronze *nao* bell from Zhejiang Changxing. Probably late second millennium BCE (though usually dated to the Spring and Autumn period).

Source. After *Zhongguo qingtongqi quanji,* vol. 11 (Plate 146) by Zhongguo Qingtongqi quanji bianji weiyuanhui, 1999, Beijing: Wenwu Chubanshe.

Figure 95c. Diagram drawn by Tan Derui to illustrate the insurmountable difficulty of withdrawing mold pieces from the model, for both cases of the Xin'gan tiger (Figure 95a) and Changxing *nao* (Figure 95b).

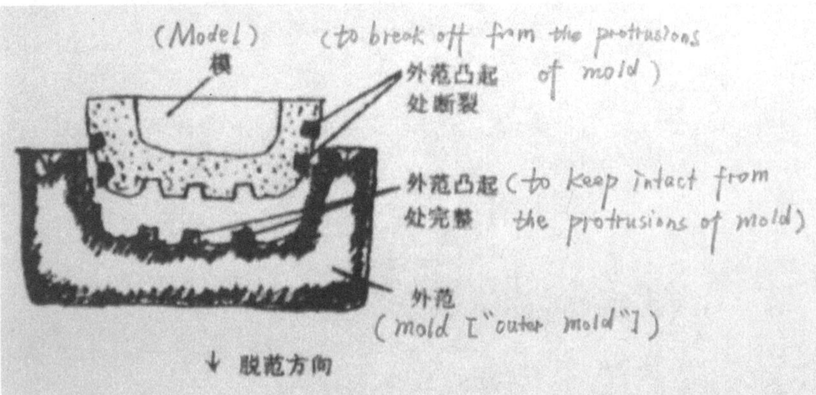

Source. After "Zhongguo Zaoqi Shila Zhuzao Wenti de Guancha yu Sikao" (fig. 2) by Tan Derui, *Nanfang Wenwu* 2 (2007): 36–40.

Figure 96a. Bronze *gui* from Zhejiang Changxing. Probably late second millennium BCE (though usually dated to the Spring and Autumn period). H. 9.9cm. Side view.

Source. After "Yunwen Gui—Wenshi Zhizuo Gongyi Qite de Nanfang Qingtongqi" (p. 1175, Color Plate), *Tezhong Zhuzao ji Youse Hejin* 12 (2009): 1175–1176.

Figure 96b. Top view of the bronze *gui* of Figure 96a.

Source. After "Yunwen Gui—Wenshi Zhizuo Gongyi Qite de Nanfang Qingtongqi" (p. 1176, fig. 1), *Tezhong Zhuzao ji Youse Hejin* 12 (2009): 1175–1176.

Figure 96c. Diagram drawn by Tan Derui to illustrate the detail of decoration from the bronze *gui* of Figure 96a.

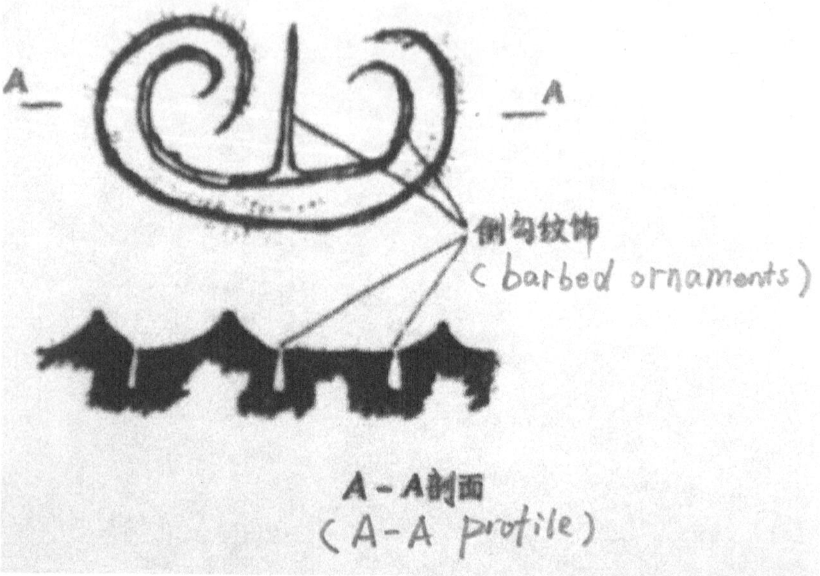

Source. After "Zhongguo Zaoqi Shila Zhuzao Wenti de Guancha yu Sikao" (fig. 3) by Tan Derui, *Nanfang Wenwu* 2 (2007): 36–40.

Figure 97a. The double-owl bronze *you* in the Winthrop collection, Fogg Art Museum, Harvard University. H.19 cm. Anyang period.

Source. Artifact in the Winthrop Collection, Fogg Art Museum (acc. 1943. 52. 102), Harvard Art Museums. After *Shang Ritual Bronzes in the Arthur M. Sackler Collections* (p. 115, fig.153) by Robert Bagley, 1987, Washington, D.C.: Arthur M. Sackler Foundation; Arthur M. Sackler Museum, Harvard University; distributed by Harvard University Press, Cambridge, MA.

Figure 97b. Swing handle of the bronze *you* shown in Figure 97a.

Source. Artifact in the Winthrop Collection, Fogg Art Museum (acc. 1943. 52. 102), Harvard Art Museums. After *Shang Ritual Bronzes in the Arthur M. Sackler Collections* (p. 115, fig.153) by Robert Bagley, 1987, Washington, D.C.: Arthur M. Sackler Foundation; Arthur M. Sackler Museum, Harvard University; distributed by Harvard University Press, Cambridge, MA.

Figure 97c. Swing handle of the bronze *you* shown in Figure 97a.

Source. Artifact in the Winthrop Collection, Fogg Art Museum (acc. 1943. 52. 102), Harvard Art Museums. After *Shang Ritual Bronzes in the Arthur M. Sackler Collections* (p. 115, fig.153) by Robert Bagley, 1987, Washington, D.C.: Arthur M. Sackler Foundation; Arthur M. Sackler Museum, Harvard University; distributed by Harvard University Press, Cambridge, MA.

Figure 98a. Bronze plaque from Chenggong Tianzimiao M41, Yunnan. Circa 350 BCE.

Source. After *Zhongguo qingtongqi quanji*, vol. 14 (Plate 124), edited by Zhongguo qingtongqi quanji bianji weiyuanhui, 1996, Beijing: Wenwu Chubanshe.

Figure 98b. Bronze plaque from Jiangchuan Lijiashan M13, Yunnan. Circa 3rd century BCE.

Source. After *Zhongguo qingtongqi quanji*, vol. 14 (Plate 126), edited by Zhongguo qingtongqi quanji bianji weiyuanhui, 1996, Beijing: Wenwu Chubanshe.

Figure 99a. Bronze mask with trunk-like projection from Sichuan Guanghan Sanxingdui K2. H.84.3 cm; height of mask 31.7 cm. Late second millennium BCE.

Source. Artifact in the Sanxingdui Museum, Guanghan (Sichuan). After *Zhongguo qingtongqi quanji*, vol. 13 (Plate 30), edited by Zhongguo qingtongqi quanji bianji weiyuanhui, 1994, Beijing: Wenwu Chubanshe.

Figure 99b. Detail of the rear of the mask shown in Figure 99a.

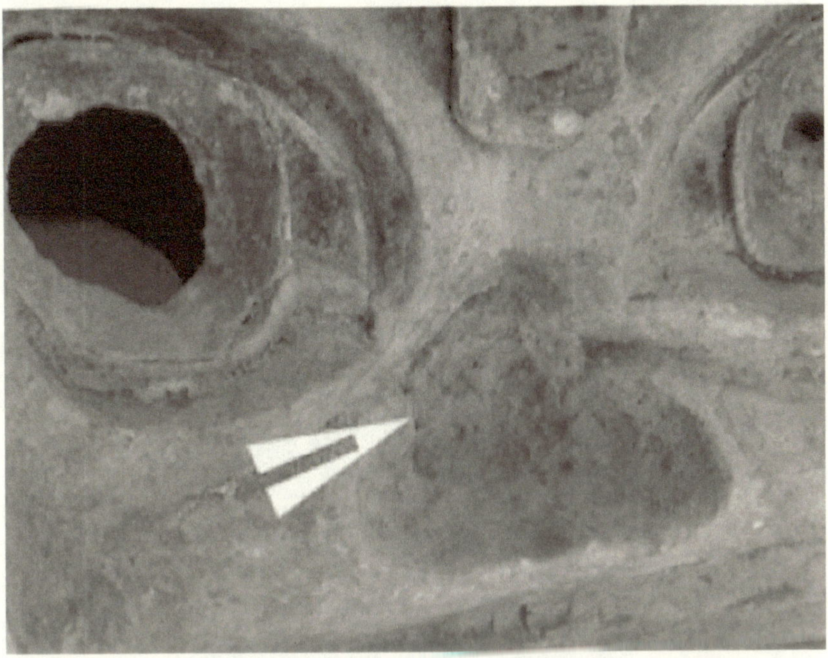

Source. After "The Early History of Lost-wax Casting" (p. 152, fig. 5: a) by Christopher Davey,2009, in *Metallurgy and Civilisation: Eurasia and Beyond,* edited by Mei Jianjun and Thilo Rehren, 147–154. London: Archetype.
Note. This trace, according to Christopher Davey's *problematic* explanation, shows mold remains in the nose cavity.

Figure 99c. Detail of the rear of the mask shown in Figure 99a.

Source. After "The Early History of Lost-wax Casting" (p. 152, fig. 5: b) by Christopher Davey,2009, in *Metallurgy and Civilisation: Eurasia and Beyond*, edited by Mei Jianjun and Thilo Rehren, 147–154. London: Archetype.
Note. This trace, according to Christopher Davey's *problematic* explanation, shows a "false cut" originally made in the wax model.

Figure 100. Bronze dagger from Hebei Qinglong Chaodaogou with an intricately formed animal-head pommel. L. 30.2 cm. Late second millennium BCE.

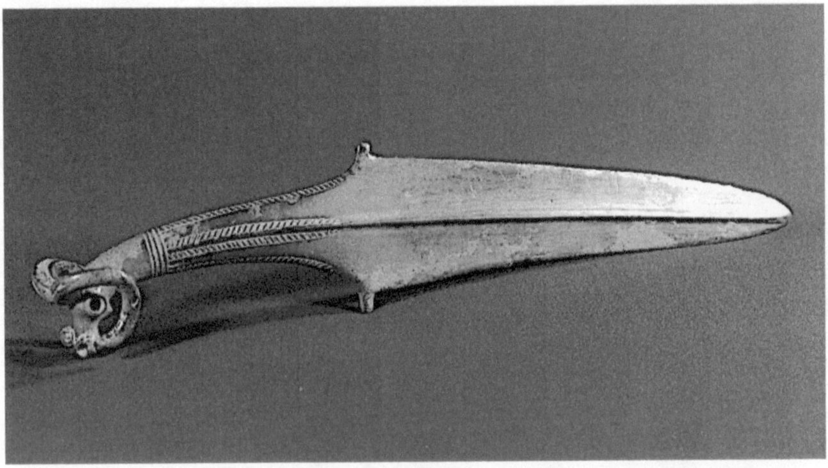

Source. After *Zhongguo qingtongqi quanji*, vol. 15 (Plate 3), edited by Zhongguo qingtongqi quanji bianji weiyuanhui, 1995, Beijing: Wenwu Chubanshe.

Figure 101a. Bronze hair ornament formerly in the Arthur M. Sackler Foundation collection, probably from Western Mongolia, circa 13th–11th century BCE. L. 9cm.

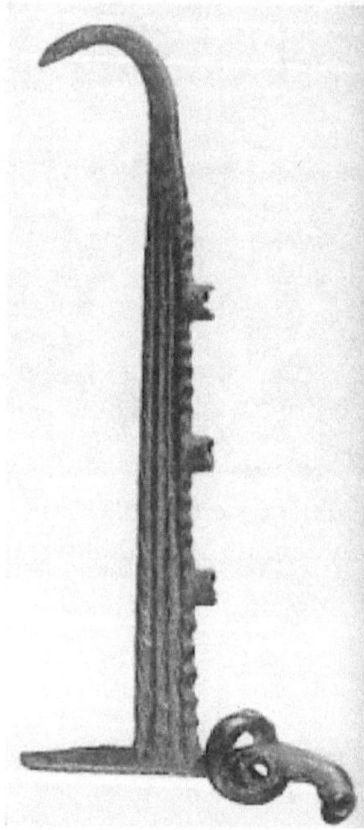

Source. After *Ancient Bronzes of the Eastern Eurasian Steppes from the Arthur M. Sackler Collections* (p. 142, no. 32) by Emma Bunker et al., 1997, New York: Arthur M. Sackler Foundation; distributed by Abrams.

Figure 101b. The ram head of the bronze hair ornament shown in Figure 101a.

Source. After *Ancient Bronzes of the Eastern Eurasian Steppes from the Arthur M. Sackler Collections* (p. 142, no. 32) by Emma Bunker et al., 1997, New York: Arthur M. Sackler Foundation; distributed by Abrams.

Figure 102a. Bronze harness jingle with stag from Northeast China, circa 7th–6th century BCE. H. 13.3 cm

Source. After *Nomadic Art of the Eastern Eurasian Steppes: The Eugene V. Thaw and other New York Collections* (Plate 29) by Emma Bunker, 2002, with contributions by James C.Y. Watt and Zhixin Sun. New York: Metropolitan Museum of Art, and New Haven, CT: Yale University Press.
Note. Emma Bunker, *questionably*, considers it as cast by the direct lost-wax process.

Figure 102b. Bronze harness jingle with an animal head from Northeast China, circa 7th–6th century BCE. H.7.3 cm.

Source. After *Nomadic Art of the Eastern Eurasian Steppes: The Eugene V. Thaw and other New York Collections* (Plate 30) by Emma Bunker, 2002, with contributions by James C.Y. Watt and Zhixin Sun. New York: Metropolitan Museum of Art, and New Haven, CT: Yale University Press.
Note. Emma Bunker, *questionably*, considers it as cast by the direct lost-wax process .

Figure 102c. Bronze harness ornament with tigers from Northeast China, circa 7th–6th century BCE. L. 10.6 cm.

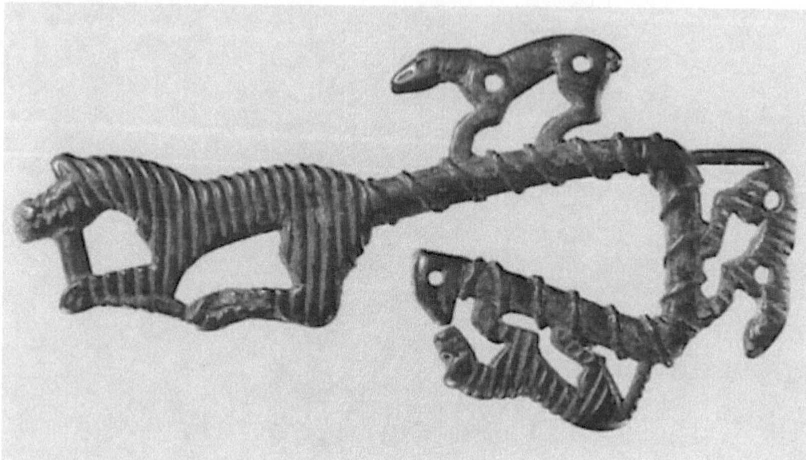

Source. After *Nomadic Art of the Eastern Eurasian Steppes: The Eugene V. Thaw and other New York Collections* (Plate 27) by Emma Bunker, 2002, with contributions by James C.Y. Watt and Zhixin Sun. New York: Metropolitan Museum of Art, and New Haven, CT: Yale University Press.

Note. Emma Bunker, *questionably*, considers it as cast by the indirect lost-wax process (with the wax model formed in a two-piece mold).

Figure 102d. Bronze harness jingle with mounted hunters from Northeast China, circa 7th–6th century BCE. H.13 cm.

Source. After *Nomadic Art of the Eastern Eurasian Steppes: The Eugene V. Thaw and other New York Collections* (Plate 26) by Emma Bunker, 2002, with contributions by James C.Y. Watt and Zhixin Sun. New York: Metropolitan Museum of Art, and New Haven, CT: Yale University Press.

Note. Emma Bunker, *questionably*, considers it as cast by the indirect lost-wax process (with the wax model formed in a two-piece mold).

Figure 103a. A set of four bronze finials with single ram from regions in the center of northern Hebei, circa 7th-6th century BCE. H.16.2 cm.

Source. After *Nomadic Art of the Eastern Eurasian Steppes: The Eugene V. Thaw and other New York Collections* (Plate 24) by Emma Bunker, 2002, with contributions by James C.Y. Watt and Zhixin Sun. New York: Metropolitan Museum of Art, and New Haven, CT: Yale University Press.

Figure 103b. Bronze finial with single ram (ram horns are different from those shown in Figure 103a) from regions in the center of northern Hebei, circa 7th-6th century BCE. H. 15.5 cm.

Source. After *Nomadic Art of the Eastern Eurasian Steppes: The Eugene V. Thaw and other New York Collections* (Plate 25) by Emma Bunker, 2002, with contributions by James C.Y. Watt and Zhixin Sun. New York: Metropolitan Museum of Art, and New Haven, CT: Yale University Press.

Figure 103c. Bronze finial with two rams from Hebei Chicheng, circa 7th-6th century BCE.

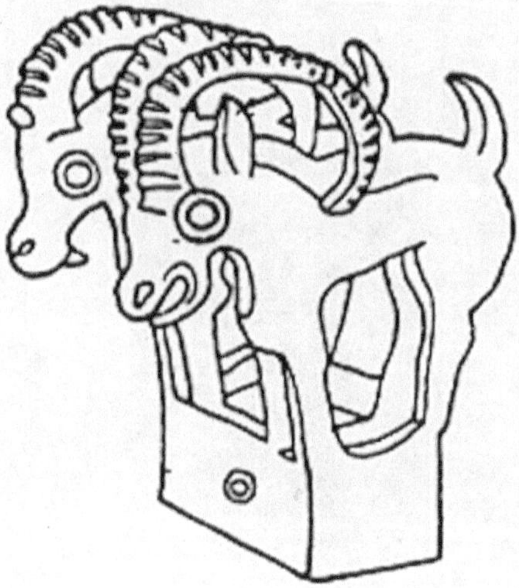

Source. After *Ancient Bronzes of the Eastern Eurasian Steppes from the Arthur M. Sackler Collections* (p. 177, fig. 92.3) by Emma Bunker et al., 1997, New York: Arthur M. Sackler Foundation; distributed by Abrams.

Figure 103d. Bronze finial with single horse from regions in the center of northern Hebei, circa 7th-6th century BCE. H. 10.9 cm.

Source. After *Ancient Bronzes of the Eastern Eurasian Steppes from the Arthur M. Sackler Collections* (p. 177, fig. 92) by Emma Bunker et al., 1997, New York: Arthur M. Sackler Foundation; distributed by Abrams.

Figure 103e. Bronze finial with single horse/ass from Hebei Zhangjiakou, circa 7th-6th century BCE.

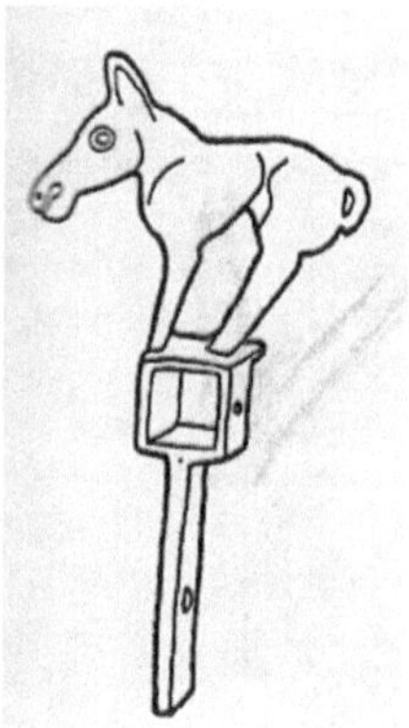

Source. After *Ancient Bronzes of the Eastern Eurasian Steppes from the Arthur M. Sackler Collections* (p. 177, fig. 92.1) by Emma Bunker et al., 1997, New York: Arthur M. Sackler Foundation; distributed by Abrams.

Figure 103f. Bronze finial with single water bird from Hebei Kangbao, circa 7th-6th century BCE.

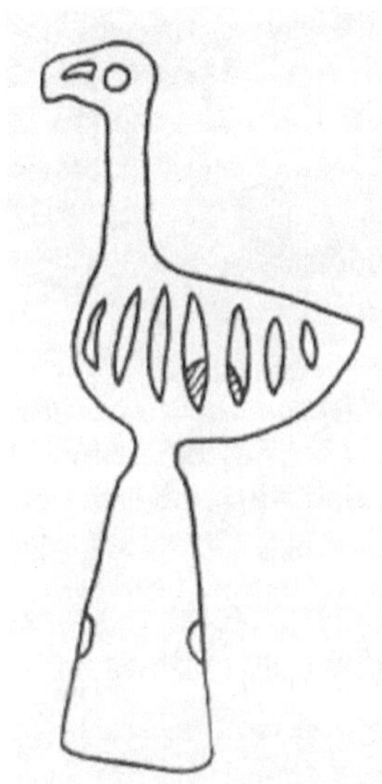

Source. After *Ancient Bronzes of the Eastern Eurasian Steppes from the Arthur M. Sackler Collections* (p. 177, fig. 92.2) by Emma Bunker et al., 1997, New York: Arthur M. Sackler Foundation; distributed by Abrams.

Figure 104a. Crouching feline pectoral in gold from Beijing Yanqing Jundushan, dimensions not specified.

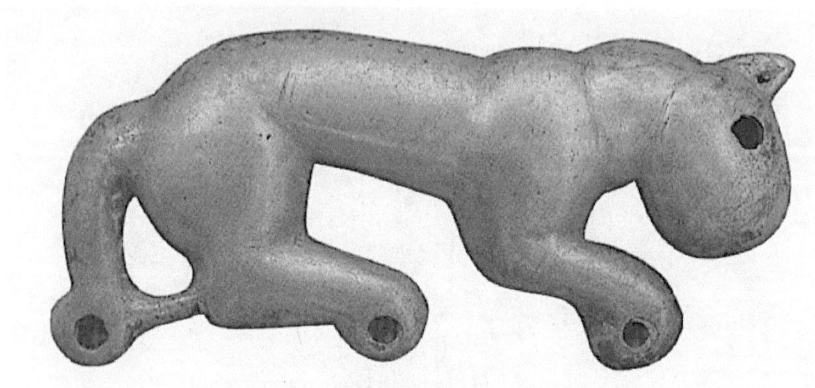

Source. After *Traders and Raiders on China's Northern Frontier* (p. 50, fig. 18) by Jenny So and Emma Bunker, 1995, Seattle: Arthur M. Sackler Gallery, Smithsonian Institution, in association with University of Washington Press.

Figure 104b. Crouching feline pectoral in bronze (in the Levy and White collection), L. 9.4cm, probably from northwest China/northern Hebei.

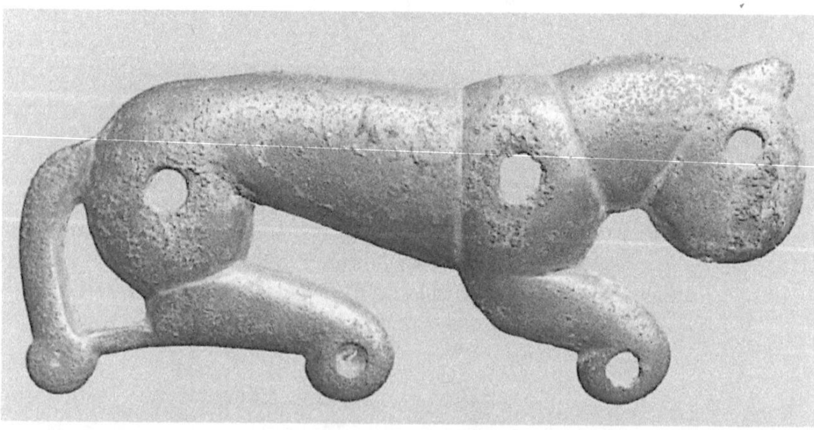

Source. After *Traders and Raiders on China's Northern Frontier* (p. 112, Plate 27) by Jenny So and Emma Bunker, 1995, Seattle: Arthur M. Sackler Gallery, Smithsonian Institution, in association with University of Washington Press.

Figure 105a. No. 21 bronze crane (broken) from the mausoleum of the First Emperor of Qin (r. 247–210 BCE, mausoleum constructed from 247 to 208 BCE).

Source. The image of No. 21 bronze crane (heavily broken and unpublished) is directly provided by Dr. Shao Anding.

Figure 105b. Multi-layer feathers from the tail of the bronze crane shown in Figure 105a.

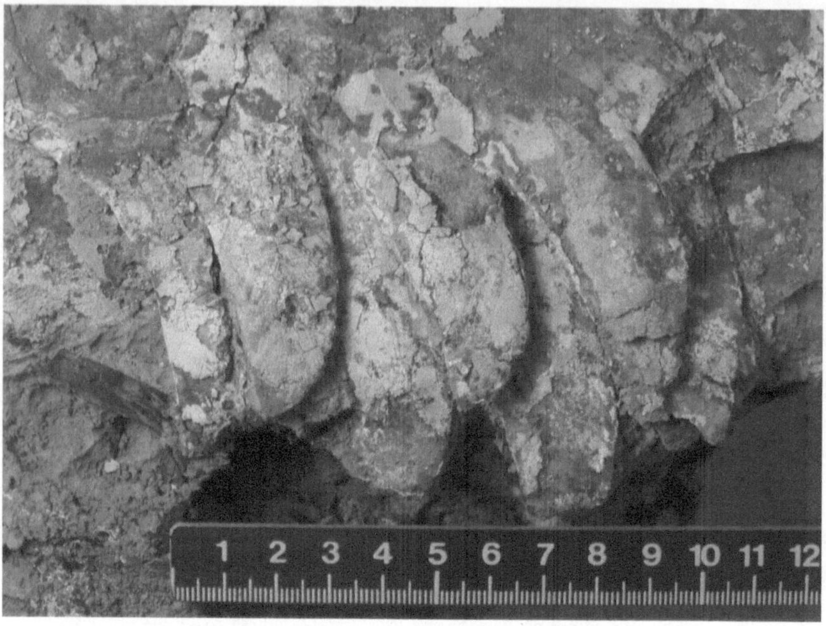

Source. After *Qinshihuang Lingyuan Chutu Caihui Qingtong Shuiqin Zhizuo Gongyi ji Xiangguan Wenti Yanjiu* (p. 63, fig. 2-168) by Shao Anding, 2019, Beijing: Kexue chubanshe.

Figure 106a. No. 10 bronze swan (broken) from the mausoleum of the First Emperor of Qin (r. 247–210 BCE, mausoleum constructed from 247 to 208 BCE).

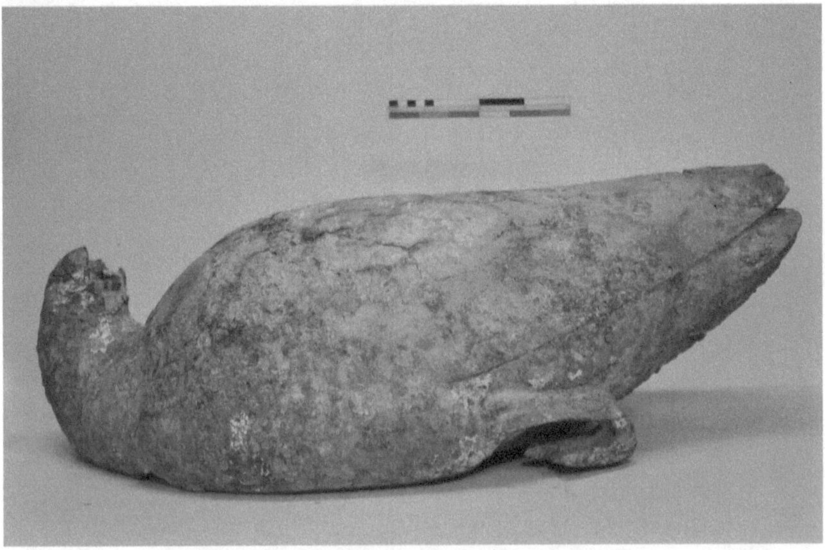

Source. After "Qinshihuang Lingyuan Chutu Caihui Qingtong Shuiqin Zhizuo Gongyi ji Xiangguan Wenti Yanjiu" (p. 40, fig. 3-26) by Shao Anding, Ph.D. dissertation, University of Science and Technology Beijing, 2013.

Figure 106b. Brush marks left on the inner surface of the bronze swan shown in Figure 106a. .

Source. After "Qinshihuang Lingyuan Chutu Caihui Qingtong Shuiqin Zhizuo Gongyi ji Xiangguan Wenti Yanjiu" (p. 41, fig. 3-35) by Shao Anding, Ph.D. dissertation, University of Science and Technology Beijing, 2013.

Figure 107. Belt hook from Jiangsu Lianshui Sanlidun. Cast Gold. Warring States period. L. 12cm.

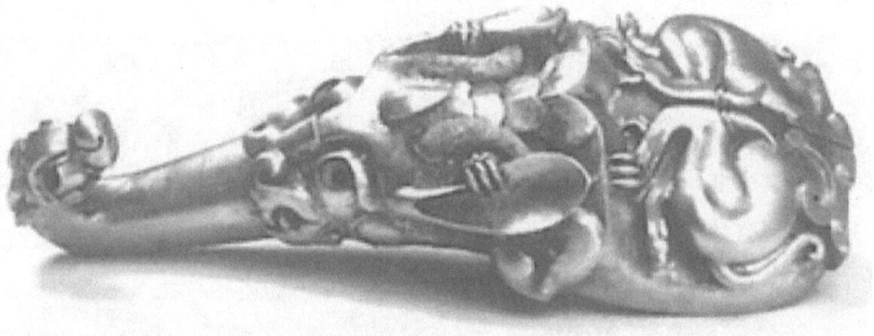

Source. After "Jiangsu Lianshui Sanlidun Xihan mu" (Plate 12.3) by Nanjing Bowuyuan, *Kaogu* 2 (1973): 80–87, 89.

Figure 108. Gilded bronze belt buckle plaque. Originally one of a pair, stylistically associated with the State of Qin. Circa 3rd century BCE. L. 11.0 cm, H. 7.5 cm.

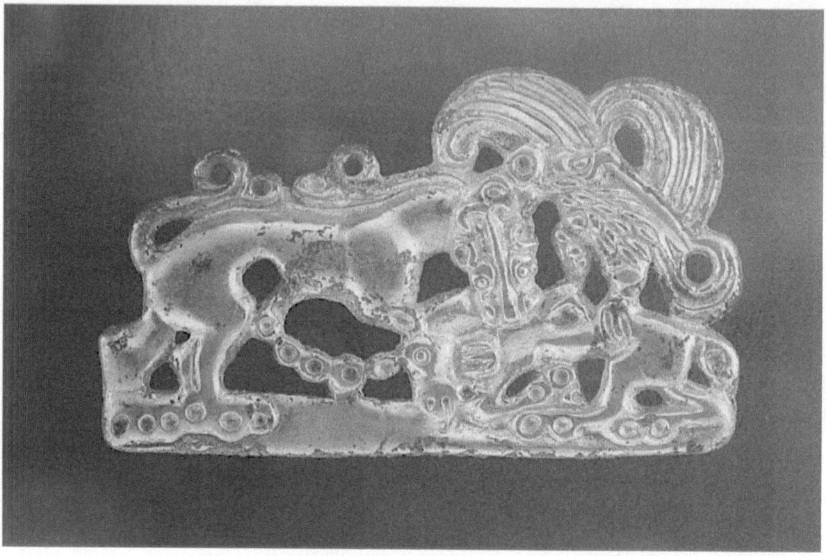

Source. Artifact in the Mengdiexuan collection, Hong Kong (acc. GI-022, published in *Ancient Chinese and Ordos Bronzes* by Jessica Rawson and Emma Bunker, no. 217). After *Adornment for Eternity: Status and Rank in Chinese Ornament* (p. 109, Plate 28) by Julia White and Emma Bunker, 1994, Denver: Denver Art Museum in association with the Woods Pub. Co.

Figure 109a. Bronze miniature house from Zhejiang Shaoxing Shizishan M306. H. 17 cm, W.13 cm, D.11.5 cm.

Source. Image courtesy of Yu Shanying and the Zhejiang Provincial Museum.

Figure 109b. Bottom of the bronze miniature house shown in Figure 109a.

Source. Author's photograph, courtesy of Zhejiang Provincial Museum.

Figure 109c. An ensemble of six musicians within the bronze miniature house of Figure 109a.

Source. Author's photograph, courtesy of Zhejiang Provincial Museum.

Figure 109d. Clay core exposed in the left ear of one figurine within the bronze miniature house of Figure 109a.

Source. Author's photograph, courtesy of Zhejiang Provincial Museum.

Figure 110a. Bronze *he* from Zhejiang Shaoxing Shizishan M306. L. 29 cm.

Source. Author's photograph, courtesy of Zhejiang Provincial Museum.

Figure 110b. Vessel spout of the bronze *he* shown in Figure 110a.

Source. Author's photograph, courtesy of Zhejiang Provincial Museum.

Figure 110c. Lid of the bronze *he* shown in Figure 110a.

Source. Author's photograph, courtesy of Zhejiang Provincial Museum.

Figure 111a. Iron dagger with interlaced gold hilt from Yimencun Tomb 2, Shaanxi Baoji. L. 35.2 cm (Length of hilt: 12.8 cm). Middle/Late Spring and Autumn period.

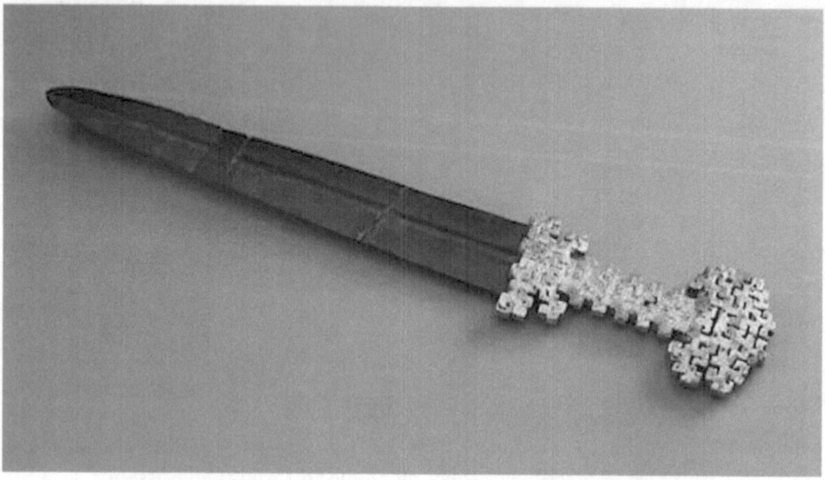

Source. After https://read01.com/zh-hk/dE4MLEm.html#.Xfk6HS2B3EY (Accessed December18, 2019). For further images and information, see *Chunqiu Zhanguo: Zhengba Tuqiang de Shidai* (*p.* 92) by Liu Wei and He Hong, 2001, Hong Kong: Commercial Press.

Figure 111b. Interlace of the iron dagger shown in Figure 111a.

Source. After https://read01.com/zh-hk/dE4MLEm.html#.Xfk6HS2B3EY (Accessed December18, 2019). For further images and information, see *Chunqiu Zhanguo: Zhengba Tuqiang de Shidai* (*p.* 92) by Liu Wei and He Hong, 2001, Hong Kong: Commercial Press.

Figure 112a. Bronze *lei* with copper inlays from Hougudui Tomb 1, Henan Gushi. H. 45 cm.

Source. After *Gushi Hougudui Yihao Mu* (Color Plate 15.1), edited by Henan Sheng Wenwu Kaogu Yanjiusuo, 2004, Zhengzhou: Daxiang Chubanshe.

Figure 112b. Illustration showing how each copper inlay of the bronze *lei* of Figure 112a was fixed with a tenon-like device.

Source. After *Gushi Hougudui Yihao Mu* (p. 136, fig. 3), edited by Henan Sheng Wenwu Kaogu Yanjiusuo, 2004, Zhengzhou: Daxiang Chubanshe.

Figure 113a. Bronze *gui* (M1:6) with cast copper inlays from Xulou Tomb1, Shandong Zaozhuang. H.14.6 cm. Probably during the first quarter of the 6th century BCE.

Source. After "Zaozhuang Xulou Chutu Zhuxiang Hongtong Qingtongqi Tanlun—Jianji Hongtong Zhuxiang Wenshi Qingtongqi de Shidai yu Chandi Wenti" (p. 3, fig. 3) by Su Rongyu and Wang Lihua, 2015, presented in the academic conference "Qingtongqi yu Shandong Guguo Xueshu Yantaohui" (Dec. 10–11, 2015; conference proceedings published in 2017).

Figure 113b. Tenon-like device behind the "王"-shaped copper ornamentation of the bronze *gui* shown in Figure 113a.

Source. After "Zaozhuang Xulou Chutu Zhuxiang Hongtong Qingtongqi Tanlun—Jianji Hongtong Zhuxiang Wenshi Qingtongqi de Shidai yu Chandi Wenti" (p. 8, fig. 15) by Su Rongyu and Wang Lihua, 2015, presented in the academic conference "Qingtongqi yu Shandong Guguo Xueshu Yantaohui" (Dec. 10–11, 2015; conference proceedings published in 2017).

Figure 114a. Bronze *gui* (M1:44) with cast copper inlays from Xulou Tomb1, Shandong Zaozhuang. H.14.6 cm. Probably during the first quarter of the 6th century BCE.

Source. After "Zaozhuang Xulou Chutu Zhuxiang Hongtong Qingtongqi Tanlun—Jianji Hongtong Zhuxiang Wenshi Qingtongqi de Shidai yu Chandi Wenti" (p. 2, fig. 1) by Su Rongyu and Wang Lihua, 2015, presented in the academic conference "Qingtongqi yu Shandong Guguo Xueshu Yantaohui" (Dec. 10–11, 2015; conference proceedings published in 2017).

Figure 114b. Scar left by a metal spacer, detached from an animal-shaped copper ornament of the bronze *gui* shown in Figure 114a.

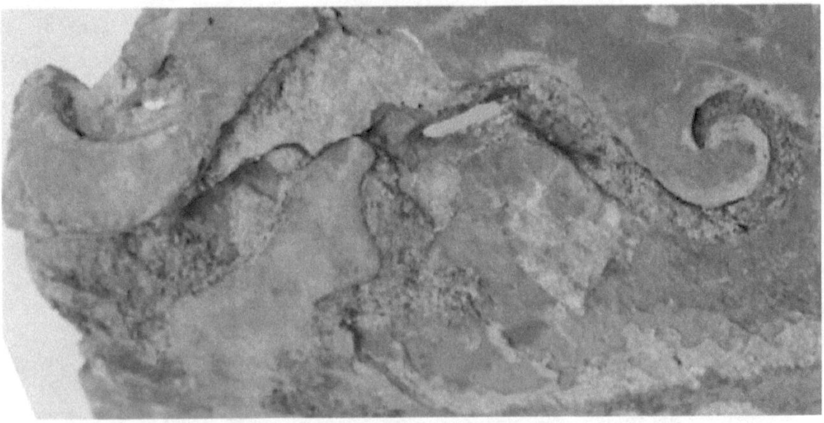

Source. After "Zaozhuang Xulou Chutu Zhuxiang Hongtong Qingtongqi Tanlun—Jianji Hongtong Zhuxiang Wenshi Qingtongqi de Shidai yu Chandi Wenti" (p. 9, fig. 17) by Su Rongyu and Wang Lihua, 2015, presented in the academic conference "Qingtongqi yu Shandong Guguo Xueshu Yantaohui" (Dec. 10–11, 2015; conference proceedings published in 2017).

Figure 115a. Bronze drum stand from Leigudun M1, Hubei Suizhou. H.54 cm. Tomb dated 433 BCE.

Source. After *Zhongguo qingtongqi quanji*, vol. 10 (Plate 167), edited by Zhongguo qingtongqi quanji bianji weiyuanhui, 1996, Beijing: Wenwu Chubanshe.

Figure 115b. Detail of the bronze drum stand shown in Figure 115a.

Source. After *Zhongguo qingtongqi quanji*, vol. 10 (Plate 168), edited by Zhongguo qingtongqi quanji bianji weiyuanhui, 1996, Beijing: Wenwu Chubanshe.

Figure 116. Bronze *zun* from Ningxiang County, Hunan province. Circa 12th century BCE. H. 58.3 cm.

Source. After *Zhongguo qingtongqi quanji*, vol. 4 (Plate 115), edited by Zhongguo qingtongqi quanji bianji weiyuanhui, 1998, Beijing: Wenwu Chubanshe.

Figure 117a. Gold hook buckle, L. 7.8 cm. 3rd–2nd century BCE. Front side.

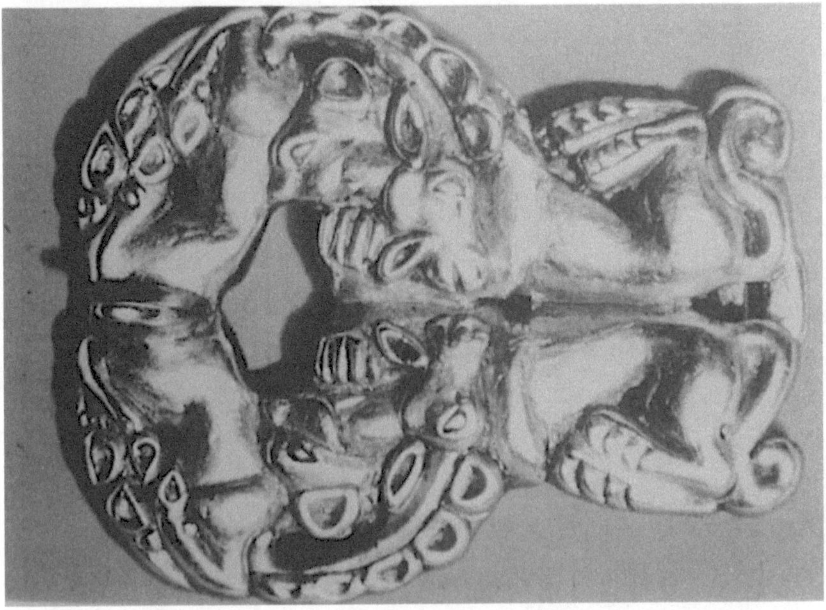

Source. After "Lost Wax and Lost Textile: An Unusual Ancient Technique for Casting Gold Belt Plaques" (p. 223, fig. 20.2) by Emma Bunker, 1988, in *The Beginning of the Use of Metals and Alloys: Papers from the Second International Conference on the Beginning of the Use of Metals and Alloys, Zhengzhou, China, 21-26 October 1986*, edited by R. Maddin, 222–225. Cambridge, MA: MIT Press.

Figure 117b. Back side of the gold hook buckle shown in Figure 117a.

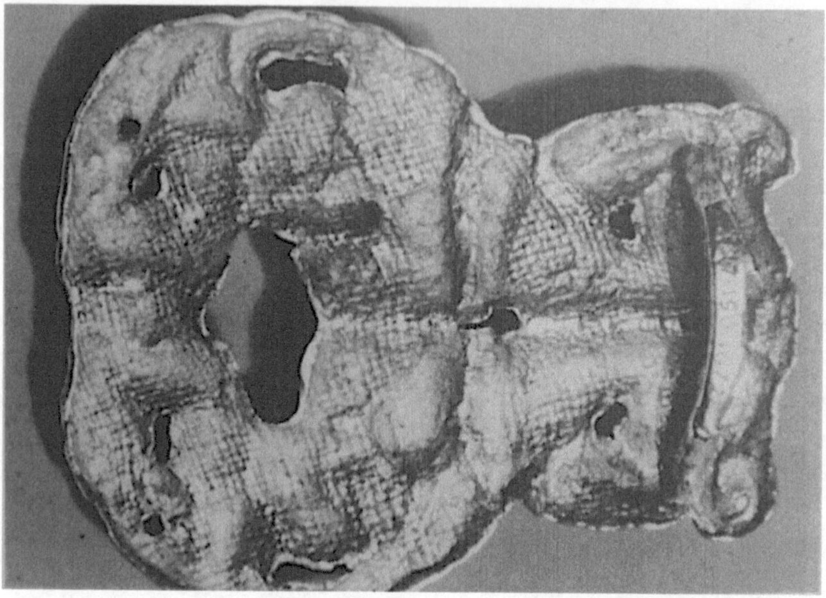

Source. After "Lost Wax and Lost Textile: An Unusual Ancient Technique for Casting Gold Belt Plaques" (p. 223, fig. 20.1) by Emma Bunker, 1988, in *The Beginning of the Use of Metals and Alloys: Papers from the Second International Conference on the Beginning of the Use of Metals and Alloys, Zhengzhou, China, 21-26 October 1986*, edited by R. Maddin, 222–225. Cambridge, MA: MIT Press.

Character Glossary

Anyang 安陽

An Zhimin 安志敏

Baicaopo 白草坡

baihu 白虎

Baoji 寶雞

bi 匕

Bu [illegible]

bu 瓿

Caomenwan 曹門灣

Chang'an 長安

Changsha 長沙

Changxing 長興

Chaodaogou 抄道溝

Chen 陳

Chen Mengjia 陳夢家

Cheng, Duke 成［公］

cheng geng 撐梗

Chenggong 呈貢

Chen Peifen 陳佩芬

Chen Zhang 陳璋

Chu 楚

"Chu Wang Yin Shen" 楚王酓審

Chuxiong 楚雄

Cuo 譽

cuo feng 錯縫

Dadianzi 大甸子

Dahekou 大河口

Dangyang 當陽

Dayangzhou 大洋州

Dian 滇

ding 鼎

Dongbei 東北

Dongxiafeng 東下馮

Dong Yawei 董亞巍

dougong 斗拱

dui 敦

Erlitou 二裡頭

Erligang 二里崗

Fangshan 房山

Fengxingzuishan 鳳形嘴山

fang hu 方壺

fan feng 範縫

fei leng 飛棱

Fengxiang 鳳翔

fou 缶

fu 簠

Fufeng 扶風

Fuhao 婦好

Gao Ming 高明

ge 戈

gu 觚

gui 簋

Gong, king 共[王]

gongxing geng 拱形梗

Guanghan 廣漢

Guo 虢

Guo Baojun 郭寶鈞

Gushi 固始

Gu Wenzi Lei Bian 古文字類編

Guyuege 古越閣

Haimenkou 海門口

Han 漢

Hancheng 韓城

Hanhuai 漢淮

he 盉

*he** 鉌

He Hao 何浩

He Yonglin 何永林

Hougudui 侯古堆

Hou Hanshu 後漢書

Houma 侯馬

hu 壺

Hua Jueming 華覺明

Hua Lin 華林

Huan, Duke 桓[公]

Huang 黃

huang 璜

Huang Fengchun 黃鳳春

Huangpi 黃陂

Huang Wei 黃維

"Huang Zi" 黃子

"Huang Zi Lou" 黃子婁

huawen geng 花紋梗

Hunyuan 渾源

jia 斝

Jiangchuan 江川

Jiangling 江陵

Jin 晉

jin 禁

Jingmen 荊門

Jinguo Gudu Bowuguan 晉國古都博物館

"Jin Jiang" 晉姜

Jinning 晉寧

jue 爵

Jundushan 軍都山

Ke *bo* 克鎛

Kexingzhuang 客省莊

kuang geng 框梗

kui 夔

lei 罍

Leigudun 擂鼓墩

Lian Haiping 廉海萍

lian he fang hu 蓮鶴方壺

Liangdaicun 梁帶村

Liao, River 遼[河]

Li Chaoyuan 李朝遠

Li Ji 李濟

Lijialou 李家樓

Lijiashan 李家山

Li Jinghua 李京華

Lingnan 嶺南

Lingshi 靈石

Lingtai 靈台

Li Zhijia 黎志佳

Liujiadianzi 刘家店子

Liujiahe 劉家河

Liulihe 琉璃河

Liulin 劉林

Liuquan 柳泉

Li Yuxin 黎毓馨

Li Zhiwei 李志偉

loukong 鏤空

loukong gao fudiao 鏤空高浮雕

Longshan 龍山

Lu 魯

Luo Feng 羅豐

Luoyang 洛陽

Majiayuan 馬家塬

Mei Jianjun 梅建軍

Meixian 郿縣

Mengdiexuan 夢蝶軒

Mingmenhuafu 名門華府

mingqi 明器/冥器

mo 模

Nanguan 南關

Nanjing 南京

Nanshangen 南山根

Nanyang 南陽

Nanyaozhuang 南窯庄

nao 鐃

Ning, Duke 寧[公]

Ningcheng 寧城

Ningxiang 寧鄉

nitiao panzhu 泥條盤築

pan 盤

Panlongcheng 盤龍城

pan long fang hu 蟠龍方壺

Peng Zi She/Ye 彭子射

Pinggu 平穀

Pingshan 平山

Pixian 邳縣

Qi 齊

Qijia 齊家

Qilian 祁連

Qin 秦

Qinglong 青龍

Qiulin 秋林

Ren Changzhong 任常中

Renjiacun 任家村

Rong Geng 容庚

"Rong Sheng" 戎生

Rui 芮

Sanji Gongshe 三汲公社

Sanmenxia 三門峽

Sanxingdui 三星堆

Shang 商

Shangmacun 上馬村

Shang-Zhou 商周

Shanrong 山戎

Shao Anding 邵安定

Shaoxing 紹興

She/Ye 射

Shen, Duke 申[公]

"Shen Gong zhi Sun" 申公之孫

sheng 笙

sheng ding 升鼎

Shiji 史記

Shi Quan 石泉

Shizhaishan 石寨山

Shi Zhangru 石璋如

Shizishan 獅子山

Sima Qian 司馬遷

Songze 崧澤

Su Rongyu 蘇榮譽

Suizhou 隨州

Tai, Lake 太[湖]

Taihang 太行

Tan Derui 譚德睿

Tang Wenxing 湯文興

Taosi 陶寺

taotie 饕餮

Tianzimiao 天子廟

Tongling 銅陵

Tunxi 屯溪

toukong 透空

tusun 突榫

Wanjiaba 萬家壩

Wan Jiabao 萬家保

Wan Li 萬俐

Wang Changqing 王長青

Wang Jinchao 王金潮

Wang Ningsheng 汪寧生

Wang Yong 王勇

Wei, River 渭[河]

Wenfengta 文峰塔

Wuhan 武漢

Wu Laiming 吳來明

Xiang, Duke 襄[公]

Xi'an 西安

Xiasi 下寺

Xibei 西北

xincheng 芯撑

Xichuan 淅川

Xichuan Xiasi Chunqiu Chumu 淅川下寺春秋楚墓

xindian 芯墊

xinfan 芯范

Xin'gan 新幹

Xinjiang 新絳

Xinzhai 新寨

Xinzheng 新鄭

xinfan 芯范

Xu 許

Xuchang 許昌

Xu Huikang 徐惠康

Xulou 徐樓

Xuyi 盱眙

Yan 燕

yan 甗

Yangjiacun 楊家村

Yanqing 延慶

Yanshi 偃師

Yaojiagang 姚家崗

Ye 葉

Yexian 葉縣

Yi 乙

yi 匜

Yichang 宜昌

Yicheng 翼城

Yiluo, River 伊洛[河]

Yingcheng 郢稱

Yimencun 益門村

Yin 殷

Yinxu 殷墟

Yishui 沂水

Yixing 宜興

you 卣

yu 盂

yuan leng 圓棱

yue 鉞

Yueyang 岳陽

Yuhuangmiao 玉皇廟

"Yun Er" 慍兒

Yutaishan 雨台山

Zaoyang 棗陽

Zaozhuang 棗莊

Zeng 曾

zhan 盞

Zhang Changping 張昌平

Zhang Guangyuan 張光遠

Zhangjiakou 張家口

Zhangjiapo 張家坡

zhan-yu 盞盂

Zhaojiahu 趙家湖

Zhao Shigang 趙世綱

Zhao Xihu 趙希鵠

Zhengzhou 鄭州

zhi geng 直梗

Zhongshan 中山

Zhongtiao 中條

Zhou 周

Zhou Weirong 周衛榮

Zhuang Qiao 莊蹻

Zhuyuangou 竹園溝

Zilinggang 子陵崗

zisha 紫砂

zun 尊

zun-pan 尊盤

Bibliography

Agrawal, Dharma P. *Ancient Metal Technology and Archaeology of South Asia: A Pan-Asian Perspective.* New Delhi: Aryan Books International, 2000.

An Zhimin 安誌敏. "On Early Copper and Bronze Objects in Ancient China." In *The Beginnings of Metallurgy in China*, edited by Katheryn M. Linduff, Han Rubin, and Sun Shuyun, 29–46. Lewiston and Queenston: The Edwin Mellen Press, 2000. (Originally published in *Kaogu Xuebao* 考古學報 12 [1993]: 1110–1119).

An Zhimin 安誌敏. "Some Problems Concerning Early Chinese Copper and Bronze Artifacts." In *The Beginnings of Metallurgy in China*, edited by Katheryn M. Linduff, Han Rubin, and Sun Shuyun. Lewiston and Queenston: The Edwin Mellen Press, 2000, 63–86. (Originally published in *Kaogu Xuebao* 考古學報 3 [1981]: 269–285).

Anthony, David W. *The Horse, the Wheel, and Language: How Bronze-Age Riders from the Eurasian Steppes Shaped the Modern World.* Princeton, NJ: Princeton University Press, 2007.

———. "The Sintashta Genesis: The Roles of Climate Change, Warfare, and Long-distance Trade." In *Social Complexity in Prehistoric Eurasia: Monuments, Metals, and Mobility*, edited by Bryan K. Hanks and Katheryn M. Linduff, 47–73. Cambridge and New York: Cambridge University Press, 2009.

Bai Lifan 白黎璠. "Xia Shang Xizhou Jinqi Yanjiu 夏商西周金器研究." *Zhongyuan Wenwu* 中原文物5 (2006): 39–49, 93.

Bagley, Robert W. "Anyang Mold-making and the Decorated Model." *Artibus Asiae* 69 (2009): 39–90.

———. "Debris from the Houma Foundry." *Orientations* 27, no. 9 (October 1996): 50–58.

———. *Gombrich Among the Egyptians, and Other Essays in the History of Art.* Seattle: Marquand Books, Inc., 2015.

———. "Replication Techniques in Eastern Zhou Bronze Casting." In *History from Things: Essays on Material Culture*, edited by Steven Lubar and W. David Kingery, 231–241. Washington and London: Smithsonian Institution Press, 1993.

———. "Shang Archaeology." In *The Cambridge History of Ancient China*, edited by Michael Loewe and Edward L. Shaughnessy, 124–231. Cambridge and New York: Cambridge University Press, 1999.

———. *Shang Ritual Bronzes in the Arthur M. Sackler Collections.* Washington, D.C.: Arthur M. Sackler Foundation; Arthur M. Sackler Museum, Harvard University; distributed by Harvard University Press, Cambridge, MA, 1987.

———. "Shang Ritual Bronzes: Casting Technique and Vessel Design." *Archives of Asian Art* 43 (1990): 7–20.

———. "The Beginnings of the Bronze Age," and "The Zhengzhou Phase (The Erligang Period)." In *The Great Bronze Age of China: An Exhibition from the People's Republic of China*, edited by Fong Wen, 67–108. New York: Metropolitan Museum of Art: Knopf, 1980.

———. "What the Bronzes from Hunyuan Tell Us about the Foundry at Houma." *Orientations* 26, no. 1 (January 1995): 214–222.

Baines, John. "Civilizations and Empires: A Perspective on Erligang from Early Egypt." In *Art and Archaeology of the Erligang Civilization*, edited by Kyle Steinke and Dora C.Y. Ching, 99–120. Princeton, NJ: Princeton University Press, 2014.

Baoji, shi Kaogu Yanjiusuo 寶雞市考古研究所. *Qinmu Yizhen: Baoji Yimen Erhao Chunqiumu* 秦墓遺珍:寶雞益門二號春秋墓. Beijing: Kexue Chubanshe科學出版社, 2016.

Barnard, Noel. *Bronze Casting and Bronze Alloys in Ancient China.* Tokyo: Monumenta Serica Monograph XIV, 1961.

———. "The Entry of Cire-Perdue Investment Casting, and Certain Other Metallurgical Techniques (Mainly Metalworking) into South China and Their Progress Northwards." Paper presented at the Kiola Conference, N.S.W., Ancient Chinese and Southeast Asian Bronze Cultures, February 8–12, 1988 (pre-print in June 1987). Part 1, translated by Huang Jianhua 黃劍華 as "Shilafa he Qita Yejin Jishu zai Zhongguo de Chuanbo"失蠟法和其它金屬技術在中國的傳播. *Sichuan Wenwu* 四川文物 4 (1996): 70–76.

———. "The Entry of Cire-Perdue Investment Casting, and Certain Other Metallurgical Techniques (Mainly Metalworking) into South China and Their Progress Northwards." In *Ancient Chinese and Southeast Asian Bronze Age Cultures*, vol. 1, edited by Bulbeck, F. David, and Noel Barnard, 1–94. Taipei: SMC, 1996.

———. "Chinese Bronze Vessels with Copper Inlaid Decor and Pseudo-Copper Inlay of Ch'un-Ch'iu and Chan-Kuo Times - Part Two." In *Ancient Chinese and Southeast Asian Bronze Age Cultures*, vol. I, edited by Bulbeck, F. David, and Noel Barnard, 177–272. Taipei: SMC, 1996.

———. "Origin of Bronze Casting in Ancient China." In *Metallurgical Remains of Ancient China*, edited by Noel Barnard and Sato Tamotsu, 1–75. Tokyo: Nichiosha, 1975.

———. "Thoughts on the Emergence of Metallurgy in Pre-Shang and Early Shang China and a Technical Appraisal of Relevant Bronze Artifacts of the Time." Unpublished Paper presented at the Hsia Culture Conference, UCLA, May 23–25, 1990.

Bunker, Emma C. "The Beginning of Metallurgy in Ancient China." *Asian Art Articles*, Denver Art Museum,http://www.denverartmuseumasian.org/asianart/articles/metalwork/art_li_mat.html (Accessed January 5, 2015).

———. "First Millennium BCE Beifang Artifacts as Historical Documents." In *Social Complexity in Prehistoric Eurasia: Monuments, Metals, and Mobility*, edited by Bryan K. Hanks and Katheryn M. Linduff, 272–95. Cambridge and New York: Cambridge University Press, 2009.

———. "Lost Wax and Lost Textile: An Unusual Ancient Technique for Casting Gold Belt Plaques." In *The Beginning of the Use of Metals and Alloys: Papers from the Second International Conference on the Beginning of the Use of Metals and Alloys, Zhengzhou, China, 21-26 October 1986*, edited by R. Maddin, 222–225. Cambridge, MA: MIT Press, 1988.

———. *Nomadic Art of the Eastern Eurasian Steppes: The Eugene V. Thaw and other New York Collections*, with contributions by James C.Y. Watt and Zhixin Sun. New York: Metropolitan Museum of Art; New Haven, CT: Yale University Press, 2002.

Bunker, Emma C., Trudy S. Kawami, and Katheryn M. Linduff. *Ancient Bronzes of the Eastern Eurasian Steppes from the Arthur M. Sackler Collections*. New York: Arthur M. Sackler Foundation; distributed by Abrams, 1997.

Bunker, Emma C., and Douglas Latchford. *Khmer Bronzes: New Interpretations of the Past*. Chicago: Art Media Resources, 2011.

Bunker, Emma C., and Joseph Ternbach. "A Variation of the 'Lost-wax' Process." *Expedition* 12, no. 3 (1970): 41–43.

Bushell, Stephen W. *Chinese Art*, vol. 1. London: H. M. Stationery Office, 1904.

Cao, Dazhi. "The Loess Highland in a Trading Network (1300–1050 BC)." Ph.D. dissertation, Princeton University, 2014.

Chang, Kwang-chih. "The Chinese Bronze Age: A Modern Synthesis." In *The Great Bronze Age of China: An Exhibition from the People's Republic of China*, edited by Fong Wen, 35–50. New York: Metropolitan Museum of Art and Knopf, 1980.

Chase, W. Thomas. *Ancient Chinese Bronze Art: Casting the Precious Sacral Vessel*. New York City: China House Gallery: China Institute in America, 1991.

Chen, Peifen 陳佩芬. "Luelun Chu Jing de Zhuzao Jishu" 略論楚鏡的鑄造技術. In *Lianxing Shenye, Yingzhi Lianggong: Shanghai Bowuguan cang Tongjing Jingpin* 练形神冶, 瑩質良工: 上海博物館藏銅鏡精品, edited by Shanghai Bowuguan上海博物館, 30–35. Shanghai: Shanghai Shuhua Chubanshe, 2005.

———. *Xia Shang Zhou Qingtongqi Yanjiu: Shanghai Bowuguan Cangpin* 夏商周青銅器研究:上海博物館藏品. Shanghai: Shanghai Guji Chubanshe, 2004.

Chen, Hongliang 陳洪良, and Xu Guoping 許國平. "Xu Gong Ning Panhuiwen Toukong Shijian Fuzhi Yanjiu" 許公寧蟠虺紋透空飾件復

制研究. In *Qingtong Wenhua Yanjiu*. Vol. 6 青銅文化研究(第六輯), 298–305. Tongling: Huangshan Shushe, 2009.

Chen, Hongliang 陳洪良, Xu Guoping 許國平, Li Yuanzhi 李元芝, and Zhang Fangtao 張方濤. "Xu Gong Ning Panhuiwen Toukong Shijian Fuzhi Yanjiu" 許公寧墓蟠虺紋透空飾件複製研究. *Nanfang Wenwu* 南方文物 4 (2009): 46–50.

Chen, Mengjia 陳夢家. "Yindai Tongqi Sanpian" 殷代銅器三篇. *Kaogu Xuebao* 考古學報 7 (1954): 15–59.

Chen, Peifen. "Early Chinese Bronze Mirrors." *Orientations* (March 1997): 88–93. Reprinted in "*Chinese Bronzes: Selected Articles from Orientations 1983–2000*," 260–265. Hong Kong: Orientations Magazine Ltd., 2001.

Chernykh, EvgeniI Nikolaevich. "Ancient Metallurgy in the Eurasian Steppes and China: Problems of Interactions." In *Metallurgy and Civilisation: Eurasia and Beyond*, edited by Mei Jianjun and Thilo Rehren, 3–8. London: Archetype, 2009.

———. *Ancient Metallurgy in the USSR: The Early Metal Age*. New York: Cambridge University Press, 1992.

———. "Formation of the Eurasian Steppe Belt Cultures: Viewed through the Lens of Archaeometallurgy and Radiocarbon Dating." In *Social Complexity in Prehistoric Eurasia*, edited by Bryan Hanks and Katheryn Linduff, 115–145. Cambridge and New York: Cambridge University Press, 2009.

Childe, V. Gordon. "Archaeological Ages as Technological Stages." *The Journal of the Royal Anthropological Institute of Great Britain and Ireland* 74, no. 1/2 (1944): 7–24.

Chiou-Peng, Tzehuey. "Incipient Metallurgy in Yunnan: New Data for Old Debates." In *Metallurgy and Civilisation: Eurasia and Beyond*, edited by Mei Jianjun and Thilo Rehren, 79–84. London: Archetype, 2009.

———. "The Technical History of Early Asian Kettledrums." In *Khmer Bronzes: New Interpretations of the Past*, edited by Emma C. Bunker and Douglas Latchford, 17–26. Chicago: Art Media Resources, 2011.

Collins, William F. "The Corrosion of Early Chinese Bronzes." *Journal of the Institute of Metals* 45 (1931): 23–55.

Creel, Herrlee G. "On the Origins of the Manufacture and Decoration of Bronze in the Shang Period." *Monumenta Serica* 1, no. 1 (1935): 39–69.

Daniel, Glyn Edmund. *The Origins and Growth of Archaeology.* Baltimore: Penguin Books, 1967.

Davey, Christopher J. "The Early History of Lost-wax Casting." In *Metallurgy and Civilisation: Eurasia and Beyond*, edited by Mei Jianjun and Thilo Rehren, 147–154. London: Archetype, 2009.

Davidson, J. Leroy. "Toward a Grouping of Early Chinese Bronzes." *Parnassus* 9, no. (1937): 29–34, 51.

Doeringer, Suzannah, David Gordon Mitten, and Arthur Steinberg, eds. *Art and Technology: A Symposium on Classical Bronzes.* Cambridge, MA: MIT Press, 1970.

Dong, Yawei 董亞巍. *Fanzhu Qingtong* 範鑄青銅. Beijing: Beijing Yishu yu Kexue Dianzi Chubanshe, 2007.

Dong, Yawei 董亞巍, Pan Lu 潘路, Wan Quanwen 萬全文, Zhou Weirong 周衛榮, Ma Juncai 馬俊才, and Wang Changsui 王昌燧. "Zaitan Zenghou Yi Zun-Pan de Zhuzao gongyi" 再探曾侯乙尊盤的鑄造工藝. *Zhongyuan Wenwu* 中原文物1 (2008): 97–107.

Económica, Fondo de Cultura. *The Art of Gold, the Legacy of Pre-Hispanic Colombia: Collection of the Gold Museum in Bogotá.* México, D.F.: Fondo de Cultura Económica; Bogotá: Banco de la República; Milano: Skira, 2007.

Fairbank, Wilma. "Piece-mold Craftsmanship and Shang Bronze Design." *Archives of the Chinese Art Society of America* 16 (1962): 18–32.

———. "Review of Noel Barnard's 'Bronze Casting and Bronze Alloys in Ancient China.'" *Technology and Culture* 3, no. 2 (1962): 178–180.

Falkenhausen, Lothar von. *Chinese Society in the Age of Confucius (1000–250 BC): The Archaeological Evidence.* Los Angeles: Cotsen Institute of Archaeology, UCLA, 2006.

———. *Suspended Music: Chime-Bells in the Culture of Bronze Age China.* Berkeley and Los Angeles: University of California Press, 1993.

Fan, Taofeng 範陶峰, and Wan Li 萬俐. “Chen Zhang Hu Zhizuo Jishu de Chubu Tantao”陳璋壺製作技術的初步探討. *Zhuzao* 鑄造 61, no. 10 (2012): 1163–1167.

Fitzgerald-Huber, Louisa G. “Qijia and Erlitou: The Question of Contact with Distant Cultures.” *Early China* 20 (1995): 17–67.

———. “The Relationship of the Painted Pottery and Lung-shan Cultures.” In *The Origins of Chinese Civilization*, edited by David N. Keightley, 177–216. Berkeley: University of California Press, 1983.

Fong, Wen, ed. *The Great Bronze Age of China: An Exhibition from the People's Republic of China.* New York: Metropolitan Museum of Art and Knopf, 1980.

Franklin, Ursula M. “The Beginning of Metallurgy in China: A Comparative Approach.” In *The Great Bronze Age of China: A Symposium*, edited by George Kuwayama, 94–99. Los Angeles, CA: Los Angeles County Museum of Art; Seattle: distributed by University of Washington Press, 1983.

———. *The Real World of Technology.* Concord, ON: House of Anansi Press Limited, 1992.

Gao, Ming. *Gu Wenzi Lei Bian.* 古文字類編. Beijing: Zhonghua Shuju, 1980.

Gettens, Rutherford J. *The Freer Chinese Bronzes*, vol II: Technical Studies. Washington: Smithsonian Institution, 1969.

———. “Joining Methods in the Fabrication of Ancient Chinese Bronze Ceremonial Vessels.” In *Application of Science in Examination of Works of Art,* 205–217. Boston: Museum of Fine Arts, 1967,

Guo, Baojun 郭寶鈞. *Ruixian Yiqi* 濬縣彝器. Edited by Sun Haibo 孫海波. Henan Tongzhiguan 河南通誌館, 1937.

———. *Shangzhou Tongqiqun Zonghe Yanjiu* 商周銅器群綜合研究. Beijing: Wenwu Chubanshe, 1981.

Han, Rubin, and Li Xiaocen. "The Bronze-using Cultures in the Northern Frontier of Ancient China and the Metallurgies of Ancient Dian Area in Yunnan Province." In *Social Complexity in Prehistoric Eurasia: Monuments, Metals, and Mobility*, edited by Bryan K. Hanks and Katheryn M. Linduff, 168–186. Cambridge and New York: Cambridge University Press, 2009.

He, Hao 何浩. "Chunqiu shi Chu Mieguo Xintan" 春秋時楚滅國新探. *Jianghan Luntan* 江漢論壇 4 (1982): 55–63.

Hemingway, Seán. "Bronze Sculpture." In *Making Classical Art: Process & Practice*, edited by Roger Ling, 37–46. Stroud: Tempus; Charleston, SC: Arcadia Pub, 2000.

Henan Sheng Wenwu Kaogu Yanjiusuo 河南省文物考古研究所, ed. *Gushi Hougudui Yihao Mu* 固始侯古堆一號墓. Zhengzhou: Daxiang Chubanshe, 2004.

Henan Sheng Wenwu Yanjiusuo 河南省文物研究所, Henan Sheng Danjiang Kuqu Kaogu Fajuedui 河南省丹江庫區考古發掘隊, and Xichuan Xian Bowuguan 淅川縣博物館. *Xichuan Xiasi Chunqiu Chumu* 淅川下寺春秋楚墓. Beijing: Wenwu Chubanshe, 1991.

Henan Bowuyuan 河南博物院, and Taibei Guoli Lishi Bowuguan 臺北國立歷史博物館. *Xinzheng Zhenggong Damu Qingtongqi* 新鄭鄭公大墓青銅器. Zhengzhou: Daxiang Chubanshe, 2001.

Higham, Charles. "Prehistoric Metallurgy in Southeast Asia: Some New Information from the Excavation of Ban Na Di." In *The Beginning of the Use of Metals and Alloys: Papers from the Second International Conference on the Beginning of the Use of Metals and Alloys, Zhengzhou, China, 21–26 October 1986*, edited by R. Maddin, 130–155. Cambridge, MA: MIT Press.

———. *The Bronze Age of Southeast Asia*. Cambridge: Cambridge University Press, 1996.

Higham, Charles, and Amphan Kijngam. "Prehistoric Investigations in Northeastern Thailand." *BAR International Series* 231. Oxford: British Archaeological Reports, 1984.

Higham, Charles, and Thomas Higham. "A New Chronological Framework for Prehistoric Southeast Asia, based on a Bayesian Model from Ban Non Wat." *Antiquity* 83 (2009): 125–44.

Hu, Gang 胡鋼, Liu Baige 劉百舸, Zhang Xia 張夏, and Wang Lihua 王麗華. "Zaozhuang Shi Xulou Cun Hongtong Zhuxiang Qingtongqi Wenshi Xiangzhu Tezheng" 棗莊市徐樓村紅銅鑄鑲青銅器紋飾鑲鑄特徵. *Wenwu Baohu yu Kaogu Kexue* 文物保護與考古科學, 1 (2017): 51–56.

Hua, Jueming 華覺明. "Niuyue Daduhui Bowuguan Suocang Shangzhou Qingtongqi de Jishu Kaocha 紐約大都會博物館所藏商周青銅器的技術考察." In *Qingtong Wenhua Yanjiu*,Vol. 3 青銅文化研究(第三輯). Tongling: Huangshan Shushe, 2003, 53–56.

———. "Shilafa de Qiyuan he Fazhan" 失蠟法的起源和發展. In *Kejishi Wenji: Jinshushi Zhuanji* 科技史文集: 金屬史專輯. Shanghai: Shanghai Keji Chubanshe, 1985, 63–81.

———. *Zhongguo Gudai Jinshu Jishu: Tong he Tie Zaojiu de Wenming* 中國古代金屬技術: 銅和鐵造就的文明. Zhengzhou: Daxiang Chubanshe, 1999.

———. *Zhongguo Yezhu shi Lunji* 中國冶鑄史論集. Beijing: Wenwu Chubanshe, 1986.

———. "Zhongxifang Shilafa zhi Tongyi: Jianping 'Xianqin bu Cunzai Shilafa' yi Shuo" 中西方失蠟法之同異——兼評'先秦不存在失蠟法'一說. *Kaogu* 考古4 (2010): 87–96.

Hua, Jueming 華覺明, and Jia Yunfu 賈雲福. "Zenghou Yu zun-pan he shilafa de qiyuan" 曾侯遇尊盤和失蠟法的起源. Ziran Kexucshi Yanjiu 自然科學史研究4 (1983): 352–359.

Hua, Jueming 華覺明, and Xiao Huifang 蕭蕙芳. "Chu Wang Yu: Yizhi Zhongguo Zuizao de Shila Zhujian, Jianlun Shimofa Jiqi Gongyi Yanbian" 楚王盂——已知中國最早的失蠟鑄件, 兼論失模法及其工藝演變. In *Qingtong Wenhua Yanjiu*, Vol. 4 青銅文化研究(第四輯). Tongling: Huangshan Shushe, 2005, 80–89.

Hua, Lin 華林. "Tongjing Zonghe Yanjiu 銅鏡綜合研究." Research published online in 2014: https://tongjing.jimdo.com/%E9%93%9C%E9%95%9C%E7%BB%BC%E5%90%88%E7%A0%94%E7%A9%B6/ (Accessed March 6, 2018).

———. "'Toukong Jiaolongwen Jing' yanjiu" '透空交龍紋鏡'研究. Research Series published online from 2013 to 2015: https://tongjing.jimdo.com/%E5%9B%9B%E5%8F%B6%E7%BA%B9%E9%95%9C/ (Accessed March 6, 2018).

Huang, Cuimei 黃翠梅, and Li Jianwei 李建緯. "Jinyu Tongmeng: Dongzhou Jinqi he Yuqi zhi Zhuangshi Fengge yu Juese Yanbian" 金玉同盟——東周金器和玉器之裝飾風格與角色演變. *Zhongyuan Wenwu* 中原文物 1 (2007): 42–58.

Huang, Jinzhou 黃金洲. "Zenghou Yi Zun-Pan Caiyong Shilafa Gongyi Zhuzao Wuyongzhiyi: yu 'Zhongguo Qingtong Shidai bu Cunzai Shilafa Zhuzao Gongyi' Taolun" 曾侯乙尊盤采用失蠟法工藝鑄造毋庸置疑——與《中國青銅時代不存在失蠟法鑄造工藝》討論. *Jianghan Kaogu* 江漢考古 4 (2008): 109–117, 101.

Huang, Shengzhang 黃盛璋. "Xuyi Xinchu Tongqi, Jinqi ji xiangguang Wenti Kaobian" 盱眙新出銅器, 金器及相關問題考辨. *Wenwu* 文物 10 (1984): 59–64.

Huang, Wei 黃維, Wu Xiaohong 吳小紅, Chen Jianli 陳建立, and Wang Hui 王輝. "Zhangjiachuan Majiayuan Mudi Chutu Jin Guanshi de Yanjiu" 張家川馬家塬墓地出土金管飾的研究. *Wenwu* 文物 10 (2009): 78–85.

Huang, Wei 黃維, Chen Jianli 陳建立, Wu Xiaohong 吳小紅, Wang Hui 王輝, and Zhou Guangji 周廣濟. "Majiayuan Mudi Jinzhipin de Chengfen yu Zhizuo Jishu Chubu Yanjiu" 馬家塬墓地金製品的成分與製作技術初步研究. *Bianjiang Kaogu Yanjiu* 邊疆考古研究, Vol. 11 (2012): 405–422.

Huang, Fengchun 黃鳳春, Hu Gang 胡剛, and Guo Changjiang 郭長江. "Hubei Suizhou shi Wenfengta Dongzhou Mudi" 湖北隨州市文峰塔東周墓地. *Kaogu* 考古 7 (2014): 18–33.

Hubei Sheng Bowuguan 湖北省博物館. *Zeng hou Yi Mu* 曾侯乙墓. Beijing: Wenwu Chubanshe, 1989.

Hubei Sheng Yichang Diqu Bowuguan 湖北省宜昌地區博物館, and Beijing Daxue Kaoguxi 北京大學考古系. *Dangyang Zhaojiahu Chumu* 當陽趙家湖楚墓. Beijing: Wenwu Chubanshe, 1992.

Hughes, Thomas P. *Networks of Power: Electrification in Western Society, 1880–1930.* Baltimore: The Johns Hopkins University Press, 1983.

Hwang, Ming-chorng 黃銘崇. "Maixiang Zhongqi Shidai: Zhutong Jishu de Shuruyu Zhongguo Qingtong Jishu de Xingcheng" 邁向重器時代——鑄銅技術的輸入與中國青銅技術的形成. *Zhongyang Yanjiuyuan Lishi Yuyan Yanjiusuo Jikan* 中央研究院歷史語言研究所集刊 85, no. 4 (2014): 557–678.

Ivanova, Mariya. *The Black Sea and the Early Civilizations of Europe, the Near East and Asia.* New York: Cambridge University Press, 2013.

Jia, Yunfu 賈雲福, and Hu Caibin 胡才彬. "Dui Gudai Qingtongqi Hongtong Qianxiang de Yanjiu" 對古代青銅器紅銅嵌鑲的研究. *Wuhan Hongxueyuan Xuebao* 武漢工學院學報 1 (1980): 27–34.

Jia, Yunfu 賈雲福, Hu Caibin 胡才彬, and Hua Jueming 華覺明. "Zeng Hou Yi Qingtongqi Hongtong Huawen Zhuxiangfa de Yanjiu" 曾侯乙青銅器紅銅花紋鑄鑲法的研究. In *Kejishi Wenji* 科技史文集, Vol. 13. Shanghai: Shanghai Kexue Jishu Chubanshe, 1985, 82–87.

Jiang, Suhua 蔣素華. "Zhongguo Gudai Qingtongqi de Guibao: Zhong Jin Luo Hu" 中國古代青銅器的瑰寶——重金絡壺. *Tezhong Zhuzao ji Youse Hejin* 特種鑄造及有色合金2 (2000): 57–8.

Jiangxisheng Wenwu Kaogu Yanjiusuo 江西省文物考古研究所, Jiangxisheng Bowuguan 江西省博物館, *Xin'gan xian Bowuguan* 新幹縣博物館. *Xin'gan Shangdai Da Mu* 新幹商代大墓. Beijing: Wenwu Chubanshe, 1997.

Jing, Hongwei 景宏偉, and Wang Zhouying 王周應 "Fengxiang Faxian Zhanguo Fengniao Xianhuan Tong Xunlu" 鳳翔發現戰國鳳鳥銜環銅熏爐. *Wenbo* 文博1 (1996): 57.

Jingmen shi Bowuguan 荊門市博物館. *Jingmen Zilinggang* 荊門子陵崗. Beijing: Wenwu Chubanshe, 2008.

Karlbeck, Orvar. "Anyang Moulds." *Bulletin of the Museum of Far Eastern Antiquities* 7 (1935): 39–60.

Keyser, Barbara W. "Decor Replication in Two Late Chou Bronze Chien," *Ars Orientalis* 11 (1979): 127–62.

Knauth, Percy. *The Metalsmiths.* New York: Time-Life Books, 1974.

Kunming, Shi Wenwu Guanli Weiyuanhui 昆明市文物管理委員會. "Chenggong Tianzimiao Dian Mu" 呈貢天子廟滇墓. *Kaogu Xuebao* 考古學報 4 (1985): 507–544.

Kuz'mina, Elena E. *The Origin of the Indo-Iranians.* Leiden, The Netherlands; Boston: Brill, 2007.

Lawton, Thomas. *Chinese Art of the Warring States Period: Change and Continuity, 480–222 B.C.* Washington, D.C.: Published for the Freer Gallery of Art by the Smithsonian Institution Press, 1982.

Le Bas, Anthony, Noel Kennon, Ross Smith, and Noel Barnard. "Chinese Bronze Vessels with Copper Inlaid Decor and Pseudo-Copper Inlay of Ch'un-Ch'iu and Chan-Kuo Times–Part One." In *Ancient Chinese and Southeast Asian Bronze Age Cultures,* vol. 1, edited by Bulbeck, F. David and Noel Barnard, 123-176. Taipei: SMC, 1996.

Ledderose, Lothar. *Ten Thousand Things: Module and Mass Production in Chinese Art.* Princeton, NJ: Princeton University Press, 2000.

Li, Chaoyuan 李朝遠. "Xinhuo Zhanguo Toukong Fuhejing Yanjiu" 新獲戰國透空復合鏡研究. In *Lianxing Shenye, Yingzhi Lianggong: Shanghai Bowuguan cang Tongjing Jingpin* 練形神冶, 瑩質良工: 上海博物館藏銅鏡精品, edited by Shanghai Bowuguan 上海博物館, 36–53. Shanghai: Shanghai Shuhua Chubanshe, 2005.

Lee, Chienwei 李建緯. "Chengqi zhi Dao—Zhongguo Xianqin zhi Handai dui Huangjin de Renshi yu Gongyi Jishu Yanjiu"成器之道－中國先秦至漢代對黃金的認識與工藝技術研究. *Meishu Xuebao* 美術學報 (Journal of Fine Arts), edited by Taipei National University of the Arts, no. 4 (2011): 1–52.

Li, Ji 李濟. "Fushen Zang" 俯身葬. *Anyang Fajue Baogao* 安陽發掘報告 3 (1931): 447–480.

———. "Yinxu Chutu Qingtong Liqi zhi Zong Jiantao" 殷墟出土青銅禮器之總檢討. In Li Ji李濟, *Yinxu Qingtongqi Yanjiu* 殷墟青銅器研究. Shanghai: Shanghai Shiji Chuban Jituan, 2008, 521–570. (Originally published in *Lishi Yuyan Yanjiusuo Jikan* 歷史語言研究所集刊, vol. 47, 1976).

Li, Jinghua 李京華. "Gushi Hougudui Qingtong Zhuxiang Hongtong Huawen Gongyi Tantao" 固始侯古堆青銅鑄鑲紅銅花紋工藝探討. In *Gushi Hougudui Yihao Mu* 固始侯古堆一號墓, edited by Henansheng Wenwu Kaogu Yanjiusuo 河南省文物考古研究所, 134–146. Zhengzhou: Daxiang Chubanshe.

———. "Xichuan Chunqiu Chumu Tongjin Shila Zhuzaofa de Gongyi Tantao" 淅川春秋楚墓銅禁失蠟鑄造法的工藝探討. *Wenwu Baohu yu Kaogu Kexue*文物保護與考古科學 6, no. 1 (1994): 39–45.

———. "Xichuan Chunqiu Chumu Tongjin Shila Zhuzaofa de Gongyi Tantao" 淅川春秋楚墓銅禁失蠟鑄造法的工藝探討. In *Li Jinghua* 李京華, *Zhongyuan Gudai Yejin Jishu Yanjiu* 中原古代冶金技術研究. Zhengzhou: Zhongzhou Guji Chubanshe, 1994, 48–52.

Li, Kin Sum. "Mirrors from 500–200 BC Middle Yangzi Region: Design and Manufacture." Ph.D. dissertation, Princeton University, 2015.

Li, Xiaocen 李曉岑, and Han Rubin 韓汝玢. *Gu Dianguo Jinshu Jishu Yanjiu* 古滇國金屬技術研究. Beijing: Kexue Chubanshe, 2011.

Li, Xueqin. "Chu Bronzes and Chu Culture." In *New Perspectives on Chu Culture during the Eastern Zhou Period*, edited by Thomas Lawton, 1–47. Washington, D.C.: Arthur M. Sackler Gallery, Smithsonian Institution; Princeton, NJ: distributed by Princeton University Press, 1991.

Li, Xueqin 李學勤. "Guyuege suo cang Qingtong Bingqi Xuancui" 古越閣所藏青銅兵器選萃. *Wenwu* 文物 4 (1993): 18–28.

———. "Rongsheng Bianzhong Lunsh" 戎生編鐘論釋. *Wenwu* 文物 9 (1999): 75–82.

———. *Sihai Xunzhen* 四海尋珍. Beijing: Qinghua Daxue Chubanshe, 1998.

———. "Yimencun Jin, Yu Qi Wenshi Yanjiu" 益門村金、玉器紋飾研究. *Wenwu* 文物 10 (1993): 15–19.

Li, Xueqin 李學勤, and Zhu Minshen 祝敏申. "Xuyi Hu Ming yu Qi Po Yan Niandai" 盱眙壺銘與齊破燕年代. *Wenwu Chunqiu* 文物春秋 1 (1989): 3–17.

Li, Yuanzhi 李元芝, Zhang Fangtao 張方濤, Chen Hongliang 陳洪良, Xu Guoping 許國平, Hua Jueming 華覺明, and Tan Derui 譚德睿. "Zhongguo Zaoqi Shila Zhujian—Xu Gong Ning Panhui Wen Toukong Shijian de Fuzhi Yanjiu" 中國早期失蠟鑄件——許公寧蟠虺紋透空飾件的複制研究. *Tezhong Zhuzao ji Youse Hejin* 特種鑄造及有色合金 2 (2010): 192–194.

Li, Yuanzhi 李元芝, Zhang Fangtao 張方濤, Tan Derui 譚德睿, Hua Jueming 華覺明, and Zhang Mingwu 張明悟. "Xu Gong Ning Toukong Panhui Wen Qingtong Shijian—Xian Qin Shilafa zhi yi Qili 許公寧透空蟠虺紋青銅飾件—先秦失蠟法之一器例." *Zhongyuan Wenwu* 中原文物1 (2007): 96–103.

Li, Yung-ti. "The Anyang Bronze Foundries: Archaeological Remains, Casting Technology, and Production Organization." Ph.D. dissertation, Harvard University, 2003.

Li, Zhiwei 李誌偉. "Gudai Qingtong Zhuzao Fangfa de Zai Tantao" 古代青銅鑄造方法的再探討. *Jianghan Kaogu* 江漢考古 4 (1986): 94–100.

———. "Youguan Zeng Hou Yi Zun Pan Zhuzao Fangfa de Zhengming—Lun Zhongguo Qingtong Shidai de Rongmo Jishu" 有關曾侯乙尊盤鑄造方法的證明—論中國青銅時代的熔模技術. *Nanfang Wenwu* 南方文物 2 (2008): 39–46.

———. "Zeng Hou Yi Bianzhong Zhuzao Fangfa zhi Wojian—Jian Lun Xianqin Zhuzao Gongyi" 曾侯乙編鐘鑄造方法之我見—兼論先秦青銅鑄造工藝. *Gugong Xueshu Jikan* 故宮學術季刊 20, no. 2 (2002): 109–27, 208.

———. "Zeng Hou Yi Bianzhong ji Zun, Zunzuo Zhuzao Fangfa Xintan—Jian Lun Xianqin Qingtong Zhuzao Gongyi" 曾侯乙墓編鐘及尊、尊座鑄造方法新探—兼論先秦青銅鑄造工藝. In *Chushi Luncong* 楚史論叢. Wuhan: Hubei Renmin Chubanshe, 1984, 73–92.

Lin, Meicun 林梅村. "Saiyima-Tuerbinnuo Wenhua yu Shiqian Sichou zhi Lu" 塞伊瑪-圖爾賓諾文化與史前絲綢之路. *Wenwu* 文物 10 (2015): 49–63.

Lin, Yun 林沄. "A Reexamination of the Relationship Between Bronzes of the Shang Culture and of the Northern Zone." In *Studies in Shang Archaeology*, edited by K.C. Chang, 237–273. New Haven, CT: Yale University Press, 1986.

Linduff, Katheryn M. "Production of Signature Artifacts for the Nomad Market in the State of Qin during the Late Warring States Period in China (4th–3rd Century BCE)." In *Metallurgy and Civilisation: Eurasia and Beyond*, edited by Mei Jianjun and Thilo Rehren, 90–96. London: Archetype, 2009.

Liu, Li, and Xingcan Chen. *The Archaeology of China: From the Late Paleolithic to the Early Bronze Age*. New York: Cambridge University Press, 2012.

Liu, Wei 劉煒, and He Hong 何洪. *Chunqiu Zhanguo: Zhengba Tuqiang de Shidai* 春秋戰國—爭霸圖強的時代 (Chinese Civilization in a New Light, vol.3), Hong Kong: Commercial Press, 2001.

Liu, Yu 劉煜, Zhang Changping 張昌平, Hu Dongbo 胡東波, and Yue Zhanwei 岳佔偉. "Jishu Xuanze he Jishu Fengge de Xingcheng: Yi Ding Weili Kaocha Erlitou Shiqi dao Wanshang Qingtongqi de Jishu Yanjin" 技術選擇和技術風格的形成: 以鼎為例考察二裏頭時期到晚商青銅器的技術演進. In *Jinyu Jiaohui: Shangzhou Kaogu, Yishu yu Wenhua Lunwenji* 金玉交輝: 商周考古、藝術與文化論文集, edited by Chen Guangzu陳光祖, 191–218. Taibei: Zhongyang Yanjiuyuan Lishi Yuyan Yanjiusuo, 2013.

Liu, Zhao 劉釗. *Chutu Jianbo Wenzi Congkao* 出土簡帛文字叢考. Taibei: Taiwan Guji Chuban Youxian Gongsi, 2004.

Loehr, Max. "Weapons and Tools from Anyang, and Siberian Analogies." *American Journal of Archaeology* 53 (1949): 126–144.

Loveday, Helen. "Diversity in Eastern Zhou Bronze Casting: A Look at a Group of Openwork Vessels." *Journal of East Asian Archaeology* 4, no. 1–4 (2002): 101–142.

Luo, Feng 羅豐. "Zhongyuan Zhizao—Guanyu Beifang Dongwu Wen Jinshu Paishi" 中原製造——關於北方動物紋金屬牌飾. *Wenwu* 文物 3 (2010): 56–63, 96.

Luo, Wugan, Tao Li, Changsui Wang, and Fengchun Huang. "Discovery of Beeswax as Binding Agent on a 6th-century BC Chinese Turquoise-inlaid Bronze Sword." *Journal of Archaeological Science* 39, no. 5 (2012): 1227–1237.

Luo, Xunzhang 羅勛章. "Shandong Yishui Liujiadianzi Chunqiu Mu Fajue Jianbao" 山東沂水劉家店子春秋墓發掘簡報. *Wenwu* 文物 9 (1984), 1–10.

Mackenzie, Colin. "Chu Bronze Work: a Unilinear Tradition or a Synthesis of Diverse Sources?" In *New Perspectives on Chu Culture during the Eastern Zhou Period*, edited by Thomas Lawton, 107–157. Washington, D.C.: Arthur M. Sackler Gallery, Smithsonian Institution; Princeton, N.J.: Distributed by Princeton University Press, 1991.

———. "The Chu Tradition of Wood Carving." In *Style in the East Asian Tradition*, edited by Rosemary E. Scott and Graham Hutt, 82–102. London: University of London, School of Oriental and African Studies, Percival David Foundation of Chinese Art, 1987.

Maryon, Herbert, and Harold Plenderleith. "Fine Metal-work." In *A History of Technology*, edited by Charles Singer et al., 623–662. Oxford: Oxford University Press, 1954.

Masson, Vadim M. *Altyn-depe: Raskopki goroda bronzovogo veka v Yuzhnom Turkmenistane.* Leningrad: "Nauka," Leningradskoe otd-nie, 1981.

Mattusch, Carol C. *Classical Bronzes: The Art and Craft of Greek and Roman Statuary.* Ithaca, NY: Cornell University Press, 1996.

———. *Enduring Bronze: Ancient Art, Modern Views.* Los Angeles: J. Paul Getty Museum, 2014.

———. *Greek Bronze Statuary: From the Beginnings through the Fifth Century B.C.* Ithaca, NY: Cornell University Press, 1988.

McGovern, Patrick E. et al. "Fermented Beverages of Pre- and Proto-Historic China." *Proceedings of the National Academy of Sciences (PNAS)* 101, no. 51 (2004): 17593–17598.

Mei, Jianjun 梅建軍. "Qijia and Seima-Turbino: The Question of Early Contacts between Northwest China and the Eurasian Steppe." *Bulletin of the Museum of Far Eastern Antiquities* 75 (2003): 31–54.

Mei, Jianjun, Pu Wang, Kunlong Chen, Lu Wang, Yingchen Wang, and Yaxiong Liu. "Archaeometallurgical Studies in China: Some Recent Developments and Challenging Issues." *Journal of Archaeological Science* 56 (2015): 221–232.

Meyers, Pieter. "Applications of X-Ray Radiography in the Study of Archaeological Objects." In *Archaeological Chemistry-II*, edited by Giles F Carter, 79-96. Washington, D.C.: American Chemical Society, 1978.

Meyers, Pieter. "Casting Technology in Cambodia and Related Southeast Asian Civilizations." In *Khmer Bronzes: New Interpretations of the Past*, edited by Emma C. Bunker and Douglas Latchford, 27–41. Chicago: Art Media Resources 2011.

Mou, Yongkang 牟永抗. "Shaoxing 306 hao Zhanguo mu Fajue Jianbao" 紹興 306 號戰國墓發掘簡報. *Wenwu* 文物 1 (1984): 10–6.

Murowchick, Robert E. "The Development of Early Bronze Metallurgy in Vietnam and Kampuchea: A Reexamination of Recent Work." In *The Beginning of the Use of Metals and Alloys: Papers from the Second International Conference on the Beginning of the Use of Metals and Alloys, Zhengzhou, China*, 21-26 October 1986, edited by R. Maddin, 182–199. Cambridge, MA: MIT Press, 1988.

Namba, Junko 難波純子. *Chugoku ōchō no sui* 中國王朝の粋. Osaka: Osaka Yoshi Karabu 大阪美術倶楽部, 2004.

Nanjing Bowuyuan 南京博物院. "1959 nian dong Xuzhou diqu kaogu diaocha 1959 年冬徐州地區考古調查." *Kaogu* 考古 3 (1960): 25–29.

———. “Jiangsu Lianshui Sanlidun Xihan mu江蘇漣水三裏墩西漢墓.” *Kaogu*考古2 (1973): 80–87, 89.

Nanyang shi Wenwu Kaogu Yanjiusuo 南陽市文物考古研究所. “Henan Nanyang Chunqiu Chu Pengshe Mu Fajue Jianbao” 河南南陽春秋楚彭射墓發掘簡報. *Wenwu* 文物3 (2011): 4–31.

Nanyang Municipal Institute of Cultural Relics and Archaeology. “Peng She’s Tomb of the Chu State of the Spring-and-Autumn Period in Nanyang, Henan,” translated by Paul Nicholas Vogt. *Chinese Archaeology* 12 (2012): 48–60.

Nickel, Lucas. “The First Emperor and Sculpture in China.” *Bulletin of SOAS* 76, no. 3 (2013): 413–447.

———. “Imperfect Symmetry: Re-thinking Bronze Casting Technology in Ancient China.” *Artibus Asiae* 66, no. 1 (2006): 5–39.

Notis, Michael R., and Dongning Wang. “Ancient Chinese Bronze Casting Methods: The Dilemma of Choice.” *MRS (Materials Research Society) Advances* 2, no. 33–34 (2017): 1743–1768.

Odagi, Harutaro 小田木治太郎. “Hoppō-keichōhōkei obikazariban no tenkai: Shīan hokkō shin-bo shutsudo igata no bunseki kara”北方系長方形帯飾板の展開-西安北郊秦墓出土鋳型の分析から. *Chūgoku kōkogaku* 中國考古學 5 (2005): 79–94.

Peng, Peng. “The Lost-wax Casting in Bronze Age China: Art, Technology, and Social Agency.” Ph.D. dissertation, Princeton University, 2018.

———. “Was Lost-wax Casting Adopted in Pre-Qin Chinese Bronze Art?” *Orientations* January/February 2018: 103–111.

Piggott, Stuart. Wagon*, Chariot, and Carriage: Symbol and Status in the History of Transport.* London: Thames and Hudson, 1992.

Pirazzoli, Michèle (皮拉左裏, 米歇爾). “Dian Wenhua de Niandai Wenti 滇文化的年代問題.” *Kaogu* 考古 1 (1990): 78–86.

Pittman, Holly. *Art of the Bronze Age: Southeastern Iran, Western Central Asia, and the Indus Valley.* New York: Metropolitan Museum of Art, 1984.

Pope, John, Rutherford Gettens, James Cahill, and Noel Barnard, eds. *The Freer Chinese Bronzes.* Washington, D.C.: Smithsonian Institution, 1967–69.

Possehl, Gregory L. *The Indus Civilization: A Contemporary Perspective.* Lanham, MD: Altamira Press, 2002.

———. "The Middle Asian Interaction Sphere." *Expedition* 49, no. 1 (2007): 40–42.

Qiao, Baotong 喬保同. "Zuixin Kaogu Faxian: Nanyang Chu Peng Shi Jiazu Mu" 最新考古發現: 南陽楚彭氏家族墓. *Wenwu yu Kaogu* 文物與考古6 (2009): 148–49.

Qiu, Shihua 仇士華et al. "Youguan Suowei 'Xia Wenhua' de Tanshisi Niandai Ceding de Chubu Baogao" 有關所謂'夏文化'的碳十四年代測定的初步報告. *Kaogu* 考古10 (1983): 923–928.

Qiu, Zihui 邱兹惠. "Yunnan Qingtong Wenhua de Qima Wenyang" 雲南青銅文化的騎馬紋樣. In *Xingbie Yanjiu yu Zhongguo Kaoguxue* 性別研究與中國考古學, edited by Katheryn M. Linduff and Sun Yan, 235–250. Beijing: Kexue Chubanshe, 2006.

Rao, P. N. *Zhizao Jishu: Zhuzao, Chengxing he Hanjie* 制造技術: 鑄造、成形和焊接. Published in China following the author's book *Manufacturing Technology (Volume 1): Foundry, Forming and Welding.* Beijing: Jixie Gongye Chubanshe, 2010.

Rawson, Jessica. "A Bronze-Casting Revolution in the Western Zhou and Its Impact on Provincial Industries." In *The Beginning of the Use of Metals and Alloys: Papers from the Second International Conference on the Beginning of the Use of Metals and Alloys, Zhengzhou, China,* 21-26 October 1986, edited by R. Maddin, 228–238. Cambridge, MA: MIT Press.

———. "Carnelian Beads, Animal Figures and Exotic Vessels: Traces of Contact between the Chinese States and Inner Asia, c. 1000–650 BC." In *Archäologie in China,* vol. 1, *Bridging Eurasia,* 1–42. Berlin: Deutsches Archäeologisches Institut, 2010.

———. *Chinese Bronzes: Art and Ritual.* London: published for the Trustees of the British Museum in association with the Sainsbury

Centre for Visual Arts, University of East Anglia by British Museum Publications, 1987.

———. *Western Zhou Ritual Bronzes from the Arthur M. Sackler Collections.* Washington, D.C.: Arthur M. Sackler Foundation; Cambridge, MA: Arthur M. Sackler Museum, Harvard University; Distributed by Harvard University Press, 1990.

———. "Western Zhou Sources of Interlaced Motifs." In *Style in the East Asian Tradition*, edited by Rosemary E. Scott and Graham Hutt, 38–64. London: University of London, School of Oriental and African Studies, Percival David Foundation of Chinese Art, 1987.

Ren, Changzhong 任常中, and Wang Changqing 王長青. "Henan Xichuan Xiasi Chunqiu Yunwen Tong Jin de Zhuzao yu Xiufu" 河南淅川下寺春秋雲紋銅禁的鑄造與修復. *Kaogu* 考古 5 (1987): 474–478.

Rong, Geng 容庚. *Shangzhou Yiqi Tongkao* 商周彝器通考. Beijing: Zhonghua Shuju, 1941.

Sandars, Nancy K. *Prehistoric Art in Europe.* Harmondsworth: Penguin Books, 1968.

Schaberg, David. "Texts and Artifacts: A Review of the Cambridge History of Ancient China." *Monumenta Serica* 49 (2001): 463–515.

Shaanxi Sheng Kaogu Yanjiusuo 陕西省考古研究所 et al. "Shaanxi Meixian Yangjiacun Xi Zhou Qingtongqi Jiaocang Fajue Jianbao" 陕西郿縣楊家村西周青銅器窖藏發掘簡報, *Wenwu* 文物 6 (2003): 4–42.

Shanghai Bowuguan 上海博物館, ed. *Lianxing Shenye, Yingzhi Lianggong: Shanghai Bowuguan cang Tongjing Jingpin* 練形神冶, 瑩質良工: 上海博物館藏銅鏡精品. Shanghai: Shanghai Shuhua Chubanshe, 2005.

Shao, Anding 邵安定. "Qinshihuang Lingyuan Chutu Caihui Qingtong Shuiqin Zhizuo Gongyi ji Xiangguan Wenti Yanjiu" 秦始皇陵園出土彩繪青銅水禽製作工藝及相關問題研究. Ph.D. dissertation, University of Science and Technology Beijing, 2013.

———. *Qinshihuang Lingyuan Chutu Caihui Qingtong Shuiqin Zhizuo Gongyi ji Xiangguan Wenti Yanjiu* 秦始皇帝陵園出土彩繪青銅水禽製作工藝及相關問題研究. Beijing: Kexue chubanshe, 2019.

Shaw, Ian, and Robert Jameson, eds. *A Dictionary of Archaeology.* Malden, MA: Blackwell Publishers, 1999.

Shi, Zhangru 石璋如. "Yin dai de Zhutong Gongyi" 殷代的鑄銅工藝. *Zhongyang Yanjiuyuan Lishi Yuyan Yanjiusuo Jikan* 中央研究院歷史語言研究所集刊 26 (1955): 95–129.

Simpson, Bruce L. *Development of the Metal Castings Industry.* Technical Development Program, American Foundrymen's Assn., 1948.

Singer, Paul. "The 'Unique' Object in Chinese Art." *Oriental Art* 7 (1961).

Smith, Cyril Stanley. *A Search for Structure: Selected Essays on Science, Art, and History.* Cambridge, MA: MIT Press, 1981.

So, Jenny F. *Eastern Zhou Ritual Bronzes from the Arthur M. Sackler Collections.* New York: Published by the Arthur M. Sackler Foundation in association with the Arthur M. Sackler Gallery, Smithsonian Institution: Distributed by Abrams, 1995.

———. "Hu Vessels from Xinzheng: Toward a Definition of Chu Style." In *The Great Bronze Age of China: A Symposium*, edited by George Kuwayama, 64–71. Los Angeles, CA: Los Angeles County Museum of Art; Seattle: Distributed by University of Washington Press, 1983.

———. "New Departures in Eastern Zhou Bronze Designs: The Spring and Autumn Period." In *The Great Bronze Age of China: An Exhibition from the People's Republic of China*, edited by Fong Wen, 251–269. New York: Metropolitan Museum of Art: Knopf, 1980.

So, Jenny F., and Emma C. Bunker. *Traders and Raiders on China's Northern Frontier.* Seattle: Arthur M. Sackler Gallery, Smithsonian Institution, in association with University of Washington Press, 1995.

Steinberg, Arthur. "Technology and Culture: Technological Styles in the Bronzes of Shang China, Phrygia and Urnfield Central Europe." In *Material Culture: Styles, Organization, and Dynamics of Technology*, edited by Heather Lechtman and Robert S. Merrill, 53–86. St. Paul et al.: West Publishing Co., 1977.

Steinke, Kyle. "Erligang and the Southern Bronze Industries." In *Art and Archaeology of the Erligang Civilization*, edited byKyle Steinke and Dora C.Y. Ching, 151–170. Princeton, NJ: Princeton University Press, 2014.

Su, Rongyu 蘇榮譽. "Anyang Yinxu Qingtong Jishu Yuanyuan de Shangdai Nanfang Yinsu" 安陽殷墟青銅技術淵源的商代南方因素. In *Quanwu Toushang: Quanwu Boguguan cang Qingtongqi Toushe Saomiao Jiexi* 泉屋透賞: 泉屋博古館藏青銅器透射掃描解析, edited by Quanwu Boguguan泉屋博古館 and Jiuzhou Bowuguan九州國立博物館, 352–386. Beijing: Zhongguo Keji Chubanshe, 2015.

———. "Ershi Shiji dui Xianqin Qingtong Liqi Zhuzao Jishu de Yanjiu" 二十世紀對先秦青銅禮器鑄造技術的研究. In *Quanwu Toushang: Quanwu Boguguan cang Qingtongqi Toushe Saomiao Jiexi* 泉屋透賞: 泉屋博古館藏青銅器透射掃描解析, edited by Quanwu Boguguan泉屋博古館 and Jiuzhou Bowuguan九州國立博物館, 387–445. Beijing: Zhongguo Keji Chubanshe, 2015.

———. "Hubei Zaoyang Caomenwan Zengguo Mudi Yihao Mu Chutu Jinshuqi Keji Jiazhi" 湖北棗陽曹門灣曾國墓地一號墓出土金屬器科技價值. *Zhongguo Wenwubao* 中國文物報, March 31, 2015.

———. "Zhanguo Zhongshan Wang Cuo Mu Qingtong Qiqun Zhuzao Gongyi Yanjiu" 戰國中山王嚳墓青銅器群鑄造工藝研究. In *Moji: Su Rongyu Zixuanji* 磨戟: 蘇榮譽自選集. Shanghai: Shanghai Renmin Chubanshe, 2012, 226–268. (Originally published as the appendix of Cuo Mu–Zhanguo Zhongshanguo Guowang zhi Mu嚳墓——戰國中山國國王之墓, Beijing: Wenwu Chubanshe, 1995, 548–577).

Su, Rongyu 蘇榮譽, Hua Jueming 華覺明, Li Kemin 李克敏, and Lu Benshan 盧本珊. *Zhongguo Shanggu Jinshu Jishu* 中國上古金屬技術. Jinan: Shandong Kexue Jishu Chubanshe, 1995.

Su, Rongyu 蘇榮譽, and Wang Lihua 王麗華. "Zaozhuang Xulou Chutu Zhuxiang Hongtong Qingtongqi Tanlun–Jianji Hongtong Zhuxiang Wenshi Qingtongqi de Shidai yu Chandi Wenti" 棗莊徐樓出土鑄鑲紅銅青銅器探論——兼及紅銅鑄鑲紋飾青銅器的時代與產地問題. Paper presented in the conference,

"Qingtongqi yu Shandong Guguo Xueshu Yantaohui" 青銅器與山東古國學術研討會 (Dec. 10–11, 2015; conference proceedings published as *Qingtongqi yu Shandong Guguo Xueshu Yantaohui Lunwenji* 青銅器與山東古國學術研討會論文集. Shanghai: Shanghai Guji Chubanshe, 2017).

Sun, Haibo 孫海波. *Xinzheng Yiqi* 新鄭彝器. Beijing, 1937.

Sun, Yan. "Colonizing China's Northern Frontier: Yan and Her Neighbors during the Early Western Zhou Period." *International Journal of Historical Archaeology* 10, no. 2 (2006): 159–177.

Tan, Derui 譚德睿. "Bianxing Kuiwen Fenti Yan—Cunzai Fenshifa de you yi Shengdong Lizheng" 變形夔紋分體甗–存在焚失法的又一生動例證. *Tezhong Zhuzao ji Youse Hejin* 特種鑄造及有色合金 3 (2010): 293–294.

———. *Canlan de Zhongguo Gudai Shila Zhuzao* 燦爛的中國古代失蠟鑄造. Shanghai: Shanghai Kexue Jishu Wenxian Chubanshe, 1989.

———. "Qianyi Gudai Shila Zhujian de Bianren—Jianping Zhou Weirong deng zhi Wupan Shuo" 淺議古代失蠟鑄件的辨認——兼評周衛榮等之誤判說. In *Qingtong Wenhua Yanjiu*. Vol. 5 青銅文化研究(第五輯). Tongling: Huangshan Shushe, 2008, 118–124.

———. "Shang Zhongqi de Tiliang Hu" 商中期的提梁壺. *Tezhong Zhuzao ji Youse Hejin* 特種鑄造及有色合金6 (2007): 491–492.

———. "Yunwen Gui—Wenshi Zhizuo Gongyi Qite de Nanfang Qingtongqi" 雲紋簋–紋飾製作工藝奇特的南方青銅器. *Tezhong Zhuzao ji Youse Hejin* 特種鑄造及有色合金 12 (2009): 1175–1176.

———. "Zhongguo Gudai Shila Zhuzao Qiyuan Wenti de Sikao" 中國古代失蠟鑄造起源問題的思考. *Wenwu Baohu yu Kaogu Kexue* 文物保護與考古科學2 (1994): 43–48.

———. "Zhongguo Zaoqi Shila Zhuzao Wenti de Guancha yu Sikao" 中國早期失蠟鑄造問題的考察與思考. *Nanfang Wenwu* 南方文物 2 (2007): 36–40.

Tang, Lan 唐蘭. "Zhongguo Qingtongqi de Qiyuan yu Fazhan" 中国青铜器的起源与发展. *Gugong Bowuyuan Yuankan* 故宫博物院院刊 1 (1979): 4–10.

Tang, Wenxing 湯文興. "Xichuan Xiasi Yihao Mu Qingtongqi de Zhuzao Jishu" 淅川下寺一號墓青銅器的鑄造技術. *Kaogu* 考古 2 (1981): 174–176.

Tian, Renxiao田仁孝. "Baojishi Yimencun Erhao Chunqiumu Fajue Jianbao" 寶雞市益門村二號春秋墓發掘簡報. *Wenwu* 文物 10 (1993): 1–14.

Wan, Li 萬俐. "Cong Chenzhang Hu kan Chunqiu Zhanguo de Yezhu Jishu" 從陳璋壺看春秋戰國的冶鑄技術. *Wenwu Jianding yu Jianshang* 文物鑒定與鑒賞3 (2010): 36–40.

———. "Nanjing Bowuyuan cang Zhanguo Toukong Panlongwen Qingtongqi" 南京博物院藏戰國透空蟠龍紋青銅器. *Dongnan Wenhua* 東南文化 6 (2013): 72–76.

———. "Shixi Zhongguo Xianqin Shiqi Toukong Panlongwen Qingtongqi Zhuzao Jishu de Fazhan yu Chuanbo" 試析中國先秦時期透空蟠龍紋青銅器鑄造技術的發展與傳播. *Wenwu Baohu yu Kaogu Kexue* 文物保護與考古科學 1 (2014): 110–116.

Wang, Jinchao 王金潮. "Nifanfa Fuyuan Xugong Ning Loukong Panhuiwen Qingtong Shijian Yanjiu" 泥範法復原許公寧鏤空蟠虺紋青銅飾件研究. *Nanfang Wenwu* 南方文物 4 (2008): 34–36.

———. "Tan Zeng Hou Yi Zun Pan de Zhuzao Gongyi" 談曾侯乙尊盤的鑄造工藝. *Dongnan Wenhua* 東南文化 1 (2002): 70–74.

———. "Zhongguo Qingtong Shidai Loukong Qingtongqi Nifan Zhuzuo Gongyi Yanjiu" 中國青銅時代鏤空青銅器泥範鑄作工藝研究. In *Qingtong Wenhua Yanjiu*, Vol. 6 青銅文化研究(第六輯). Tongling: Huangshan Shushe, 2009, 278–287, 190.

Wang, Ningsheng 汪寧生. *Yunnan Kaogu* 雲南考古. Kunming: Yunnan Renmin Chubanshe, 1980.

Wang, Wei 王瑋. “Zhongqiu Hua Yue Ye, Jiajing Gong Shang Shi” 中秋花月夜,佳鏡共賞時. *Wenwu Jianding yu Jianshang* 文物鑒定與鑒賞 9 (2011): 26–29.

Wang, Wei 王瑋, Wang Jinchao 王金潮, and Sheng You 盛友. “Sanwei Donghua li de Gudai Chuantong Gongyi” 三維動畫裏的古代傳統工藝. *Jianghan Kaogu* 江漢考古3 (2010): 113–122.

White, Joyce C. (Translated as 喬伊斯.J.懷特). “Banqing yu Taiguo Dongbeibu Qingtong Wenhua de Biannian Wenti” 班清與泰國東北部青銅文化的編年問題. In *Nanfang Minzu Kaogu*, Vol. 3 南方民族考古(第三輯). Chengdu, 1990: 15–22.

White, Joyce C. “Dating Early Bronze at Ban Chiang, Thailand.” In *From Homo Erectus to the Living Traditions: Choice of Papers from the 11th International Conference of the European Association of Southeast Asian Archaeologists, Bougon, 25th-29th September 2006*, edited by Jean-Pierre Pautreau, Anne-Sophie Coupey, and Valéry Zeitoun, 91–104. 11th International Conference of the European Association of Southeast Asian Archaeologists, 2008.

——— “Early East Asian Metallurgy: The Southern Tradition.” In *The Beginning of the Use of Metals and Alloys: Papers from the Second International Conference on the Beginning of the Use of Metals and Alloys, Zhengzhou, China, 21-26 October 1986*, edited by R. Maddin, 175–181. Cambridge, MA: MIT Press.

White, Julia M., and Emma C. Bunker. *Adornment for Eternity: Status and Rank in Chinese Ornament.* Denver: Denver Art Museum in association with the Woods Pub. Co., 1994.

Wu, Laiming 吳來明, and Xu Huikang 徐惠康. “Shanghai Bowuguan Wenwu Baohu yu Kaogu Kexue Shiyanshi Guance yu Jiance Baogao”上海博物館文物保護與考古科學實驗室觀測與檢測報告. Appendix one to Li Chaoyuan李朝遠. “Xinhuo Zhanguo Toukong Fuhejing Yanjiu” 新獲戰國透空復合鏡研究. In *Lianxing Shenye, Yingzhi Lianggong: Shanghai Bowuguan cang Tongjing Jingpin* 練形神冶, 瑩質良工: 上海博物館藏銅鏡精品, edited by Shanghai Bowuguan上海博物館, 50–52. Shanghai: Shanghai Shuhua Chubanshe, 2005.

Wu, Zhenfeng 吳鎮烽. *Shang Zhou Qingtongqi Mingwen ji Tuxiang Jicheng Xubian* 商周青銅器銘文暨圖像集成續編. Shanghai: Shanghai Fuji Chubanshe, 2016.

Wu'en, Yuesitu 烏恩岳斯圖. *Beifang Caoyuan Kaoguxue Wenhua Yanjiu: Qingtong Shidai zhi Zaoqi Tieqi Shidai* 北方草原考古學文化研究: 青銅時代至早期鐵器時代. Beijing: Kexue Chubanshe, 2007

Xu, Jay. "Bronze at Sanxingdui." In *Ancient Sichuan: Treasures from a Lost Civilization*, edited by Robert Bagley, 59–151. Seattle, WA: Seattle Art Museum; Princeton, NJ: Princeton University Press, 2001.

Yang, Jindong 楊金東. "Handai Tong Xunlu Yanjiu" 漢代銅熏爐研究. *Huaxia Kaogu* 華夏考古 2 (2014): 83–94.

Yetts, Walter P. *The George Eumorfopoulos Collection; Catalogue of the Chinese & Corean Bronzes, Sculpture, Jades, Jewellery and Miscellaneous Objects*, Vol. 1. London, E. Benn, 1929.

———. "Recent Finds near An-Yang." *The Journal of the Royal Asiatic Society of Great Britain and Ireland* 3 (1935): 467–474.

Yueyang shi Wenwu Gongzuodui 岳陽市文物工作隊. "Hunan sheng Yueyang xian Fengxingzuishan yi hao mu Fajue Jianbao" 湖南省岳陽縣鳳形嘴山一號墓發掘簡報. *Wenwu* 文物 1 (1993): 1–8.

Yunnan Sheng Bowuguan Kaogu Fajue Gongzuodui 雲南省博物館考古發掘工作隊. "Yunnan Jinning Shizhaishan Guyizhi ji Muzang" 雲南晉寧石寨山古遺址及墓葬. *Kaogu Xuebao* 考古學報 1 (1956): 43–63.

Yunnan Sheng Bowuguan 雲南省博物館. *Yunnan Jinning Shizhaishan Gu Muqun Fajue Baogao* 雲南晉寧石寨山古墓群發掘報告. Beijing: Wenwu Chubanshe, 1959.

———. "Yunnan Jinning Shizhaishan di san ci Fajue Jianbao 雲南晉寧石寨山第三次發掘簡報." *Kaogu* 考古 9 (1959): 459–461.

———. "Yunnan Jinning Shizhaishan Gumu di si ci Fajue Jianba" 雲南晉寧石寨山古墓第四次發掘簡報. *Kaogu* 考古 9 (1963): 480–485.

———. “Yunnan Jinning Shizhaishan Gumuqun Chutu Tongtieqi Buyi” 雲南晉寧石寨山古墓群出土銅鐵器補遺. *Wenwu* 文物 12 (1964): 41–49.

———. “Yunnan Jiangchuan Lijiashan Gumuqun Fajue Baoga” 雲南江川李家山古墓群發掘報告. *Kaogu Xuebao* 考古學報 2 (1975): 97–156.

Zaozhuang Shi Bowuguan 棗莊市博物館, Zaozhuang Shi Wenwu Guanli Weiyuanhui Bangongshi 棗莊市文物管理委員會辦公室, and Zaozhuang Shi Yicheng Qu Wenguang Xinju 棗莊市嶧城區文廣新局. “Shandong Zaozhuang Xulou Dongzhou Mu Fajue Jianbao” 山東棗莊徐樓東周墓發掘簡報. *Wenwu* 文物 1 (2014): 4–27.

Zhang, Changping 張昌平. “Guanyu Zeng Hou Yi Zun Pan” 關於曾侯乙尊盤. Shanghai *Wenbo Luncong* 上海文博論叢 3 (2007): 36–44.

———. “Guanyu Zeng Hou Yi Zun Pan Shifou Caiyong Shilafa Zhuzao de Shuping” 關於曾侯乙尊盤是否采用失蠟法鑄造的述評. *Jianghan Kaogu* 江漢考古 4 (2007): 85–90, 52.

———. “Panlongcheng Shangdai Qingtong Rongqi de Chubu Kaocha” 盤龍城商代青銅容器的初步考察. *Jianghan Kaogu* 江漢考古 1 (2003): 45–51.

———. *Zeng Guo Qingtongqi Yanjiu* 曾國青銅器研究. Beijing: Wenwu chubanshe, 2009.

Zhang, Guangyuan 張光遠. “Zhongguo Zuizao ‘Shilafa’: Chunqiu Zhongqi ‘She Wang Gaiguan Long Hu Fang Hu’ de Zhufa Lunzheng” 中國最早‘失蠟法’:春秋中期‘蛇網蓋冠龍虎方壺’的鑄法論證. *Dongnan Wenhua* 東南文化 1 (2002): 63–69.

Zhang, Linsheng 張臨生. “Guoli Gugong Bowuyuan suocang Dongzhou Xiangqianqi Yanjiu” 國立故宮博物院所藏東周鑲嵌器研究. *Gugong Xueshu Jikan* 故宮學術季刊7, no. 2 (1989): 1–78.

Zhang, Tian’en 張天恩. “Qinqi Sanlun—Yimen Chunqiumu Jige Wenti Qiantan” 秦器三論——益門春秋墓幾個問題淺談. *Wenwu* 文物 10 (1993): 20–27.

Zhao, Shigang 趙世綱. "Chunqiu Shiqi Shilafa Zhuzao Gongyi Wenti Tantao" 春秋時期失蠟法鑄造工藝問題探討. *Zhongyuan Wenwu* 中原文物 6 (2006): 84–91.

Zhongguo Shehui Kexueyuan Kaogu Yanjiusuo 中國社會科學院考古研究所. *Zhangjiapo Xizhou Mudi* 張家坡西周墓地. Beijing: Zhongguo Dabaike Quanshu Chubanshe, 1999.

Zhongguo Wenwu Jinghua Bianji Weiyuanhui《中國文物精華》編輯委員會. *Zhongguo Wenwu Jinghua* 中國文物精華. Beijing: Wenwu Chubanshe, 1992.

Zhou, Weirong 周衛榮, Dong Yawei 董亞巍, Wan Quanwen 萬全文, and Wang Changsui 王昌遂. "Zhongguo Qingtong Shidai Bu Cunzai Shilafa Zhuzao Gongyi." *Jianghan Kaogu* 江漢考古 2 (2006): 80–85.

———. "Zai Lun 'Shila Gongyi Bushi Zhongguo Qingtong Shidai de Xuanze'" 再論'失蠟工藝不是中國青銅時代的選擇. *Nanfang Wenwu* 南方文物2 (2007): 41–48, 132.

Zhou, Weirong 周衛榮, and Huang Wei 黃維. "Shi Lun Qingtong Shidai Toukong Qingtongqi de Gongyi Tese—Jian Tan Shila Zhuzao Wenti" 試論青銅時代透空青銅器的工藝特色——兼談失蠟鑄造問題. In *Zhongguo Guojia Bowuguan Guankan* 中國國家博物館館刊1 (2015): 147–155.

———. "Lost-wax Casting in Ancient China: New Discussion on Old Debates." *The Journal of the Minerals, Metals & Materials* 7 (2015): 1629–1636.

———. "'Shila Shizhi Fa' Shangque" '失蠟失織法'商榷. *Nanfang Wenwu* 南方文物 1 (2016): 59–63.

Zhou, Xiaolu 周曉陸. "Xuyi Suochu Zhongjin luo ling. Chen Zhang Yuanhu Dukao" 盱眙所出重金絡[illegible]. 陳璋圓壺讀考. *Kaogu* 考古 3 (1988): 258–263.

Works Cited by Title

Beijing Wenwu Jingcui Daxi (Qingtongqi juan) 北京文物精粹大系(青銅器卷), edited by Mei Ninghua 梅寧華 and Tao Xincheng 陶信成. Beijing: Beijing Chubanshe, 2006.

Hou Hanshu 後漢書 [Book of the Later Han], by Fan Ye范曄 (398–445 CE). Beijing: Zhonghua Shuju, 1965.

Kaogu 考古 [Archaeology], journal issued by the Institute of Archaeology, Chinese Academy of Social Sciences 中國社會科學院考古研究所.

Shiji 史記 [Records of the Historian], authored by Sima Qian 司馬遷 (145 or 135–c. 86/85 BCE). Beijing: Zhonghua Shuju, 1959.

Zhongguo qingtongqi quanji 中國青銅器全集, edited by Zhongguo qingtongqi quanji bianji weiyuanhui 中國青銅器全集編輯委員會, vols. 1–16. Beijing: Wenwu Chubanshe, 1993–1998.

Zhongguo Wenwubao 中國文物報 [China Cultural Relics Newspaper], issued by the State Administration of Cultural Heritage 國家文物局, China.

Zuo Zhuan 左傳 [The Transmission/Commentary of Zuo], attributed to Zuo Qiuming左丘明 (during 6th–5th centuries BCE?) and probably compiled later. See Yang Bojun楊伯峻. *Chunqiu Zuozhuan Zhu* 春秋左傳註. Beijing: Zhonghua Shuju, 1990.

Index

About the Author

Peng Peng is an assistant professor at the Chinese University of Hong Kong (CUHK). Prior to this appointment, Professor Peng was a visiting assistant professor of Asian art history at the University of Minnesota, Twin Cities. He holds a PhD and an MA from Princeton University, an MA and a BA from Peking University. Professor Peng's research centers on the art, archaeology, visual and material cultures of ancient East Asia; and his work has been published in *Orientations* and *Sino-Platonic Papers.*

Cambria Sinophone World Series

General Editor: Victor H. Mair (University of Pennsylvania)

Books in the Cambria Sinophone World Series

Metalworking in Bronze Age China: The Lost-Wax Process by Peng Peng

The Chinese Lyric Sequence: Poems, Paintings, Anthologies by Joseph R. Allen

The Great Leap Backward: Forgetting and Representing the Mao Years by Lingchei Letty Chen

Rethinking the Sinosphere: Poetics, Aesthetics, and Identity Formation edited by Nanxiu Qian, Richard J. Smith, and Bowei Zhang

Reexamining the Sinosphere: Transmissions and Transformations in East Asia edited by Nanxiu Qian, Richard J. Smith, and Bowei Zhang

Insects in Chinese Literature: A Study and Anthology by Wilt L. Idema

The Poetics and Politics of Sensuality in China: The "Fragrant and Bedazzling" Movement (1600-1930) by Xiaorong Li

Spatial Imaginaries in Mid-Tang China: Geography, Cartography, and Literature by Ao Wang

Texts and Transformations: Essays in Honor of the 75th Birthday of Victor H. Mair edited by Haun Saussy

Chinese Women Writers and Modern Print Culture by Megan M. Ferry

Reading Lu Xun Through Carl Jung by Carolyn Brown

Gao Xingjian and Transmedia Aesthetics edited by Mabel Lee and Liu Jianmei

Imperfect Understanding: Intimate Portraits of Chinese Celebrities edited by Christopher Rea

Zhang Yimou: Globalization and the Subject of Culture by Wendy Larson

The Borderlands of Asia: Culture, Place, Poetry by Mark Bender

Buddhist Transformations and Interactions: Essays in Honor of Antonino Forte edited by Victor H. Mair

Chinese Avant-garde Fiction: Quest for Historicity and Transcendent Truth by Zhansui Yu

Eroticism and Other Literary Conventions In Chinese Literature: Intertextuality In The Story Of The Stone by I-Hsien Wu

The Sinophone Cinema of Hou Hsiao-hsien: Culture, Style, Voice and Motion by Christopher Lupke

Supernatural Sinophone Taiwan and Beyond by Chia-rong Wu

Cosmopolitanism in China edited by Minghui Hu and Johan Elverskog

"The Immortal Maiden Equal to Heaven" and Other Precious Scrolls from Western Gansu by Wilt L. Idema

Chinese Ethnic Minority Oral Traditions: A Recovered Text of Bai Folk Songs in a Sinoxenic Script by Jingqi Fu and Zhao Min with Xu Lin and Duan Ling

China and Beyond in the Mediaeval Period: Cultural Crossings and Inter-Regional Connections edited by Dorothy C. Wong and Gustav Heldt

Anglophone Literatures in the Asian Diaspora: Literary Transnationalism and Translingual Migrations by Karen An-hwei Lee

Modern Poetry in China: A Visual-Verbal Dynamic by Paul Manfredi

Sinophone Malaysian Literature: Not Made in China by Alison M. Groppe

Infected Korean Language, Purity versus Hybridity: From the Sinographic Cosmopolis to Japanese Colonialism to Global English by Koh Jongsok (translated by Ross King)

The Chinese Prose Poem: A Study of Lu Xun's Wild Grass (Yecao) by Nicholas A. Kaldis

Gao Xingjian: Aesthetics and Creation by Gao Xingjian (translated by Mabel Lee)

Rethinking Chineseness: Translational Sinophone Identities in the Nanyang Literary World by E. K. Tan

A Study of Two Classics: A Cultural Critique of The Romance of the Three Kingdoms and The Water Margin by Liu Zaifu (translated by Shu Yunzhong)

Confucian Prophet: Political Thought in Du Fu's Poetry (752–757) by David K. Schneider

The Classic of Changes in Cultural Context: A Textual Archaeology of the Yi jing by Scott Davis

www.ingramcontent.com/pod-product-compliance
Lightning Source LLC
LaVergne TN
LVHW091522200926
848483LV00002B/281